TRANSLATION

OF THE

SURYA-SIDDHANTA,

A TEXT-BOOK OF HINDU ASTRONOMY;
WITH NOTES AND AN APPENDIX,

CONTAINING ADDITIONAL NOTES AND TABLES, CALCULATIONS OF ECLIPSES, A STELLAR MAP, AND INDEXES.

TRANSLATION

OF THE

SURYA-SIDDHANTA,

A TEXT-BOOK OF HINDU ASTRONOMY;
WITH NOTES AND AN APPENDIX,

CONTAINING ADDITIONAL NOTES AND TABLES, CALCULATIONS OF ECLIPSES, A STELLAR MAP, AND INDEXES.

BY

EBENEZER BURGESS

GYAN

Published by

GYAN

Gyan Publishing House
5, Ansari Road
Daryaganj, New Delhi-110002
Phone: 011-47034999, 9811692060
E-mail: books@gyanbooks.com

Distribution Network
gyanbooks.com
India, USA, Canada, UK, Australia

ISBN: 978-93-6280-303-0 (HB)
First Published, 1860

2nd Impression 2024

Printed at: Gyan Press, Delhi.

TRANSLATION OF THE SURYA-SIDDHANTA
AUTHOR: EBENEZER BURGESS

HINDU ASTRONOMY.

TRANSLATION

OF THE

SÛRYA-SIDDHÂNTA,

A TEXT-BOOK OF HINDU ASTRONOMY;

WITH NOTES, AND AN APPENDIX,

CONTAINING ADDITIONAL NOTES AND TABLES, CALCULATIONS OF ECLIPSES, A STELLAR MAP, AND INDEXES.

BY REV. EBENEZER BURGESS,

FORMERLY MISSIONARY OF THE A. B. C. F. M. IN INDIA;

ASSISTED BY THE

COMMITTEE OF PUBLICATION OF THE AMERICAN ORIENTAL SOCIETY.

[FROM THE JOURNAL OF THE AMERICAN ORIENTAL SOCIETY, VOL. VI, 1860.]

NEW HAVEN:
FOR THE AMERICAN ORIENTAL SOCIETY,
PRINTED BY E. HAYES, PRINTER TO YALE COLLEGE.
MDCCCLX.

SOLD BY THE SOCIETY'S AGENTS:
NEW YORK: JOHN WILEY, 56 WALKER ST.;
LONDON: TRÜBNER & CO.; PARIS: BENJ. DUPRAT;
LEIPZIG: F. A. BROCKHAUS.

TRANSLATION

OF THE

SÛRYA-SIDDHÂNTA,

WITH NOTES, AND AN APPENDIX.

[Communicated to the American Oriental Society May 17, 1858, and published in the Sixth Volume of its Journal.]

Introductory Note.

Soon after my entrance upon the missionary field, in the Marâtha country of western India, in the year 1839, my attention was directed to the preparation, in the Marâthî language, of an astronomical text-book for schools. I was thus led to a study of the Hindu science of astronomy, as exhibited in the native text-books, and to an examination of what had been written respecting it by European scholars. I at once found myself, on the one hand, highly interested by the subject itself, and, on the other, somewhat embarrassed for want of a satisfactory introduction to it. A comprehensive exhibition of the Hindu system had nowhere been made. The Astronomie Indienne of Bailly, the first extended work upon its subject, had long been acknowledged to be founded upon insufficient data, to contain a greatly exaggerated estimate of the antiquity and value of the Hindu astronomy, and to have been written for the purpose of supporting an untenable theory. The articles in the Asiatic Researches, by Davis, Colebrooke, and Bentley, which were the first, as they still remain the most important, sources of knowledge respecting the matters with which they deal, relate only to particular points in the system, of especial prominence and interest. Bentley's volume on Hindu astronomy is mainly occupied with an endeavor to ascertain the age of the principal astronomical treatises, and the epochs of astronomical discovery and progress, and is, moreover, even in these respects, an exceedingly unsafe guide. The treatment of the subject by Delambre, in his History of Ancient Astronomy, being founded only upon Bailly and the earliest of the essays in the Asiatic Researches, partakes, of course, of the incompleteness of his authorities. Works of value have been published in India also, into which more or less of Hindu astronomy enters, as Warren's Kâla Sankalita, Jervis's Weights Measures and Coins of India, Hoisington's Oriental Astronomer, and the like; but these, too, give, for the most part, hardly more than the practical processes employed in parts of the system, and they are, like many of the authorities already mentioned, only with difficulty accessible. In short, there was nothing in existence which showed the world how much and how little the Hindus know of astronomy, as also their mode of presenting the subject in its totality, the intermixture in their science of old ideas with new, of astronomy with astrology, of observation and mathematical deduction with arbitrary theory, mythology, cosmogony, and pure imagination. It seemed to me that nothing would so well supply the deficiency as the translation and detailed explication of a complete treatise of Hindu astronomy: and this work I accordingly undertook to execute.

Among the different Siddhântas, or text-books of astronomy, existing in India in the Sanskrit language, none appeared better suited to my purpose than the Sûrya-Siddhânta. That it is one of the most highly esteemed, best known, and most frequently employed, of all, must be evident to any one who has noticed how much oftener than any other it is referred to as authority in the various papers on the Hindu astronomy. In fact, the science as practised in modern India is in the greater part founded upon its data and processes. In the lists of Siddhântas given by native authorities it is almost invariably mentioned second, the Brahma-Siddhânta being placed first: the latter enjoys this preminence, perhaps, mainly on account of its name; it is, at any rate, comparatively rare and little known. For completeness, simplicity, and conciseness combined, the Sûrya-Siddhânta is believed not to be surpassed by any other. It is also more easily obtainable. In general, it is difficult, without official influence or exorbitant pay, to gain possession of texts which are rare and held in high esteem. During my stay in India, I was able to procure copies of only three astronomical treatises besides the Sûrya-Siddhânta; the Çâkalya-Sanhitâ of the Brahma-Siddhânta, the Siddhânta-Çiromaṇi of Bhâskara, and the Graha-Lâghava, of which the two latter have also been printed at Calcutta. Of the Sûrya-Siddhânta I obtained three copies, two of them giving the text alone, and the third also the commentary entitled Gûḍhârthaprakâçaka, by Ranganâtha, of which the date is unknown to me. The latter manuscript agrees in all respects with the edition of the Sûrya-Siddhânta, accompanied by the same commentary, of which the publication, in the series entitled Bibliotheca Indica, has been commenced in India by an American scholar, and a member of this Society, Prof. Fitz-Edward Hall of Benares; to this I have also had access, although not until my work was nearly completed.

My first rough draft of the translation and notes was made while I was still in India, with the aid of Brahmans who were familiar with the Sanskrit and well versed in Hindu astronomical science. In a few points also I received help from the native Professor of Mathematics in the Sanskrit College at Pûna. But notwithstanding this, there remained not a few obscure and difficult points, connected with the demonstration and application of the processes taught in the text. In the solution of these, I have received very important assistance from the Committee of Publication of the Society. They have also—the main share of the work falling to Prof. Whitney—enriched the notes with much additional matter of value. My whole collected material, in fact, was placed in their hands for revision, expansion, and reduction to the form best answering to the requirements of modern scholars, my own engrossing occupations, and distance from the place of publication, as well as my confidence in their ability and judgment, leading me to prefer to intrust this work to them rather than to undertake its execution myself.

We have also to express our acknowledgments to Mr. Hubert A. Newton, Professor of Mathematics in Yale College, for valuable aid rendered us in the more difficult demonstrations, and in the comparison of the Hindu and Greek astronomies, as well as for his constant advice and suggestions, which add not a little to the value of the work.

The Sûrya-Siddhânta, like the larger portion of the Sanskrit literature, is written in the verse commonly called the *çloka*, or in stanzas of two lines, each line being composed of two halves, or *pâdas*, of eight syllables each. With its metrical form are connected one or two peculiarities which call for notice. In the first place, for the terms used there are often many synonyms, which are employed according to the exigencies of the verse: thus, the sun has twelve different names, Mars six, the divisions of time two or three each, radius six or eight, and so on. Again, the method of expressing numbers, large or small, is by naming the figures which compose them, beginning with the last and going backward; using for each figure not only its own proper name, but that of any object associated in the Hindu mind with the number it represents. Thus, the number 1,577,917,828 (i. 37) is thus given: Vasu (a class of deities, *eight* in number) -two-eight-mountain (the *seven* mythical chains of mountains) -form-figure (the *nine* digits) -seven-mountain-lunar days (of which there are *fifteen* in the half-month). Once more, the style of expression of the treatise is, in general, excessively concise and elliptical, often to a degree that would make its meaning entirely unintelligible without a commentary, the exposition of a native teacher, or such a knowledge of the subject treated of as should show what the text must be meant to say. Some striking instances are pointed out in the notes. This over-conciseness, however, is not wholly due to the metrical form of the treatise: it is characteristic of much of the Hindu scientific literature, in its various branches; its text-books are wont to be intended as only the text for written comment or oral explication, and hint, rather than fully express, the meaning they contain. In our translation, we have not thought it worth while to indicate, by parentheses or otherwise, the words and phrases introduced by us to make the meaning of the text evident: such a course would occasion the reader much more embarrassment than satisfaction. Our endeavor is, in all cases, to hit the true mean between unintelligibility and diffuseness, altering the phraseology and construction of the original only so far as is necessary. In both the translation and the notes, moreover, we keep steadily in view the interests of the two classes of readers for whose benefit the work is undertaken: those who are orientalists without being astronomers, and those who are astronomers without being orientalists. For the sakeof the former, our explanations and demonstrations are made more elementary and full than would be necessary, were we addressing mathematicians only: for the sake of the latter, we cast the whole into a form as occidental as may be, translating every technical term which admits of translation: since to compel all those who may desire to inform themselves respecting the scientific content of the Hindu astronomy to learn the Sanskrit technical language would be highly unreasonable. To furnish no ground of complaint, however, to those who are familiar with and attached to these terms, we insert them liberally in the translation, in connection with their English equivalents. The derivation and literal signification of the greater part of the technical terms employed in the treatise are also given in the notes, since such an explanation of the history of a term is often essential to its full comprehension, and throws valuable light upon the conceptions of those by whom it was originally applied.

We adopt, as the text of our translation, the published edition of the Siddhânta, referred to above, following its readings and its order of arrangement, wherever they differ, as they do in many places, from those of the manuscripts without commentary in our possession. The discordances of the two versions, when they are of sufficient consequence to be worth notice, are mentioned in the notes.

As regards the transcription of Sanskrit words in Roman letters, we need only specify that *c* represents the sound of the English *ch* in "church," Italian *c* before *e* and *i*: that *j* is the English *j*: that *ç* is pronounced like the English *sh*, German *sch*, French *ch*, while *sh* is a sound nearly resembling it, but uttered with the tip of the tongue turned back into the top of the mouth, as are the other lingual letters, *ṭ*, *ḍ*, *ṇ*: finally, that the Sanskrit *r* used as a vowel (which value it has also in some of the Slavonic dialects) is written with a dot underneath, as *ṛ*.

The demonstrations of principles and processes given by the native commentary are made without the help of figures. The figures which we introduce are for the most part our own, although a few of them were suggested by those of a set obtained in India, from native mathematicians.

For the discussion of such general questions relating to this Siddhânta as its age, its authorship, the alterations which it may have undergone before being brought into its present form, the stage which it represents in the progress of Hindu mathematical science, the extent and character of the mathematical and astronomical knowledge displayed in it, and the relation of the same to that of other ancient nations, especially of the Greeks, the reader is referred to the notes upon the text. The form in which our publication is made does not allow us to sum up here, in a preface, the final results of our investigations into these and kindred topics. It may perhaps be found advisable to present such a summary at the end of the article, in connection with the additional notes and other matters to be there given.

SÛRYA-SIDDHÂNTA.

CHAPTER I.

OF THE MEAN MOTIONS OF THE PLANETS.

CONTENTS:—1, homage to the Deity; 2–9, revelation of the present treatise; 10–11, modes of dividing time; 11–12, subdivisions of a day; 12–14, of a year; 14–17, of the Ages; 18–19, of an Æon; 20–21, of Brahma's life; 21–23, part of it already elapsed; 24, time occupied in the work of creation; 25–27, general account of the movements of the planets; 28, subdivisions of the circle; 29–33, number of revolutions of the planets, and of the moon's apsis and node, in an Age; 34–39, number of days and months, of different kinds, in an Age; 40, in an Æon; 41–44, number of revolutions, in an Æon, of the apsides and nodes of the planets; 45–47, time elapsed from the end of creation to that of the Golden Age; 48–51, rule for the reduction to civil days of the whole time since the creation; 51–52, method of finding the lords of the day, the month, and the year; 53–54, rule for finding the mean place of a planet, and of its apsis and node; 55, to find the current year of the cycle of Jupiter; 56, simplification of the above calculations; 57–58, situation of the planets, and of the moon's apsis and node, at the end of the Golden Age; 59–60, dimensions of the earth; 60–61, correction, for difference of longitude, of the mean place of a planet as found; 62, situation of the principal meridian; 63–65, ascertainment of difference of longitude by difference between observed and computed time of a lunar eclipse; 66, difference of time owing to difference of longitude; 67, to find the mean place of a planet for any required hour of the day; 68–70, inclination of the orbits of the planets.

1. To him whose shape is inconceivable and unmanifested, who is unaffected by the qualities, whose nature is quality, whose form is the support of the entire creation—to Brahma be homage!

The usual propitiatory expression of homage to some deity, with which Hindu works are wont to commence.

2. When but little of the Golden Age (*kṛta yuga*) was left, a great demon (*asura*), named Maya, being desirous to know that mysterious, supreme, pure, and exalted science,

3. That chief auxiliary of the scripture (*vedânga*), in its entirety—the cause, namely, of the motion of the heavenly bodies (*jyotis*), performed, in propitiation of the Sun, very severe religious austerities.

According to this, the Sûrya-Siddhânta was revealed more than 2,164,960 years ago, that amount of time having elapsed, according to Hindu reckoning, since the end of the Golden Age; see below, under verse 48, for the computation of the period. As regards the actual date of the treatise, it is, like all dates in Hindu history and the history of Hindu literature, exceedingly difficult to ascertain. It is the more difficult, because, unlike most, or all, of the astronomical treatises, the Sûrya-Siddhânta attaches itself to the name of no individual as its author, but professes to be a direct revelation from the Sun (*sûrya*). A treatise of this name, however, is confessedly among the earliest text-books of the Indian science. It was one of the five earlier works upon which was founded the Pañca-siddhântika, Compendium of Five Astronomies, of Varâha-mihira, one of the earliest astronomers whose works have been, in part, preserved to us, and who is supposed to have lived about the beginning of the sixth century of our era. A Sûrya-Siddhânta is also referred to by Brahmagupta, who is assigned to the close of the same century and the commencement of the one following. The arguments by which Mr. Bentley (Hindu Astronomy, p. 158, etc.) attempts to prove Varâha-mihira to have lived in the sixteenth century, and his professed works to be forgeries and impositions, are sufficiently refuted by the testimony of al-Bîrûnî (the same person as the Abu-r-Raiḥân, so often quoted in the first article of this volume), who visited India under Mahmûd of Ghazna, and wrote in A.D. 1031 an account of the country: he speaks of Varâha-mihira and of his Pañca-siddhântika, assigning to both nearly the same age as is attributed to them by the modern Hindus (see Reinaud in the Journal Asiatique for Sept.–Oct. 1844, ivme Série, iv. 286; and also his Mémoire sur l'Inde). He also speaks of the Sûrya-Siddhânta itself, and ascribes its authorship to Lâta (Mémoire sur l'Inde, pp. 331, 332), whom Weber (Vorlesungen über Indische Literaturgeschichte, p. 229) conjecturally identifies with a Lâḍha who is cited by Brahmagupta. Bentley has endeavored to show by internal evidence that the Sûrya-Siddhânta belongs to the end of the eleventh century: see below, under verses 29–34, where his method and results are explained, and their value estimated.

Of the six Vedângas, "limbs of the Veda," sciences auxiliary to the sacred scriptures, astronomy is claimed to be the first and chief, as representing the eyes; grammar being the mouth, ceremonial the hands, prosody the feet, etc. (see Siddhânta-Çiromaṇi, i. 12–14). The importance of astronomy to the system of religious observance lies in the fact that by it are determined the proper times of sacrifice and the like. There is a special treatise, the Jyotisha of Lagadha, or Lagata, which, attaching itself to the Vedic texts, and representing a more primitive phase of Hindu science, claims to be the astronomical Vedânga; but it is said to be of late date and of small importance.

The word *jyotis*, "heavenly body," literally "light," although the current names for astronomy and astronomers are derived from it, does not elsewhere occur in this treatise.

4. Gratified by these austerities, and rendered propitious, the Sun himself delivered unto that Maya, who besought a boon, the system of the planets.

The blessed Sun spoke:

5. Thine intent is known to me; I am gratified by thine austerities; I will give thee the science upon which time is founded, the grand system of the planets.

6. No one is able to endure my brilliancy; for communication I have no leisure; this person, who is a part of me, shall relate to thee the whole.

The manuscripts without commentary insert here the following verse: "Go therefore to Romaka-city, thine own residence; there, undergoing incarnation as a barbarian, owing to a curse of Brahma, I will impart to thee this science."

If this verse really formed a part of the text, it would be as clear an acknowledgment as the author could well convey indirectly, that the science displayed in his treatise was derived from the Greeks. Romaka-city is Rome, the great metropolis of the West; its situation is given in a following chapter (see xii. 39) as upon the equator, ninety degrees to the west of India. The incarnation of the sun there as a barbarian, for the purpose of revealing astronomy to a demon of the Hindu Pantheon, is but a transparent artifice for referring the foreign science, after all, to a Hindu origin. But the verse is clearly out of place here; it is inconsistent with the other verses among which it occurs, which give a different version of the method of revelation. How comes it here then? It can hardly have been gratuitously devised and introduced. The verse itself is found in many of the manuscripts of this Siddhânta; and the incarnation of the Sun at Romaka-city, among the Yavanas, or Greeks, and his revelation of the science of astronomy there, are variously alluded to in later works; as, for instance, in the Jñâna-bhâskara (see Weber's Catalogue of the Berlin Sanskrit Manuscripts, p. 287, etc.), where he is asserted to have revealed also the Romaka-Siddhânta. Is this verse, then, a fragment of a different, and perhaps more ancient, account of the origin of the treatise, for which, as conveying too ingenuous a confession of the source of the Hindu astronomy, another has been substituted later? Such a supposition, certainly, does not lack plausibility. There is something which looks the same way in the selection of a demon, an Asura, to be the medium of the sun's revelation; as if, while the essential truth and value of the system was acknowledged, it were sought to affix a stigma to the source whence the Hindus derived it. Weber (Ind. Stud. ii. 243; Ind. Lit. p. 225), noticing that the name of the Egyptian sovereign Ptolemaios occurs in Indian inscriptions in the form *Turumaya*, conjectures that Asura Maya is an alteration of that name, and that the demon Maya accordingly represents the author of the Almagest himself; and the conjecture is powerfully supported by the fact that al-Bîrûnî (see Reinaud, as above) ascribes the Pâuliça-Siddhânta, which the later Hindus attribute to a Puliça, to Paulus al-Yûnânî, Paulus the Greek, and that another of the astronomical treatises, alluded to above, is called the Romaka-Siddhânta.

It would be premature to discuss here the relation of the Hindu astronomy to the Greek; we propose to sum up, at the end of this work, the evidence upon the subject which it contains.

7. Thus having spoken, the god disappeared, having given directions unto the part of himself. This latter person thus addressed Maya, as he stood bowed forward, his hands suppliantly joined before him:

8. Listen with concentrated attention to the ancient and exalted science, which has been spoken, in each successive Age, to the Great Sages (*maharshi*), by the Sun himself.

9. This is that very same original text-book which the Sun of old promulgated: only, by reason of the revolution of the Ages, there is here a difference of times.

According to the commentary, the meaning of these last verses is that, in the successive Great Ages, or periods of 4,320,000 years (see below, under vv. 15–17), there are slight differences in the motions of the heavenly bodies, which render necessary a new revelation from time to time on the part of the Sun, suited to the altered conditions of things; and that when, moreover, even during the continuance of the same Age, differences of motion are noticed, owing to a difference of period, it is customary to apply to the data given a correction, which is called *bîja*. All this is very suitable for the commentator to say, but it seems not a little curious to find the Sun's superhuman representative himself insisting that this his revelation is the same one as had formerly been made by the Sun, only with different data. We cannot help suspecting in the ninth verse, rather, a virtual confession on the part of the promulgators of this treatise, that there was another, or that there were others, in existence, claiming to be the sun's revelation, or else that the data presented in this were different from those which had been previously current as revealed by the Sun. We shall have more to say hereafter (see below, under vv. 29–34) of the probable existence of more than one version of the Sûrya-Siddhânta, of the correction called *bîja*, and of its incorporation into the text of the treatise itself. The repeated revelation of the system in each successive Great Age, as stated in verse 8, presents no difficulty. It is the Puranic doctrine (see Wilson's Vishṇu Purâṇa, p. 269, etc.) that during the Iron Age the sources of knowledge become either corrupted or lost, so that a new revelation of scripture, law, and science becomes necessary during the Age succeeding.

10. Time is the destroyer of the worlds; another Time has for its nature to bring to pass. This latter, according as it is gross or minute, is called by two names, real (*mûrta*) and unreal (*amûrta*).

There is in this verse a curious mingling together of the poetical, the theoretical, and the practical. To the Hindus, as to us, Time is, in a metaphorical sense, the great destroyer of all things; as such, he is identified with Death, and with Yama, the ruler of the dead. Time, again, in the ordinary acceptation of the word, has both its imaginary, and its appreciable and practically useful divisions: the former are called real (*mûrta*, literally "embodied"), the latter unreal (*amûrta*, literally "unembodied"). The following verse explains these divisions more fully.

The epithet *kalanâtmaka*, applied to actual time in the first half of the verse, is not easy of interpretation. The commentary translates it "is an object of knowledge, is capable of being known," which does not seem satisfactory. It evidently contains a suggested etymology (*kâla*, "time," from *kalana*), and in translating it as above we have seen in it also an antithesis to the epithet bestowed upon Time the divinity. Perhaps it should be rather "has for its office enumeration."

11. That which begins with respirations (*prâṇa*) is called real; that which begins with atoms (*truṭi*) is called unreal. Six respirations make a *vinâḍî*, sixty of these a *nâḍî*;

12. And sixty nâḍîs make a sidereal day and night. . . .

The manuscripts without commentary insert, as the first half of v. 11, the usual definition of the length of a respiration: "the time occupied in pronouncing ten long syllables is called a respiration."

The table of the divisions of sidereal time is then as follows:

10 long syllables (*gurvakshara*)	=	1 respiration (*prâṇa*, period of four seconds);
6 respirations	=	1 vinâḍî (period of twenty-four seconds);
60 vinâḍîs	=	1 nâḍî (period of twenty-four minutes);
60 nâḍîs	=	1 day.

This is the method of division usually adopted in the astronomical text-books: it possesses the convenient property that its lowest subdivision, the respiration, is the same part of the day as the minute is of the circle, so that a respiration of time is equivalent to a minute of revolution of the heavenly bodies about the earth. The respiration is much more frequently called *asu*, in the text both of this and of the other Siddhântas. The vinâḍî is practically of small consequence, and is only two or three times made use of in the treatise: its usual modern name is *pala*, but as this term nowhere occurs in our text, we have not felt justified in substituting it for vinâḍî. For nâḍî also, the more common name is *daṇḍa*, but this, too, the Sûrya-Siddhânta nowhere employs, although it uses instead of nâḍî, and quite as often, *nâḍikâ* and *ghaṭikâ*. We shall uniformly make use in our translation of the terms presented above, since there are no English equivalents which admit of being substituted for them.

The ordinary Puranic division of the day is slightly different from the astronomical, viz:

15 twinklings (*nimesha*)	=	1 bit (*kâshṭhâ*);
30 bits	=	1 minute (*kalâ*);
30 minutes	=	1 hour (*muhûrta*);
30 hours	=	1 day.

Manu (i. 64) gives the same, excepting that he makes the bit to consist of 18 twinklings. Other authorities assign different values to the lesser measures of time, but all agree in the main fact of the division of the day into thirty hours, which, being perhaps an imitation of the division of the month into thirty days, is unquestionably the ancient and original Hindu method of reckoning time.

The Sûrya-Siddhânta, with commendable moderation, refrains from giving the imaginary subdivisions of the respiration which make up “unreal” time. They are thus stated in Bhâskara's Siddhânta-Çiromaṇi (i. 19, 20), along with the other, the astronomical, table:

100 atoms (*truṭi*)	= 1 speck (*tatpara*);
30 specks	= 1 twinkling (*nimesha*);
18 twinklings	= 1 bit (*kâshṭhâ*);
30 bits	= 1 minute (*kalâ*);
30 minutes	= 1 half-hour (*ghaṭikâ*);
2 half-hours	= 1 hour (*kshaṇa*);
30 hours	= 1 day.

This makes the atom equal to $\frac{1}{2\,916\,000\,000}$th of a day, or $\frac{1}{33\,750}$th of a second. Some of the Purâṇas (see Wilson's Vish. Pur. p. 22) give a different division, which makes the atom about $\frac{1}{2110}$th of a second; but they carry the division three steps farther, to the subtilissima (*paramâṇu*), which equals $\frac{1}{3\,280\,500\,000}$th of a day, or very nearly $\frac{1}{38\,000}$th of a second.

We have introduced here a statement of these minute subdivisions, because they form a natural counterpart to the immense periods which we shall soon have to consider, and are, with the latter, curiously illustrative of a fundamental trait of Hindu character: a fantastic imaginativeness, which delights itself with arbitrary theorizings, and is unrestrained by, and careless of, actual realities. Thus, having no instruments by which they could measure even seconds with any tolerable precision, they vied with one another in dividing the second down to the farthest conceivable limit of minuteness; thus, seeking infinity in the other direction also, while they were almost destitute of a chronology or a history, and could hardly fix with accuracy the date of any event beyond the memory of the living generation, they devised, and put forth as actual, a framework of chronology reaching for millions of millions of years back into the past and forward into the future.

12. . . . Of thirty of these sidereal days is composed a month; a civil (*sâvana*) month consists of as many sunrises;

13. A lunar month, of as many lunar days (*tithi*); a solar (*sâura*) month is determined by the entrance of the sun into a sign of the zodiac: twelve months make a year. . . .

We have here described days of three different kinds, and months and years of four; since, according to the commentary, the last clause translated means that twelve months of each denomination make up a year of the same denomination. Of some of these, the practical use and value will be made to appear later; but as others are not elsewhere referred to in this treatise, and as several are merely arbitrary divisions of time, of which, so far as we can discover, no use has ever been made, it may not be amiss briefly to characterize them here.

Of the measures of time referred to in the twelfth verse, the day is evidently the starting-point and standard. The sidereal day is the time of the earth's revolution on its axis; data for determining its length are given below, in v. 34, but it does not enter as an element into the later processes. Nor is a sidereal month of thirty sidereal days, or a sidereal year of three hundred and sixty such days (being less than the true sidereal year by about six and a quarter sidereal days), elsewhere mentioned in this work, or, so far as we know, made account of in any Hindu method of reckoning time. The civil (*sâvana*) day is the natural day: it is counted, in India, from sunrise to sunrise (see below, v. 36), and is accordingly of variable length: it is, of course, an important element in all computations of time. A month of thirty, and a year of three hundred and sixty, such days, are supposed to have formed the basis of the earliest Hindu chronology, an intercalary month being added once in five years. This method is long since out of use, however, and the month and year referred to here in the text, of thirty and three hundred and sixty natural days respectively, without intercalations, are elsewhere assumed and made use of only in determining, for astrological purposes, the lords of the month and year (see below, v. 52).

The standard of the lunar measure of time is the lunar month, the period of the moon's synodical revolution. It is reckoned either from new-moon to new-moon, or from full-moon to full-moon; generally, the former is called *mukhya*, "primary," and the latter *gâuṇa*, "secondary": but, according to our commentator, either of them may be denominated primary, although in fact, in this treatise, only the first of them is so regarded; and the secondary lunar month is that which is reckoned from any given lunar day to the next of the same name. This natural month, containing about twenty-nine and a half days, mean solar time, is then divided into thirty lunar days (*tithi*), and this division, although of so unnatural and arbitrary a character, the lunar days beginning and ending at any moment of the natural day and night, is, to the Hindu, of the most prominent practical importance, since by it are regulated the performance of many religious ceremonies (see below, xiv. 13), and upon it depend the chief considerations of propitious and unpropitious times, and the like. Of the lunar year of twelve lunar months, however, we know of no use made in India, either formerly or now, except as it has been introduced and employed by the Mohammedans.

Finally, the year last mentioned, the solar year, is that by which time is ordinarily reckoned in India. It is, however, not the tropical solar year, which we employ, but the sidereal, no account being made of the precession of the equinoxes. The solar month is measured by the continuance of the sun in each successive sign, and varies, according to the rapidity of his motion, from about twenty-nine and a third, to a little more than thirty-one and a half, days. There is no day corresponding to this measure of the month and of the year.

In the ordinary reckoning of time, these elements are variously combined. Throughout Southern India (see Warren's Kâla Sankalita, Madras: 1825, p. 4, etc.), the year and month made use of are the solar, and the day the civil; the beginning of each month and year being counted, in practice, from the sunrise nearest to the moment of their actual commencement. In all Northern India the year is luni-solar; the month is lunar, and is divided into both lunar and civil days; the year is composed of a variable number of months, either twelve or thirteen, beginning always with the lunar month of which the commencement next precedes the true commencement of the sidereal year. But, underneath this division, the division of the actual sidereal year into twelve solar months is likewise kept up, and to maintain the concurrence of the civil and lunar days, and the lunar and solar months, is a process of great complexity, into the details of which we need not enter here (see Warren, as above, p. 57, etc.). It will be seen later in this chapter (vv. 48–51) that the Sûrya-Siddhânta reckons time by this latter system, by the combination of civil, lunar, and sidereal elements.

13. . . . This is called a day of the gods.

14. The day and night of the gods and of the demons are mutually opposed to one another. Six times sixty of them are a year of the gods, and likewise of the demons.

"This is called," etc.: that is, as the commentary explains, the year composed of twelve solar months, as being those last mentioned; the sidereal year. It appears to us very questionable whether, in the first instance, anything more was meant by calling the year a day of the gods than to intimate that those beings of a higher order reckoned time upon a grander scale: just as the month was said to be a day of the Fathers, or Manes (xiv. 14), the Patriarchate (v. 18), a day of the Patriarchs (xiv. 21), and the Æon (v. 20), a day of Brahma; all these being familiar Puranic designations. In the astronomical reconstruction of the Puranic system, however, a physical meaning has been given to this day of the gods: the gods are made to reside at the north pole, and the demons at the south; and then, of course, during the half-year when the sun is north of the equator, it is day to the gods and night to the demons; and during the other half-year, the contrary. The subject is dwelt upon at some length in the twelfth chapter (xii. 45, etc.). To make such a division accurate, the year ought to be the tropical, and not the sidereal; but the author of the Sûrya-Siddhânta has not yet begun to take into account the precession. See what is said upon this subject in the third chapter (vv. 9–10).

The year of the gods, or the divine year, is employed only in describing the immense periods of which the statement now follows.

15. Twelve thousand of these divine years are denominated a Quadruple Age (*caturyuga*); of ten thousand times four hundred and thirty-two solar years

16. Is composed that Quadruple Age, with its dawn and twilight. The difference of the Golden and the other Ages, as measured by the difference in the number of the feet of Virtue in each, is as follows:

17. The tenth part of an Age, multiplied successively by four, three, two, and one, gives the length of the Golden and the other Ages, in order: the sixth part of each belongs to its dawn and twilight.

The period of 4,320,000 years is ordinarily styled Great Age (*mahâyuga*), or, as above in two instances, Quadruple Age (*caturyuga*). In the Sûrya-Siddhânta, however, the former term is not once found, and the latter occurs only in these verses; elsewhere, Age (*yuga*) alone is employed to denote it; and always denotes it, unless expressly limited by the name of the Golden (*kṛta*) Age.

The composition of the Age, or Great Age, is then as follows :

	Divine years.		Solar years.	
Dawn,	400		144,000	
Golden Age (*kṛta yuga*),	4000		1,440,000	
Twilight,	400		144,000	
Total duration of the Golden Age,		4,800		1,728,000
Dawn,	300		108,000	
Silver Age (*tretâ yuga*),	3000		1,080,000	
Twilight,	300		108,000	
Total duration of the Silver Age,		3,600		1,296,000
Dawn,	200		72,000	
Brazen Age (*dvâpara yuga*),	2000		720,000	
Twilight,	200		72,000	
Total duration of the Brazen Age,		2,400		864,000
Dawn,	100		36,000	
Iron Age (*kali yuga*),	1000		360,000	
Twilight,	100		36,000	
Total duration of the Iron Age,		1,200		432,000
Total duration of a Great Age,		12,000		4,320,000

Neither of the names of the last three ages is once mentioned in the Sûrya-Siddhânta. The first and last of the four are derived from the game of dice: *kṛta*, "made, won," is the side of the die marked with four dots—the lucky, or winning one; *kali* is the side marked with one dot only—the unfortunate, the losing one. In the other names, of which we do not know the original and proper meaning, the numerals *tri*, "three," and *dvâ*, "two," are plainly recognizable. The relation of the numbers four, three, two, and one, to the length of the several periods, as expressed in divine years, and also as compared with one another, is not less clearly apparent. The character attached to the different Ages by the Hindu mythological and legendary history so closely resembles that which is attributed to the Golden, Silver, Brazen, and Iron Ages, that we have not hesitated to transfer to them the latter appellations. An account of this character is given in Manu i. 81–86. During the Golden Age, Virtue stands firm upon four feet, truth and justice abound, and the life of man is four centuries; in each following Age Virtue loses a foot, and the length of life is reduced by a century, so that in the present, the Iron Age, she has but one left to hobble upon, while the extreme age attained by mortals is but a hundred years. See also Wilson's Vishṇu Purâṇa, p. 622, etc., for a description of the vices of the Iron Age.

This system of periods is not of astronomical origin, although the fixing of the commencement of the Iron Age, the only possibly historical point in it, is, as we shall see hereafter, the result of astronomical computation. Its arbitrary and artificial character is apparent. It is the system of the Purâṇas and of Manu, a part of the received Hindu cosmogony, to which astronomy was compelled to adapt itself.

We ought to remark, however, that in the text itself of Manu (i. 68–71) the duration of the Great Age, called by him Divine Age, is given as twelve thousand years simply, and that it is his commentator who, by asserting these to be divine years, brings Manu's cosmogony to an agreement with that of the Purâṇas. This is a strong indication that the divine year is an afterthought, and that the period of 4,320,000 years is an expansion of an earlier one of 12,000. Vast as this period is, however, it is far from satisfying the Hindu craving after infinity. We are next called upon to construct a new period by multiplying it by a thousand.

18. One and seventy Ages are styled here a Patriarchate (*manvantara*); at its end is said to be a twilight which has the number of years of a Golden Age, and which is a deluge.

19. In an Æon (*kalpa*) are reckoned fourteen such Patriarchs (*manu*) with their respective twilights; at the commencement of the Æon is a fifteenth dawn, having the length of a Golden Age.

The Æon is accordingly thus composed:

	Divine years.		Solar years.	
The introductory dawn,		4,800		1,728,000
Seventy-one Great Ages,	852,000		306,720,000	
A twilight,	4,800		1,728,000	
Duration of one Patriarchate,	856,800		308,448,000	
Fourteen Patriarchates,		11,995,200		4,318,272,000
Total duration of an Æon,		12,000,000		4,320,000,000

Why the factors fourteen and seventy-one were thus used in making up the Æon is not obvious; unless, indeed, in the division by fourteen is to be recognized the influence of the number seven, while at the same time such a division furnished the equal twilights, or intermediate periods of transition, which the Hindu theory demanded. The system, however, is still that of the Purâṇas (see Wilson's Vish. Pur. p. 24, etc.); and Manu (i. 72, 79) presents virtually the same, although he has not the term Æon (*kalpa*), but states simply that a thousand Divine Ages make up a day of Brahma, and seventy-one a Patriarchate. The term *manvantara*, "patriarchate," means literally "another Manu," or, "the interval of a Manu." Manu, a word identical in origin and meaning with our "man," became to the Hindus the name of a being personified as son of the Sun (*Vivasvant*) and progenitor of the human race. In each Patriarchate there arises a new Manu, who becomes for his own period the progenitor of mankind (see Wilson's Vish. Pur. p. 24).

20. The Æon, thus composed of a thousand Ages, and which brings about the destruction of all that exists, is styled a day of Brahma; his night is of the same length.

21. His extreme age is a hundred, according to this valuation of a day and a night....

We have already found indications of an assumed destruction of existing things at the termination of the lesser periods called the Age and the Patriarchate, in the necessity of a new revelation of virtue and knowledge for every Age, and of a new father of the human race for every Patriarchate. These are left, it should seem, to show us how the system of cosmical periods grew to larger and larger dimensions. The full development of it, as exhibited in the Purâṇas and here, admits only two kinds of destruction: the one occurring at the end of each Æon, or day of Brahma, when all creatures, although not the substance of the world, undergo dissolution, and remain buried in chaos during his night, to be created anew when his day begins again; the other taking place at the end of Brahma's life, when all matter even is resolved into its ultimate source.

According to the commentary, the "hundred" in verse 21 means a hundred years, each composed of three hundred and sixty days and nights, and not a hundred days and nights only, as the text might be understood to signify; since, in all statements respecting age, years are necessarily understood to be intended. The length of Brahma's life would be, then, 864,000,000,000 divine years, or 311,040,000,000,000 solar years. This period is also called in the Purâṇas a *para*, "extreme period," and its half a *parârdha* (see Wilson's Vish. Pur. p. 25); although the latter term has obtained also an independent use, as signifying a period still more enormous (ibid. p. 630). It is curious that the commentator does not seem to recognize the affinity with this period of the expression used in the text, *param âyuḥ*, "extreme age," but gives two different explanations of it, both of which are forced and unnatural.

The author of the work before us is modestly content with the number of years thus placed at his disposal, and attempts nothing farther. So is it also with the Purâṇas in general; although some of them, as the Vishṇu (Wilson, p. 637) assert that two of the greater *parârdhas* constitute only a day of Vishṇu, and others (ibid. p. 25) that Brahma's whole life is but a twinkling of the eye of Kṛshṇa or of Çiva.

21. . . . The half of his life is past; of the remainder, this is the first Æon.

22. And of this Æon, six Patriarchs (*manu*) are past, with their respective twilights; and of the Patriarch Manu son of Vivasvant, twenty-seven Ages are past;

23. Of the present, the twenty-eighth, Age, this Golden Age is past: from this point, reckoning up the time, one should compute together the whole number.

The designation of the part already elapsed of this immense period seems to be altogether arbitrary. It agrees in general with that given in the Purâṇas, and, so far as the Patriarchs and their periods are concerned, with Manu also. The name of the present Æon is *Vârâha*, "that of the boar," because Brahma, in performing anew at its commencement the act of creation, put on the form of that animal (see Wilson's Vish. Pur. p. 27, etc.). The one preceding is called the *Pâdma*, "that of the lotus." This nomenclature, however, is not universally accepted: under the word *kalpa*, in the Lexicon of Böhtlingk and Roth, may be found another system of names for these periods. Manu (i. 61, 62) gives the names of the Patriarchs of the past Patriarchates; the Purâṇas add other particulars respecting them, and also respecting those which are still to come (see Wilson's Vish. Pur. p. 259, etc.).

The end of the Golden Age of the current Great Age is the time at which the Sûrya-Siddhânta claims to have been revealed, and the epoch from which its calculations profess to commence. We will, accordingly, as the Sun directs, compute the number of years which are supposed to have elapsed before that period.

	Divine years.	Solar years.
Dawn of current Æon,	4,800	1,728,000
Six Patriarchates,	5,140,800	1,850,688,000
Twenty-seven Great Ages,	324,000	116,640,000
Total till commencement of present Great Age,	5,469,600	1,969,056,000
Golden Age of present Great Age,	4,800	1,728,000
Total time elapsed of current Æon,	5,474,400	1,970,784,000
Half Brahma's life,	432,000,000,000	155,520,000,000,000
Total time elapsed from beginning of Brahma's life to end of last Golden Age,	432,005,474,400	155,521,970,784,000

As the existing creation dates from the commencement of the current Æon, the second of the above totals is the only one with which the Sûrya-Siddhânta henceforth has any thing to do.

We are next informed that the present order of things virtually began at a period less distant than the commencement of the Æon.

24. One hundred times four hundred and seventy-four divine years passed while the All-wise was employed in creating the animate and inanimate creation, plants, stars, gods, demons, and the rest.

That is to say:

	Divine years.	Solar years.
From the total above given,	5,474,400	1,970,784,000
deduct the time occupied in creation,	47,400	17,064,000
the remainder is	5,427,000	1,953,720,000

This, then, is the time elapsed from the true commencement of the existing order of things to the epoch of this work. The deduction of this period as spent by the Deity in the work of creation is a peculiar feature of the Sûrya-Siddhânta. We shall revert to it later (see below, under vv. 29–34), as its significance cannot be shown until other data are before us.

25. The planets, moving westward with exceeding velocity, but constantly beaten by the asterisms, fall behind, at a rate precisely equal, proceeding each in its own path.

26. Hence they have an eastward motion. From the number of their revolutions is derived their daily motion, which is different according to the size of their orbits; in proportion to this daily motion they pass through the asterisms.

27. One which moves swiftly passes through them in a short time; one which moves slowly, in a long time. By their movement, the revolution is accounted complete at the end of the asterism Revatî.

We have here presented a part of the physical theory of the planetary motions, that which accounts for the mean motions: the theory is supplemented by the explanation given in the next chapter of the disturbing forces which give rise to the irregularities of movement. The earth is a sphere, and sustained immovable in the centre of the universe (xii. 32), while all the heavenly bodies, impelled by winds, or vortices, called provectors (ii. 3), revolve about it from east to west. In this general westward movement, the planets, as the commentary explains it, are, owing to their weight and the weakness of their vortices, beaten by the asterisms (*nakshatra* or *bha*, the groups of stars constituting the lunar mansions [see below, chapter viii], and used here, as in various other places, to designate the whole firmament of fixed stars), and accordingly fall behind (*lambante* = *labuntur*, *delabuntur*), as if from shame: and this is the explanation of their eastward motion, which is only apparent and relative, although wont to be regarded as real by those who do not understand the true causes of things. But now a new element is introduced into the theory, which does not seem entirely consistent with this view of the merely relative character of the eastward motion. It is asserted that the planets lag behind equally, or that each, moving in its own orbit, loses an equal amount daily, as compared with the asterisms. And we shall find farther on (xii. 73–89) that the dimensions of the planetary orbits are constructed upon this sole principle, of making the mean daily motion of each planet eastward to be the same in amount, namely 11,858.717 *yojanas:* the amount of westward motion being equal, in each case, to the difference between this amount and the whole orbit of the planet. Now if the Hindu idea of the symmetry and harmony of the universe demanded that the movements of the planets should be equal, it was certainly a very awkward and unsatisfactory way of complying with that demand to make the relative motions alone, as compared with the fixed stars, equal, and the real motions so vastly different from one another. We should rather expect that some method would have been devised for making the latter come out alike, and the former unlike, and the result of differences in the weights of the planets and the forces of the impelling currents. It looks as if this principle, and the conformity to it of the dimensions of the orbits, might have come from those who regarded the apparent daily motion as the real motion. But we know that Âryabhaṭṭa held the opinion that the earth revolved upon its axis, causing thereby the apparent westward motion of the heavenly bodies (see Colebrooke's Hindu Algebra, p. xxxviii; Essays, ii. 467), and so, of course, that the planets really moved eastward at an equal rate among the stars; and although the later astronomers are nearly unanimous against him, we cannot help surmising that the theory of the planetary orbits emanated from him or his school, or from some other of like opinion. It is not upon record, so far as we are aware, that any Hindu astronomer, of any period, held, as did some of the Greek philosophers (see Whewell's History of the Inductive Sciences, B. V. ch. i), a heliocentric theory.

The absolute motion eastward of all the planets being equal, their apparent motion is, of course, in the (inverse) ratio of their distance, or of the dimensions of their orbits.

The word translated "revolution" is *bhagaṇa,* literally "troop of asterisms;" the verbal root translated "pass through" is *bhuj,* "enjoy," from which comes also the common term for the daily motion of a planet, *bhukti,* literally "enjoyment." When a planet has "enjoyed the whole troop of asterisms," it has made a complete revolution.

The initial point of the fixed Hindu sphere, from which longitudes are reckoned, and at which the planetary motions are held by all the schools of Hindu astronomy to have commenced at the creation, is the end of the asterism Revatî, or the beginning of Açvinî (see chapter viii. for a full account of the asterisms). Its situation is most nearly marked by that of the principal star of Revatî, which, according to the Sûrya-Siddhânta, is 10′ to the west of it, but according to other authorities exactly coincides with it. That star is by all authorities identified with ζ Piscium, of which the longitude at present, as reckoned by us, from the vernal equinox, is 17° 54′. Making due allowance for the precession, we find that it coincided in position with the vernal equinox not far from the middle of the sixth century, or about A. D. 570. As such coincidence was the occasion of the point being fixed upon as the beginning of the sphere, the time of its occurrence marks approximately the era of the fixation of the sphere, and of the commencement of the history of modern Hindu astronomy. We say approximately only, because, in the first place, as will be shown in connection with the eighth chapter, the accuracy of the Hindu observations is not to be relied upon within a degree; and, in the second place, the limits of the asterisms being already long before fixed, it was necessary to take the beginning of some one of them as that of the sphere, and the Hindus may have regarded that of Açvinî as sufficiently near to the equinox for their purpose, when it was, in fact, two or three degrees, or yet more, remote from it, on either side; and each degree of removal would correspond to a difference in time of about seventy years.

In the most ancient recorded lists of the Hindu asterisms (in the texts of the Black Yajur-Veda and of the Atharva-Veda), Kṛttikâ, now the third, appears as the first. The time when the beginning of that asterism coincided with the vernal equinox would be nearly two thousand years earlier than that given above for the coincidence with it of the first point of Açvinî.

28. Sixty seconds (*vikalâ*) make a minute (*kalâ*); sixty of these, a degree (*bhâga*); of thirty of the latter is composed a sign (*râçi*); twelve of these are a revolution (*bhagaṇa*).

The Hindu divisions of the circle are thus seen to be the same with the Greek and with our own, and we shall accordingly make use, in translating, of our own familiar terms. Of the second (*vikalâ*) very little practical use is made; it is not more than two or three times alluded to in all the rest of the treatise. The minute (*kalâ*) is much more often called *liptâ* (or *liptikâ*); this is not an original Sanskrit word, but was borrowed from the Greek λεπτόν. The degree is called either *bhâga* or *ança;* both words, like the equivalent Greek word μοῖρα, mean a "part, portion." The proper signification of *râçi*, translated "sign," is simply "heap, quantity;" it is doubtless applied to designate a sign as being a certain number, or sum, of degrees, analogous to the use of *gaṇa* in *bhagaṇa* (explained above, in the last note), and of *râçi* itself in *dinarâçi*, "sum of days" (below, v. 53). In the Hindu description of an arc, the sign is as essential an element as the degree, and no arcs of greater length than thirty degrees are reckoned in degrees alone, as we are accustomed to reckon them. The Greek usage was the same. We shall hereafter see that the signs into which any circle of revolution is divided are named Aries, Taurus, etc., beginning from the point which is regarded as the starting point; so that these names are applied simply to indicate the order of succession of the arcs of thirty degrees.

29. In an Age (*yuga*), the revolutions of the sun, Mercury, and Venus, and of the conjunctions (*çîghra*) of Mars, Saturn, and Jupiter, moving eastward, are four million, three hundred and twenty thousand;

30. Of the moon, fifty-seven million, seven hundred and fifty-three thousand, three hundred and thirty-six; of Mars, two million, two hundred and ninety-six thousand, eight hundred and thirty-two;

31. Of Mercury's conjunction (*çîghra*), seventeen million, nine hundred and thirty-seven thousand, and sixty; of Jupiter, three hundred and sixty-four thousand, two hundred and twenty;

32. Of Venus's conjunction (*çîghra*), seven million, twenty-two thousand, three hundred and seventy-six; of Saturn, one hundred and forty-six thousand, five hundred and sixty-eight;

33. Of the moon's apsis (*ucca*), in an Age, four hundred and eighty-eight thousand, two hundred and three; of its node (*pâta*), in the contrary direction, two hundred and thirty-two thousand, two hundred and thirty-eight;

34. Of the asterisms, one billion, five hundred and eighty-two million, two hundred and thirty-seven thousand, eight hundred and twenty-eight. . . .

These are the fundamental and most important elements upon which is founded the astronomical system of the Sûrya-Siddhânta. We present them below in a tabular form, but must first explain the character of some of them, especially of some of those contained in verse 29, which we have omitted from the table.

The revolutions of the sun, and of Mars, Jupiter, and Saturn, require no remark, save the obvious one that those of the sun are in fact sidereal revolutions of the earth about the sun. To the sidereal revolutions of the moon we add also her synodical revolutions, anticipated from the next following passage (see v. 35). By the moon's "apsis" is to be understood her apogee; *ucca* is literally "height," i. e. "extreme distance:" the commentary explains it by *mandocca*, "apex of slowest motion:" as the same word is used to designate the aphelia of the planets, we were obliged to take in translating it the indifferent term apsis, which applies equally to both geocentric and heliocentric motion. The "node" is the ascending node (see ii. 7); the dual "nodes" is never employed in this work. But the apparent motions of the planets are greatly complicated by the fact, unknown to the Greek and the Hindu, that they are revolving about a centre about which the earth also is revolving. When any planet is on the opposite side of the sun from us, and is accordingly moving in space in a direction contrary to ours, the effect of our change of place is to increase the rate of its apparent change of place; again, when it is upon our side of the sun, and moving in the same direction with us, the effect of our motion is to retard its apparent motion, and even to cause it to seem to retrograde. This explains the "revolutions of the conjunction" of the three superior planets: their "conjunctions" revolve at the same rate with the earth, being always upon the opposite side of the sun from us; and when, by the combination of its own proper motion with that of its conjunction, the planet gets into the latter, its rate of apparent motion is greatest, becoming less in proportion as it removes from that position. The meaning of the word which we have translated "conjunction" is "swift, rapid:" a literal rendering of it would be "swift-point," or "apex of swiftest motion;" but, after much deliberation, and persevering trial of more than one term, we have concluded that "conjunction" was the least exceptionable word by which we could express it. In the case of the inferior planets, the revolution of the conjunction takes the place of the proper motion of the planet itself. By the definition given in verse 27, a planet must, in order to complete a revolution, pass through the whole zodiac; this Mercury and Venus are only able to do as they accompany the sun in his apparent annual revolution about the earth. To the Hindus, too, who had no idea of their proper movement about the sun, the annual motion must have seemed the principal one; and that by virtue of which, in their progress through the zodiac, they moved now faster and now slower, must have appeared only of secondary importance. The term "conjunction," as used in reference to these planets, must be restricted, of course, to the superior conjunction. The physical theories by which the effect of the conjunction (*çîghra*) is explained, are given in the next chapter. In the table that follows we have placed opposite each planet its own proper revolutions only.

It is farther to be observed that all the numbers of revolutions, excepting those of the moon's apsis and node, are divisible by four, so that, properly speaking, a quarter of an Age, or 1,080,000 years, rather than a whole Age, is their common period. This is a point of so much importance in the system of the Sûrya-Siddhânta, that we have added, in a second column, the number of revolutions in the lesser period.

In the third column, we add the period of revolution of each planet, as found by dividing by the number of revolutions of each the number of civil days in an Age (which is equal to the number of sidereal days, given in v. 34, diminished by the number of revolutions of the sun; see below, v. 37); they are expressed in days, nâdîs, vinâdîs, and respirations; the latter may be converted into sexagesimals of the third order by moving the decimal point one place farther to the right.

In the fourth column are given the mean daily motions.

We shall present later some comparison of these elements with those adopted in other systems of astronomy, ancient and modern.

Mean Motions of the Planets.

Planet.	Number of revolutions in 4,320,000 years.	Number of revolutions in 1,080,000 years.	Length of a revolution in mean solar time.	Mean daily motion.
			d n v p	° ′ ″ ‴ ⁗
Sun,	4,320,000	1,080,000	365 15 31 3.14	59 8 10 10.4
Mercury,	17,937,060	4,484,265	87 58 10 5.57	4 5 32 20 41.9
Venus,	7,022,376	1,755,594	224 41 54 5.06	1 36 7 43 37.3
Mars,	2,296,832	574,208	686 59 50 5.87	31 26 28 11.1
Jupiter,	364,220	91,055	4,332 19 14 2.09	4 59 8 48.6
Saturn,	146,568	36,642	10,765 46 23 0.41	2 0 22 53.4
Moon:				
sider. rev.	57,753,336	14,438,334	27 19 18 0.16	13 10 34 52 3.8
synod. rev.	53,433,336	13,358,334	29 31 50 0.70	12 11 26 41 53.4
rev. of apsis,	488,203	122,050¾	3,232 5 37 1.36	6 40 58 42.5
" " node,	232,238	58,059½	6,794 23 59 2.35	3 10 44 43.2

The arbitrary and artificial method in which the fundamental elements of the solar system are here presented is not peculiar to the Sûrya-Siddhânta; it is also adopted by all the other text-books, and is to be regarded as a characteristic feature of the general astronomical system of the Hindus. Instead of deducing the rate of motion of each planet from at least two recorded observations of its place, and establishing a genuine epoch, with the ascertained position of each at that time, they start with the assumption that, at the beginning of the present order of things, all the planets, with their apsides and nodes, commenced their movement together at that point in the heavens (near ζ Piscium, as explained above, under verse 27) fixed upon as the initial point of the sidereal sphere, and that they return, at certain fixed intervals, to a universal conjunction at the same point. As regards, however, the time when the motion commenced, the frequency of recurrence of the conjunction, and the date of that which last took place, there is discordance among the different authorities. With the Sûrya-Siddhânta, and the other treatises which adopt the same general method, the determining point of the whole system is the commencement of the current Iron Age (*kali yuga*); at that epoch the planets are assumed to have been in mean conjunction for the last time at the initial point of the sphere, the former conjunctions having taken place at intervals of 1,080,000 years previous. The instant at which the Age is made to commence is midnight on the meridian of Ujjayinî (see below, under v. 62), at the end of the 588,465th and beginning of the 588,466th day (civil reckoning) of the Julian Period, or between the 17th and 18th of February 1612 J.P., or 3102 B. C. (see below, under vv. 45–53, for the computation of the number of days since elapsed). Now, although no such conjunction as that assumed by the Hindu astronomers ever did or ever will take place, the planets were actually, at the time stated, approximating somewhat nearly to a general conjunction in the neighborhood of the initial point of the Hindu sphere; this is shown by the next table, in which we give their actual mean positions with reference to that point (including also those of the moon's apogee and node); they have been obligingly furnished us by Prof. Winlock, Superintendent of the American Ephemeris and Nautical Almanac. The positions of the primary planets are obtained by LeVerrier's times of sidereal revolution, given in the Annales de l'Observatoire, tom. ii (also in Biot's Astronomie, 3me édition, tom. v, 1857), that of the moon by Peirce's tables, and those of its apogee and node by Hansen's Tables de la Lune. The origin of the Hindu sphere is regarded as being 18° 5′ 8″ east of the vernal equinox of Jan. 1, 1860, and 50° 22′ 29″ west of that of Feb. 17, 3102 B. C., the precession in the interval being 68° 27′ 37″. We add, in a second column, the mean longitudes, as reckoned from the vernal equinox of the given date, for the sake of comparison with the similar data given by Bentley (Hind. Ast., p. 125) and by Bailly(Ast. Ind. et Or., pp. 111, 182), which we also subjoin.

Positions of the Planets, midnight, at Ujjayinî, Feb. 17–18, 3102 *B. C.*

Planet.	From beginning of Hindu sphere.			Longitude.			Bentley.			Bailly.		
	°	′	″	°	′	″	°	′	″	°	′	″
Sun,	− 7	51	48	301	45	43	301	1	1	301	5	57
Mercury,	− 41	3	26	268	34	5	267	35	26	261	14	21
Venus,	+ 24	58	59	334	36	30	333	44	37	334	22	18
Mars,	− 19	49	26	289	48	5	288	55	19	288	55	56
Jupiter,	+ 8	38	36	318	16	7	318	3	54	310	22	10
Saturn,	− 28	1	13	281	36	18	280	1	58	293	8	21
Moon,	− 1	33	41	308	3	50	306	53	42	300	51	16
do. apsis,	+ 95	19	21	44	56	42	61	12	26	61	13	33
do. node,	+198	24	45	148	2	16	144	38	32	144	37	41

The want of agreement between the results of the three different investigations illustrates the difficulty and uncertainty even yet attending inquiries into the positions of the heavenly bodies at so remote an epoch. It is very possible that the calculations of the astronomers who were the framers of the Hindu system may have led them to suppose the approach to a conjunction nearer than it actually was; but, however that may be, it seems hardly to admit of a doubt that the epoch was arrived at by astronomical calculation carried backward, and that it was fixed upon as the date of the last general conjunction, and made to determine the commencement of the present Age of the world, because the errors of the assumed positions of the planets at that time would be so small, and the number of years since elapsed so great, as to make the errors in the mean motions into which those positions entered as an element only trifling in amount.

The moon's apsis and node, however, were treated in a different manner. Their distance from the initial point of the sphere, as shown by the table, was too great to be disregarded. They were accordingly ingly exempted from the general law of a conjunction once in 1,080,000 years, and such a number of revolutions was assigned to them as should make their positions at the epoch come out, the one a quadrant, the other a half-revolution, in advance of the initial point of the sphere.

We can now see why the deduction spoken of above (v. 24), for time spent in creation, needed to be made. In order to bring all the planets to a position of mean conjunction at the epoch, the time previously elapsed must be an exact multiple of the lesser period of 1,080,000 years, or the quarter-Age; in order to give its proper position to the moon's apsis, that time must contain a certain number of whole Ages, which are the periods of conjunction of the latter with the planets, together with a remainder of three quarter-Ages; for the moon's node, in like manner, it must contain a certain number of half-Ages, with a remainder of one quarter-Age. Now the whole number of years elapsed between the beginning of the Æon and that of the current Iron Age is equal to 1826 quarter-Ages, with an odd surplus of 864,000 years: from it subtract an amount of time which shall contain this surplus, together with three, seven, eleven, fifteen, or the like (any number exceeding by three a multiple of four), quarter-Ages, and the remainder will fulfil the conditions of the problem. The deduction actually made is of fifteen periods + the surplus.

This deduction is a clear indication that, as remarked above (under v. 17), the astronomical system was compelled to adapt itself to an already established Puranic chronology. It could, indeed, fix the previously undetermined epoch of the commencement of the Iron Age, but it could not alter the arrangement of the preceding periods.

It is evident that, with whatever accuracy the mean positions of the planets may, at a given time, be ascertained by observation by the Hindu astronomers, their false assumption of a conjunction at the epoch of 3102 B. C. must introduce an element of error into their determination of the planetary motions. The annual amount of that error may indeed be small, owing to the remoteness of the epoch, and the great number of years among which the errors of assumed position are divided, yet it must in time grow to an amount not to be ignored or neglected even by observers so inaccurate, and theorists so unscrupulous, as the Hindus. This is actually the case with the elements of the Sûrya-Siddhânta; the positions of the planets, as calculated by them for the present time, are in some cases nearly 9° from the true places. The later astronomers of India, however, have known how to deal with such difficulties without abrogating their ancient text-books. As the Sûrya-Siddhânta is at present employed in astronomical calculations, there are introduced into its planetary elements certain corrections, called *bîja* (more properly *vîja;* the word means literally "seed"; we do not know how it arrived at its present significations in the mathematical language). That this was so, was known to Davis (As. Res., ii. 236), but he was unable to state the amount of the corrections, excepting in the case of the moon's apsis and node (ibid., p. 275). Bentley (Hind. Ast., p. 179) gives them in full, and upon his authority we present them in the annexed table. They are in the form, it will be noticed, of additions to, or subtractions from, the number of revolutions given for an Age, and the numbers are all divisible by four, in order not to interfere with the calculation by the lesser period of 1,080,000 years. We have added the corrected number of revolutions, for both the greater and lesser period, the corrected time of revolution, expressed in Hindu divisions of the day, and the corrected amount of mean daily motion.

These corrections were first applied, according to Mr. Bentley (As. Res., viii. 220), about the beginning of the sixteenth century; they are presented by several treatises of that as well as of later date, not having been yet superseded by others intended to secure yet greater correctness.

Mean Motions of the Planets as corrected by the bîja.

Planet.	Correction.	Corrected number of revolutions in 4,320,000 years.	Corrected number of revolutions in 1,080,000 years.	Corrected time of revolution.				Corrected daily motion.				
				d	n	v	p	°	′	″	‴	⁗
Sun,	0	4,320,000	1,080,000	365	15	31	3.14		59	8	10	10.4
Mercury,	− 16	17,937,044	4,484,261	87	58	11	1.26	4	5	32	19	54.5
Venus,	− 12	7,022,364	1,755,591	224	41	56	1.35	1	36	7	43	1.8
Mars,	0	2,296,832	574,208	686	59	50	5.87		31	26	28	11.1
Jupiter,	− 8	364,212	91,053	4,332	24	56	5.56		4	59	8	24.9
Saturn,	+ 12	146,580	36,645	10,764	53	30	1.11		2	0	23	28.9
Moon,	0	57,753,336	14,438,334	27	19	18	0.16	13	10	34	52	3.8
" apsis,	− 4	488,199	122,049¾	3,232	7	12	3.37		6	40	58	30.7
" node,	+ 4	232,242	58,060½	6,794	16	58	0.66		3	10	44	55.0

We need not, however, rely on external testimony alone for information as to the period when this correction was made. If the attempt to modify the elements in such a manner as to make them give the true positions of the planets at the time when they were so modified was in any tolerable degree successful, we ought to be able to discover by calculation the date of the alteration. If we ascertain for any given time the positions of the planets as given by the system, and compare them with the true positions as found by our best modern methods, and if we then divide the differences of position by the differences in the mean motions, we shall discover, in each separate case, when the error was or will be reduced to nothing. The results of such a calculation, made for Jan. 1, 1860, are given below, under v. 67. We see there that, if regard is had only to the absolute errors in the positions of the planets, no conclusion of value can be arrived at; the discrepancies between the dates of no error are altogether too great to allow of their being regarded as indicating any definite epoch of correction. If, on the other hand, we assume the place of the sun to have been the standard by which the positions of the other planets were tested, the dates of no error are seen to point quite distinctly to the first half of the sixteenth century as the time of the correction, their mean being A. D. 1541. Upon this assumption, also, we see why no correction of *bîja* was applied to Mars or to the moon: the former had, at the given time, only just passed his time of complete accordance with the sun, and the motion of the moon was also already so closely adjusted to that of the sun, that the difference between their errors of position is even now less than 10′. Nor is there any other supposition which will explain why the serious error in the position of the sun himself was overlooked at the time of the general correction, and why, by that correction, the absolute errors of position of more than one of the planets are made greater than they would otherwise have been, as is the case. It is, in short, clearly evident that the alteration of the elements of the Sûrya-Siddhânta which was effected early in the sixteenth century, was an adaptation of the errors of position of the other planets to that of the sun, assumed to be correct and regarded as the standard.

Now if it is possible by this method to arrive approximately at the date of a correction applied to the elements of a Siddhânta, it should be possible in like manner to arrive at the date of those elements themselves. For, owing to the false assumption of position at the epoch, there is but one point of time at which any of the periods of revolution will give the true place of its planet: if, then, as is to be presumed, the true places were nearly determined when any treatise was composed, and were made to enter as an element into the construction of its system, the comparison of the dates of no error will point to the epoch of its composition. The method, indeed, as is well known to all those who have made any studies in the history of Hindu astronomy, has already been applied to this purpose, by Mr. Bentley. It was first originated and put forth by him (in vol. vi. of the Asiatic Researches) at a time when the false estimate of the age and value of the Hindu astronomy presented by Bailly was still the prevailing one in Europe; he strenuously defended it against more than one attack (As. Res., viii, and Hind. Ast.), and finally employed it very extensively in his volume on the History of Hindu Astronomy, as a means of determining the age of the different Siddhântas. We present below the table from which, in the latter work (p. 126), he deduces the age of the Sûrya-Siddhânta; the column of approximate dates of no error we have ourselves added.

Bentley's Table of Errors in the Positions of the Planets, as calculated, for successive periods, according to the Sûrya-Siddhânta.

Planet.	Iron Age 0, B. C. 3102.	I. A. 1000, B. C. 2102.	I. A. 2000, B. C. 1102.	I. A. 3000, B. C. 102.	I. A. 3639, A. D. 538.	I. A. 4192, A. D. 1091.	When correct
	° ′ ″	° ′ ″	° ′ ″	° ′ ″	° ′ ″	° ′ ″	A. D.
Mercury,	+33 25 35	+25 9 52	+16 54 9	+8 38 26	+3 21 40	−1 12 28	945
Venus,	−32 43 36	−24 37 31	−16 31 26	−8 25 21	−3 14 45	+1 14 3	939
Mars,	+12 5 42	+ 9 26 32	+ 6 47 22	+4 8 12	+2 26 30	+0 58 29	1458
Jupiter,	−17 2 53	−12 44 16	− 8 25 39	−4 7 2	−1 21 47	+0 41 14	906
Saturn,	+20 59 3	+15 43 20	+10 27 37	+5 11 54	+1 50 10	−1 4 25	887
Moon,	− 5 52 41	− 3 50 48	− 2 9 17	−0 52 33	−0 18 30	−0 0 11	1097
" apsis,	−30 11 25	−23 9 36	−16 7 47	−9 5 58	−4 36 26	−0 43 10	1193
" node,	+23 37 31	+17 59 21	+12 31 11	+7 3 1	+3 33 19	+0 31 50	1188

From an average of the results thus obtained, Bentley draws the conclusion that the Sûrya-Siddhânta dates from the latter part of the eleventh century; or, more exactly, A. D. 1091.

The general soundness of Bentley's method will, we apprehend, be denied at the present time by few, and he is certainly entitled to not a little credit for his ingenuity in devising it, for the persevering industry shown in its application, and for the zeal and boldness with which he propounded and defended it. He succeeded in throwing not a little light upon an obscure and misapprehended subject, and his investigations have contributed very essentially to our present understanding of the Hindu systems of astronomy. But the details of his work are not to be accepted without careful testing, and his general conclusions are often unsound, and require essential modification, or are to be rejected altogether. This we will attempt to show in connection with his treatment of the Sûrya-Siddhânta.

In the first place, Bentley has made a very serious error in that part of his calculations which concerns the planet Mercury. As that planet was, at the epoch, many degrees behind its assumed place, it was necessary, of course, to assign to it a slower than its true rate of motion. But the rate actually given it by the text is not quite enough slower, and, instead of exhausting the original error of position in the tenth century of our era, as stated by Bentley, would not so dispose of it for many hundred years yet to come. Hence the correction of the *bîja*, as reported by Bentley himself, instead of giving to Mercury, as to all the rest, a more correct rate of motion, is made to have the contrary effect, in order the sooner to run out the original error of assumed position, and produce a coincidence between the calculated and the true places of the planet.

In the case of the other planets, the times of no error found by Bentley agree pretty nearly with those which we have ourselves obtained, both by calculating backward from the errors of A. D. 1860, and by calculating downward from those of B. C. 3102, and which are presented in the table given under verse 67. Upon comparing the two tables, however, it will be seen at once that Bentley's conclusions are drawn, not from the sidereal errors of position of the planets, but from the errors of their positions as compared with that of the sun, and that of the sun's own error he makes no account at all. This is a method of procedure which certainly requires a much fuller explanation and justification than he has seen fit anywhere to give of it. The Hindu sphere is a sidereal one, and in no wise bound to the movement of the sun. The sun, like the other planets, was not in the position assumed for him at the epoch of 3102 B. C., and consequently the rate of motion assigned to him by the system is palpably different from the real one: the sidereal year is about three minutes and a half too long. Why then should the sun's error be ignored, and the sidereal motions of the other planets considered only with reference to the incorrect rate of motion established for him? It is evident that Bentley ought to have taken fully into consideration the sun's position also, and to have shown either that it gave a like result with those obtained from the other planets, or, if not, what was the reason of the discrepancy. By failing to do so, he has, in our opinion, omitted the most fundamental datum of the whole calculation, and the one which leads to the most important conclusions. We have seen, in treating of the *bîja*, that it has been the aim of the modern Hindu astronomers, leaving the sun's error untouched, to amend those of the other planets to an accordance with it. Now, as things are wont to be managed in the Hindu literature, it would be no matter for surprise if such corrections were incorporated into the text itself: had not the Sûrya-Siddhânta been, at the beginning of the sixteenth century, so widely distributed, and its data so universally known, and had not the Hindu science outlived already that growing and productive period of its history when a school of astronomy might put forth a corrected text of an ancient authority, and expect to see it make its way to general acceptance, crowding out, and finally causing to disappear, the older version—such a process of alteration might, in our view, have passed upon it, and such a text might have been handed down to our time as Bentley would have pronounced, upon internal evidence, to have been composed early in the sixteenth century; while, nevertheless, the original error of the sun would remain, untouched and increasing, to indicate what was the true state of the case.

But what is the actual position of things with regard to our Siddhânta? We find that it presents us a set of planetary elements, which, when tested by the errors of position, in the manner already explained, do not appear to have been constructed so as to give the true sidereal positions at any assignable epoch, but which, on the other hand, exhibit evidences of an attempt to bring the places of the other planets into an accordance with that of the sun, made sometime in the tenth or eleventh century—the precise time is very doubtful, the discrepancies of the times of no error being far too great to give a certain result. Now it is as certain as anything in the history of Sanskrit literature can be, that there was a Sûrya-Siddhânta in existence long before that date; there is also evidence in the references and citations of other astronomical works (see Colebrooke, Essays, ii. 484; Hind. Alg., p. l) that there have been more versions than one of a treatise bearing the title; and we have seen above, in verse 9, a not very obscure intimation that the present work does not present precisely the same elements which had been accepted formerly as those of the Sûrya-Siddhânta. What can lie nearer, then, than to suppose that in the tenth or eleventh century a correction of *bîja* was calculated for application to the elements of the Siddhânta, and was then incorporated into the text, by the easy alteration of four or five of its verses; and accordingly, that while the comparative errors of the other planets betray the date of the correction, the absolute error of the sun indicates approximately the true date of the treatise?

In our table, the time of no error of the sun is given as A. D. 250. The correctness of this date, however, is not to be too strongly insisted upon, being dependent upon the correctness with which the sun's place was first determined, and then referred to the point assumed as the origin of the sphere. It was, of course, impossible to observe directly when the sun's centre, by his mean motion, was 10′ east of ζ Piscium, and there are grave errors in the determination by the Hindus of the distances from that point of the other points fixed by them in their zodiac. And a mistake of 1° in the determination of the sun's place would occasion a difference of 425 years in the resulting date of no error. We shall have occasion to recur to this subject in connection with the eighth chapter.

There is also an alternative supposition to that which we have made above, respecting the conclusion from the date of no error of the sun. If the error in the sun's motion were a fundamental feature of the whole Hindu system, appearing alike in all the different text-books of the science, that date would point to the origin rather of the whole system than of any treatise which might exhibit it. But although the different Siddhântas nearly agree with one another respecting the length of the sidereal year, they do not entirely accord, as is made evident by the following statement, in which are included all the authorities to which we have access, either in the original, or as reported by Colebrooke, Bentley, and Warren:

Authority.	Length of sidereal year.				Error.	
Sûrya-Siddhânta,	365d	6h	12m	36s.56	+ 3m	25s.81
Pâuliça-Siddhânta,	365	6	12	36	+ 3	25.25
Pârâçara-Siddhânta,	365	6	12	31.50	+ 3	20.75
Ârya-Siddhânta,	365	6	12	30.84	+ 3	20.09
Laghu-Ârya-Siddhânta,	365	6	12	30	+ 3	19.25
Siddhânta-Çiromaṇi,	365	6	12	9	+ 2	58.25

The first five of these might be regarded as unimportant variations of the same error, but it would seem that the last is an independent determination, and one of later date than the others; while, if all are independent, that of the Sûrya-Siddhânta has the appearance of being the most ancient. Such questions as these, however, are not to be too hastily decided, nor from single indications merely; they demand the most thorough investigation of each different treatise, and the careful collection of all the evidence which can be brought to bear upon them.

Here lies Bentley's chief error. He relied solely upon his method of examining the elements, applying even that, as we have seen, only partially and uncritically, and never allowing his results to be controlled or corrected by evidence of any other character. He had, in fact, no philology, and he was deficient in sound critical judgment. He thoroughly misapprehended the character of the Hindu astronomical literature, thinking it to be, in the main, a mass of forgeries framed for the purpose of deceiving the world respecting the antiquity of the Hindu people. Many of his most confident conclusions have already been overthrown by evidence of which not even he would venture to question the verity, and we are persuaded that but little of his work would stand the test of a thorough examination.

The annexed table presents a comparison of the times of mean sidereal revolution of the planets assumed by the Hindu astronomy, as represented by two of its principal text-books, with those adopted by the great Greek astronomer, and those which modern science has established. The latter are, for the primary planets, from Le Verrier; for the moon, from Nichol (Cyclopedia of the Physical Sciences, London : 1857). Those of Ptolemy are deduced from the mean daily rates of motion in longitude given by him in the Syntaxis, allowing for the movement of the equinox according to the false rate adopted by him, of 36″ yearly.

Comparative Table of the Sidereal Revolutions of the Planets.

Planet.	Sûrya-Siddhânta.				Siddhânta-Çiromani				Ptolemy.				Moderns.			
	d	h	m	s	d	h	m	s	d	h	m	s	d	h	m	s
Sun,	365	6	12	36.6	365	6	12	9.0	365	36	9	48.6	365	6	9	10.8
Mercury,	87	23	16	22.3	87	23	16	41.5	87	23	16	42.9	87	23	15	43.9
Venus,	224	16	45	56.2	224	16	45	1.9	224	16	51	56.8	224	16	49	8 0
Mars,	686	23	56	23.5	686	23	57	1.5	686	23	31	56.1	686	23	30	41.4
Jupiter,	4,332	7	41	44.4	4,332	5	45	43.7	4,332	18	9	10.5	4,332	14	2	8.6
Saturn,	10,765	18	33	13.6	10,765	19	33	56.5	10,758	17	48	14.9	10,759	5	16	32.2
Moon :																
sid. rev.	27	7	43	12.6	27	7	43	12.1	27	7	43	12.1	27	7	43	11.4
synod. rev.	29	12	44	2.8	29	12	44	2.3	29	12	44	3.3	29	12	44	2.9
rev. of apsis,	3,232	2	14	53.4	3,232	17	37	6.0	3,232	9	52	13.6	3,232	13	48	29.6
" " node,	6,794	9	35	45.4	6,792	6	5	41.9	6,799	23	18	39.4	6,798	6	41	45.6

In the additional notes at the end of the work, we shall revert to the subject of these data, and of the light thrown by them upon the origin and age of the system.

34. . . . The number of risings of the asterisms, diminished by the number of the revolutions of each planet respectively, gives the number of risings of the planets in an Age.

35. The number of lunar months is the difference between the number of revolutions of the sun and of the moon. If from it the number of solar months be subtracted, the remainder is the number of intercalary months.

36. Take the civil days from the lunar, the remainder is the number of omitted lunar days (*tithikshaya*). From rising to rising of the sun are reckoned terrestrial civil days;

37. Of these there are, in an Age, one billion, five hundred and seventy-seven million, nine hundred and seventeen thousand, eight hundred and twenty-eight; of lunar days, one billion, six hundred and three million, and eighty;

38. Of intercalary months, one million, five hundred and ninety-three thousand, three hundred and thirty-six; of omitted lunar days, twenty-five million, eighty-two thousand, two hundred and fifty-two;

39. Of solar months, fifty-one million, eight hundred and forty thousand. The number of risings of the asterisms, diminished by that of the revolutions of the sun, gives the number of terrestrial days.

40. The intercalary months, the omitted lunar days, the sidereal, lunar, and civil days—these, multiplied by a thousand, are the number of revolutions, etc., in an Æon.

The data here given are combinations of, and deductions from, those contained in the preceding passage (vv. 29–34). For convenience of reference, we present them below in a tabular form.

	In 4,320,000 years.	In 1,080,000 years.
Sidereal days,	1,582,237,828	395,559,457
deduct solar revolutions,	4,320,000	1,080,000
Natural, or civil days,	1,577,917,828	394,479,457
Sidereal solar years,	4,320,000	1,080,000
multiply by no. of solar months in a year,	12	12
Solar months,	51,840,000	12,960,000
Moon's sidereal revolutions,	57,753,336	14,438,334
deduct solar revolutions,	4,320,000	1,080,000
Synodical revolutions, lunar months,	53,433,336	13,358,334
deduct solar months,	51,840,000	12,960,000
Intercalary months,	1,593,336	398,334

4

Lunar months,	53,433,336	13,358,334
multiply by no. of lunar days in a month,	30	30
Lunar days,	1,603,000,080	400,750,020
deduct civil days,	1,577,917,828	394,479,457
Omitted lunar days,	25,082,252	6,270,563

We add a few explanatory remarks respecting some of the terms employed in this passage, or the divisions of time which they designate.

The natural day, nycthemeron, is, for astronomical purposes, reckoned in the Sûrya-Siddhânta from midnight to midnight, and is of invariable length; for the practical uses of life, the Hindus count it from sunrise to sunrise; which would cause its duration to vary, in a latitude as high as our own, sometimes as much as two or three minutes. As above noticed, the system of Brahmagupta and some others reckon the astronomical day also from sunrise.

For the lunar day, the lunar and solar month, and the general constitution of the year, see above, under verse 13. The lunar month, which is the one practically reckoned by, is named from the solar month in which it commences. An intercalation takes place when two lunar months begin in the same solar month: the former of the two is called an intercalary month (*adhimâsa*, or *adhimâsaka*, "extra month"), of the same name as that which succeeds it.

The term "omitted lunar day" (*tithikshaya*, "loss of a lunar day") is explained by the method adopted in the calendar, and in practice, of naming the days of the month. The civil day receives the name of the lunar day which ends in it; but if two lunar days end in the same solar day, the former of them is reckoned as loss (*kshaya*), and is omitted, the day being named from the other.

41. The revolutions of the sun's apsis (*manda*), moving eastward, in an Æon, are three hundred and eighty-seven; of that of Mars, two hundred and four; of that of Mercury, three hundred and sixty-eight;

42. Of that of Jupiter, nine hundred; of that of Venus, five hundred and thirty-five; of the apsis of Saturn, thirty-nine. Farther, the revolutions of the nodes, retrograde, are:

43. Of that of Mars, two hundred and fourteen; of that of Mercury, four hundred and eighty-eight; of that of Jupiter, one hundred and seventy-four; of that of Venus, nine hundred and three;

44. Of the node of Saturn, the revolutions in an Æon are six hundred and sixty-two: the revolutions of the moon's apsis and node have been given here already.

In illustration of the curious feature of the Hindu system of astronomy presented in this passage, we first give the annexed table; which shows the number of revolutions in the Æon, or period of 4,320,000,000 years, assigned by the text to the apsis and node of each planet, the resulting time of revolution, the number of years which each would require to pass through an arc of one minute, and the position of each, according to the system, in 1850; the latter being reckoned in our method, from the vernal equinox. Farther are added the actual positions for Jan. 1, 1850, as given by Biot (Traité d'Astronomie, tom. v. 529); and finally, the errors of the positions as determined by this Siddhânta.

Table of Revolutions and Present Position of the Apsides and Nodes of the Planets.

Planet.	No. of rev. in an Æon.	Time of revolution, in years.	No. of years to 1′ of motion.	Resulting position, A. D. 1850.	True position, A. D. 1850.	Error of Hindu position.
Apsides:				° ′	° ′	° ′
Sun,	387	11,162,790.7	516.8	95 4	100 22	− 5 16
Mercury,	368	11,739,130.4	543.5	238 15	255 7	− 16 52
Venus,	535	8,074,766.4	373.8	97 39	309 24	−211 45
Mars,	204	21,176,470.6	980.4	147 49	153 18	− 5 29
Jupiter,	900	4,800,000.0	222.2	189 9	191 55	− 2 46
Saturn,	39	110,769,230.8	5128.2	254 24	270 6	− 15 42
Nodes:						
Mercury,	488	8,852,459.0	409.8	38 27	46 33	− 8 6
Venus,	903	4,784,053.2	221.5	77 26	75 19	+ 2 7
Mars,	214	20,186,915.9	934.6	57 49	48 23	+ 9 26
Jupiter,	174	24,827,586.2	1149.4	97 26	98 54	− 1 28
Saturn,	662	6,525,678.2	302.1	118 7	112 22	+ 5 45

A mere inspection of this table is sufficient to show that the Hindu astronomers did not practically recognize any motion of the apsides and nodes of the planets; since, even in the case of those to which they assigned the most rapid motion, two thousand years, at the least, would be required to produce such a change of place as they, with their imperfect means of observation, would be able to detect.

This will, however, be made still more clearly apparent by the next following table, in which we give the positions of the apsides and nodes as determined by four different text-books of the Hindu science, for the commencement of the Iron Age.

Positions of the Apsides and Nodes of the Planets, according to Different Authorities, at the Commencement of the Iron Age,* 3102 *B. C.

Planet.	Sûrya-Siddhânta.	Siddhânta-Çiromaṇi.	Ârya-Siddhânta.	Pârâçara-Siddhânta.
Apsides:	(rev.) s ° ′ ″	(rev.) s ° ′ ″	(rev.) s ° ′ ″	(rev.) s ° ′ ″
Sun,	(175) 2 17 7 48	(219) 2 17 45 36	(210) 2 17 45 36	(219) 2 17 45 36
Mercury,	(166) 7 10 19 12	(151) 7 14 47 2	(154) 7 0 14 24	(162) 7 0 40 19
Venus,	(242) 2 19 39 0	(298) 2 21 2 10	(300) 0 17 16 48	(240) 2 20 42 43
Mars,	(92) 4 9 57 36	(133) 4 8 18 14	(136) 4 3 50 24	(149) 4 2 43 26
Jupiter,	(407) 5 21 0 0	(390) 5 22 15 36	(378) 5 22 48 0	(448) 5 22 35 24
Saturn,	(17) 7 26 36 36	(18) 8 20 53 31	(16) 4 29 45 36	(24) 7 28 14 52
Nodes:				
Mercury,	(221–) 0 20 52 48	(238–) 0 21 20 53	(239–) 0 20 9 36	(296–) 0 21 1 26
Venus,	(409–) 2 0 1 48	(408–) 2 0 5 2	(432–) 2 0 28 48	(408–) 2 0 5 2
Mars,	(97–) 1 10 8 24	(122–) 0 21 59 46	(136–) 1 10 19 12	(112–) 1 9 3 36
Jupiter,	(79–) 2 19 44 24	(29–) 2 22 2 38	(44–) 2 20 38 24	(87–) 2 21 43 12
Saturn,	(300–) 3 10 37 12	(267–) 3 13 23 31	(283–) 3 10 48 0	(288–) 3 10 26 24

The data of the Ârya and Pârâçara Siddhântas, from which the positions given in the table are calculated, are derived from Bentley (Hind. Ast. pp. 139, 144). To each position is prefixed the number of completed revolutions; or, in the case of the nodes, of which the motion is retrograde, the number of whole revolutions of which each falls short by the amount expressed by its position.

The almost universal disagreement of these four authorities with respect to the number of whole revolutions accomplished, and their general agreement as to the remainder, which determines the position,* prove that the Hindus had no idea of any motion of the apsides and nodes of the planets as an actual and observable phenomenon; but, knowing that the moon's apsis and node moved, they fancied that the symmetry of the universe required that those of the other planets should move also; and they constructed their systems accordingly. They held, too, as will be seen at the beginning of the second chapter, that the nodes and apsides, as well as the conjunctions (*çîghra*), were beings, stationed in the heavens, and exercising a physical influence over their respective planets, and, as the conjunctions revolved, so must these also. In framing their systems, then, they assigned to these points such a number of revolutions in an Æon as should, without attributing to them any motion which admitted of detection, make their positions what they supposed them actually to be. The differences in respect to the number of revolutions were in part rendered necessary by the differences of other features of the systems; thus, while that of the Siddhânta-Çiromaṇi makes the planetary motions commence at the beginning of the Æon, by that of the Sûrya-Siddhânta they commence 17,064,000 years later (see above, v. 24), and by that of the Ârya-Siddhânta, 3,024,000 years later (Bentley, Hind. Ast. p. 139): in part, however, they are merely arbitrary; for, although the Pârâçara-Siddhânta agrees with the Siddhânta-Çiromaṇi as to the time of the beginning of things, its numbers of revolutions correspond only in two instances with those of the latter.

It may be farther remarked, that the close accordance of the different astronomical systems in fixing the position of points which are so difficult of observation and deduction as the nodes and apsides, strongly indicates, either that the Hindus were remarkably accurate observers, and all arrived independently at a near approximation to the truth, or that some one of them was followed as an authority by the others, or that all alike derived their data from a common source, whether native or foreign. We reserve to the end of this work the discussion of these different possibilities, and the presentation of data which may tend to settle the question between them.

45. Now add together the time of the six Patriarchs (*manu*), with their respective twilights, and with the dawn at the commencement of the Æon (*kalpa*); farther, of the Patriarch Manu, son of Vivasvant,

---

* It is altogether probable that, in the two cases where the Ârya-Siddhânta seems to disagree with the others, its data were either given incorrectly by Bentley's authority, or have been incorrectly reported by him.

46. The twenty-seven Ages (*yuga*) that are past, and likewise the present Golden Age (*kṛta yuga*); from their sum subtract the time of creation, already stated in terms of divine years,

47. In solar years: the result is the time elapsed at the end of the Golden Age; namely, one billion, nine hundred and fifty-three million, seven hundred and twenty thousand solar years.

We have already presented this computation, in full, in the notes to verses 23 and 24.

48. To this, add the number of years of the time since past....

As the Sûrya-Siddhânta professes to have been revealed by the Sun about the end of the Golden Age, it is of course precluded from taking any notice of the divisions of time posterior to that period: there is nowhere in the treatise an allusion to any of the eras which are actually made use of by the inhabitants of India in reckoning time, with the exception of the cycle of sixty years, which, by its nature, is bound to no date or period (see below, v. 55). The astronomical era is the commencement of the Iron Age, the epoch, according to this Siddhânta, of the last general conjunction of the planets; this coincides, as stated above (under vv. 29–34) with Feb. 18, 1612 J. P., or 3102 B. C. From that time will have elapsed, upon the eleventh of April, 1859, the number of 4960 complete sidereal years of the Iron Age. The computation of the whole period, from the beginning of the present order of things, is then as follows:

From end of creation to end of last Golden Age,		1,953,720,000
Silver Age,	1,296,000	
Brazen Age,	864,000	
Of Iron Age,	4,960	2,164,960
Total from end of creation to April, 1859,		1,955,884,960

Since the Sûrya-Siddhânta, as will appear from the following verses, reckons by luni-solar years, it regards as the end of I. A. 4960 not the end of the solar sidereal year of that number, but that of the luni-solar year, which, by Hindu reckoning, is completed upon the third of the same month (see Ward, Kâla Sankalita, Table, p. xxxii).

48.... Reduce the sum to months, and add the months expired of the current year, beginning with the light half of Câitra.

49. Set the result down in two places; multiply it by the number of intercalary months, and divide by that of solar months, and add to the last result the number of intercalary months thus found; reduce the sum to days, and add the days expired of the current month;

50. Set the result down in two places; multiply it by the number of omitted lunar days, and divide by that of lunar days; subtract from the last result the number of omitted lunar days

thus obtained: the remainder is, at midnight, on the meridian of Lankâ,

51. The sum of days, in civil reckoning....

In these verses is taught the method of one of the most important and frequently recurring processes in Hindu Astronomy, the finding, namely, of the number of civil or natural days which have elapsed at any given date, reckoning either from the beginning of the present creation, or (see below, v. 56) from any required epoch since that time. In the modern technical language, the result is uniformly styled the *ahargaṇa*, "sum of days;" that precise term, however, does not once occur in the text of the Sûrya-Siddhânta: in the present passage we have *dyugaṇa*, which means the same thing, and in verse 53 *dinarâçi*, "heap or quantity of days."

The process will be best illustrated and explained by an example. Let it be required to find the sum of days to the beginning of Jan. 1, 1860.

It is first necessary to know what date corresponds to this in Hindu reckoning. We have remarked above that the 4960th year of the Iron Age is completed in April, 1859; in order to exhibit the place in the next following year of the date required, and, at the same time, to present the names and succession of the months, which in this treatise are assumed as known, and are nowhere stated, we have constructed the following skeleton of a Hindu calendar for the year 4961 of the Iron Age.

Solar Year.

month.	first day.
(I. A. 4960.)	
12. Câitra,	Mar. 13, 1859.
(I. A. 4961.)	
1. Vâiçâkha,	Apr. 12, do.
2. Jyâishṭha,	May 13, do.
3. Âshâḍha,	June 14, do.
4. Çrâvaṇa,	July 15, do.
5. Bhâdrapada,	Aug. 16, do.
6. Âçvina,	Sept. 16, do.
7. Kârttika,	Oct. 16, do.
8. Mârgaçîrsha,	Nov. 15, do
9. Pâusha,	Dec. 15, do.
10. Mâgha,	Jan. 13, 1860.
11. Phâlguna,	Feb. 11, do.
12. Câitra,	Mar. 12, do.

Luni-solar Year.

month.	first day.
(I. A. 4961.)	
1. Câitra,	Apr. 4, 1859.
2. Vâiçâkha,	May 3, do.
3. Jyâishṭha,	June 2, do.
4. Âshâḍha,	July 1, do.
5. Çrâvaṇa,	July 31, do.
6. Bhâdrapada,	Aug. 29, do.
7. Âçvina,	Sept. 28, do.
8. Kârttika,	Oct. 27, do.
9. Mârgaçîrsha,	Nov. 26, do.
10. Pâusha,	Dec. 25, do.
11. Mâgha,	Jan. 24, 1860.
12. Phâlguna,	Feb. 22, do.
(I. A. 4962.)	
1. Câitra,	Mar. 23, do.

The names of the solar months are derived from the names of the asterisms (see below, chap. viii.) in which, at the time of their being first so designated, the moon was full during their continuance. The same names are transferred to the lunar months. Each lunar month is divided into two parts; the first, called the light half (*çukla paksha*, "bright side"), lasts from new moon to full moon, or while the moon is waxing; the other, called the dark half (*kṛshṇa paksha*, "black side"), lasts from full moon to new moon, or while the moon is waning.

The table shows that Jan. 1, 1860, is the eighth day of the tenth month of the 4961st year of the Iron Age. The time, then, for which we have to find the sum of days, is 1,955,884,960 y., 9 m., 7 d.

Number of complete years elapsed,	1,955,884,960
multiply by number of solar months in a year,	12
Number of months,	23,470,619,520
add months elapsed of current year,	9
Whole number of months elapsed,	23,470,619,529

Now a proportion is made: as the whole number of solar months in an Age is to the number of intercalary months in the same period, so is the number of months above found to that of the corresponding intercalary months: or,

51,840,000 : 1,593,336 : : 23,470,619,529 : 721,384,703 +

Whole number of months, as above,	23,470,619,529
add intercalary months,	721,384,703
Whole number of lunar months,	24,192,004,232
multiply by number of lunar days in a month,	30
Number of lunar days,	725,760,126,960
add lunar days elapsed of current month,	7
Whole number of lunar days elapsed,	725,760,126,967

To reduce, again, the number of lunar days thus found to the corresponding number of solar days, a proportion is made, as before: as the whole number of lunar days in an Age is to the number of omitted lunar days in the same period, so is the number of lunar days in the period for which the sum of days is required to that of the corresponding omitted lunar days: or,

1,603,000,080 : 25,082,252 : : 725,760,126,967 : 11,356,018,395 +

Whole number of lunar days as above,	725,760,126,967
deduct omitted lunar days,	11,356,018,395
Total number of civil days from end of creation to beginning of Jan. 1, 1860,	714,404,108,572

This, then, is the required sum of days, for the beginning of the year A.D. 1860, at midnight, upon the Hindu prime meridian.

The first use which we are instructed to make of the result thus obtained is an astrological one.

51.... From this may be found the lords of the day, the month, and the year, counting from the sun. If the number be divided by seven, the remainder marks the lord of the day, beginning with the sun.

52. Divide the same number by the number of days in a month and in a year, multiply the one quotient by two and the

other by three, add one to each product, and divide by seven; the remainders indicate the lords of the month and of the year.

These verses explain the method of ascertaining, from the sum of days already found, the planet which is accounted to preside over the day, and also those under whose charge are placed the month and year in which that day occurs.

To find the lord of the day is to find the day of the week, since the latter derives its name from the former. The week, with the names and succession of its days, is the same in India as with us, having been derived to both from a common source. The principle upon which the assignment of the days to their respective guardians was made has been handed down by ancient authors (see Ideler, Handbuch d. math. u. tech. Chronologie, i. 178, etc.), and is well known. It depends upon the division of the day into twenty-four hours, and the assignment of each of these in succession to the planets, in their natural order; the day being regarded as under the dominion of that planet to which its first hour belongs. Thus, the planets being set down in the order of their proximity to the earth, as determined by the ancient systems of astronomy (for the Hindu, see below, xii. 84–88), beginning with the remotest, as follows: Saturn, Jupiter, Mars, sun, Venus, Mercury, moon, and the first hour of the twenty-four being assigned to the Sun, as chief of the planets, the second to Venus, etc., it will be found that the twenty-fifth hour, or the first of the second day, belongs to the moon; the forty-ninth, or the first of the third day, to Mars, and so on. Thus is obtained a new arrangement of the planets, and this is the one in which this Siddhânta, when referring to them, always assumes them to stand (see, for instance, below, v. 70; ii. 35–37): it has the convenient property that by it the sun and moon are separated from the other planets, from which they are by so many peculiarities distinguished. Upon this order depend the rules here given for ascertaining also the lords of the month and of the year. The latter, as appears both from the explanation of the commentator, and from the rules themselves, are no actual months and years, but periods of thirty and three hundred and sixty days, following one another in uniform succession, and supposed to be placed, like the day, under the guardianship of the planets to whom belong their first subdivisions: thus the lord of the day is the lord of its first hour; the lord of the month is the lord of its first day (and so of its first hour); the lord of the year is the lord of its first month (and so of its first day and hour). We give below this artificial arrangement of the planets, with the order in which they are found to succeed one another as lords of the periods of one, thirty, and three hundred and sixty days; we add their natural order of succession, as lords of the hours; and we farther prefix the ordinary names of the days, with their English equivalents. Other of the numerous names of the planets, it is to be remarked, may be put before the word *vâra* to form the name of the day: *vâra* itself means literally "successive time," or "turn," and is not used, so far as we are aware, in any other connection, to denote a day.

Name of day.		Presiding Planet.	Succession, as Lord of day,	month,	year,	hour.
Ravivâra,	Sunday,	Sun,	1	1	1	1
Somavâra,	Monday,	Moon,	2	5	6	4
Mangalavâra,	Tuesday,	Mars,	3	2	4	7
Budhavâra,	Wednesday,	Mercury,	4	6	2	3
Guruvâra,	Thursday,	Jupiter,	5	3	7	6
Çukravâra,	Friday,	Venus,	6	7	5	2
Çanivâra,	Saturday,	Saturn,	7	4	3	5

As the first day of the subsistence of the present order of things is supposed to have been a Sunday, it is only necessary to divide the sum of days by seven, and the remainder will be found, in the first column, opposite the name of the planet to which the required day belongs. Thus, taking the sum of days found above, adding to it one, for the first of January itself, and dividing by seven, we have:

$$\begin{array}{r} 7)\,714{,}404{,}108{,}573 \\ \hline 102{,}057{,}729{,}796 - 1 \end{array}$$

The first of January, 1860, accordingly, falls on a Sunday by Hindu reckoning, as by our own.

On referring to the table, it will be seen that the lords of the months follow one another at intervals of two places. To find, therefore, by a summary process, the lord of the month in which occurs any given day, first divide the sum of days by thirty; the quotient, rejecting the remainder, is the number of months elapsed; multiply this by two, that each month may push the succession forward two steps, add one for the current month, divide by seven in order to get rid of whole series, and the remainder is, in the column of lords of the day, the number of the regent of the month required. Thus:

$$\begin{array}{r} 30)\,714{,}404{,}108{,}572 \\ \hline 23{,}813{,}470{,}285+ \\ 2 \\ \hline 47{,}626{,}940{,}570 \\ 1 \\ \hline 7)\,47{,}626{,}940{,}571 \\ \hline 6{,}803{,}848{,}652 - 7 \end{array}$$

The regent of the month in question is therefore Saturn.

By a like process is found the lord of the year, saving that, as the lords of the year succeed one another at intervals of three places, the multiplication is by three instead of by two. Upon working out the process, it will be found that the final remainder is five, which designates Jupiter as the lord of the year at the given time.

Excepting here and in the parallel passage xii. 77, 78, no reference is made in the Sûrya-Siddhânta to the week, or to the names of its days. Indeed, it is not correct to speak of the week at all in connection with India, for the Hindus do not seem ever to have regarded it as a division of time, or a period to be reckoned by; they knew only of a certain order of succession, in which the days were placed under the regency of the seven planets. And since, moreover, as remarked above (under vv. 11,

12), they never made that division of the day into twenty-four hours upon which the order of regency depends, it follows that the whole system was of foreign origin, and introduced into India along with other elements of the modern sciences of astronomy and astrology, to which it belonged. Its proper foundation, the lordship of the successive hours, is shown by the other passage (xii. 78) to have been also known to the Hindus; and the name by which the hours are there called (*horâ* = ὥρα) indicates beyond a question the source whence they derived it.

53. Multiply the sum of days (*dinarâçi*) by the number of revolutions of any planet, and divide by the number of civil days; the result is the position of that planet, in virtue of its mean motion, in revolutions and parts of a revolution.

By the number of revolutions and of civil days is meant, of course, their number, as stated above, in an Age. For "position of the planet," etc., the text has, according to its usual succinct mode of expression, simply "is the planet, in revolutions, etc." There is no word for "position" or "place" in the vocabulary of this Siddhânta.

This verse gives the method of finding the mean place of the planets at any given time for which the sum of days has been ascertained, by a simple proportion: as the number of civil days in a period is to the number of revolutions during the same period, so is the sum of days to the number of revolutions and parts of a revolution accomplished down to the given time. Thus, for the sun:

1,577,917,828 : 4,320,000 : : 714,404,108,572 : 1,955,884,960rev 8s 17° 48′ 7″

The mean longitude of the sun, therefore, Jan. 1st, 1860, at midnight on the meridian of Ujjayinî, is 257° 48′ 7″. We have calculated in this manner the positions of all the planets, and of the moon's apsis and node—availing ourselves, however, of the permission given below, in verse 56, and reckoning only from the last epoch of conjunction, the beginning of the Iron Age (from which time the sum of days is 1,811,945), and also employing the numbers afforded by the lesser period of 1,080,000 years—and present the results in the following table.

Mean Places of the Planets, Jan. 1*st*, 1860, *midnight, at Ujjayinî.*

Planet.	According to the Sûrya-Siddhânta.					The same corrected by the *bîja*.			
	(rev.)	s	°	′	″	s	°	′	″
Sun,	(4,960)	8	17	48	7	8	17	48	7
Mercury,	(20,597)	4	15	13	8	4	8	36	16
Venus,	(8,063)	10	21	8	59	10	16	11	22
Mars,	(2,637)	5	24	17	36	5	24	17	36
Jupiter,	(418)	2	26	0	7	2	22	41	41
Saturn,	(168)	3	20	11	12	3	25	8	50
Moon,	(66,318)	11	15	23	24	11	15	23	24
" apsis,	(560)	10	9	42	26	10	8	3	13
" node,	(267–)	9	24	26	4	9	22	46	51

The positions are given as deduced both from the numbers of revolutions stated in the text, and from the same as corrected by the *bîja* : prefixed are the numbers of complete revolutions accomplished since the epoch. In the cases of the moon's apsis and node, however, it was necessary to employ the numbers of revolutions given for the whole Age, these not being divisible by four, and also to add to their ascertained amount of movement their longitude at the epoch (see below, under vv. 57, 58).

54. Thus also are ascertained the places of the conjunction (*çîghra*) and apsis (*mandocca*) of each planet, which have been mentioned as moving eastward; and in like manner of the nodes, which have a retrograde motion, subtracting the result from a whole circle.

The places of the apsides and nodes have already been given above (under vv. 41–44), both for the commencement of the Iron Age, and for A. D. 1850. The place of the conjunctions of the three superior planets is, of course, the mean longitude of the sun. In the case of the inferior planets, the place of the conjunction is, in fact, the mean place of the planet itself in its proper orbit, and it is this which we have given for Mercury and Venus in the preceding table : while to the Hindu apprehension, the mean place of those planets is the same with that of the sun.

55. Multiply by twelve the past revolutions of Jupiter, add the signs of the current revolution, and divide by sixty; the remainder marks the year of Jupiter's cycle, counting from Vijaya.

This is the rule for finding the current year of the cycle of sixty years, which is in use throughout all India, and which is called the cycle of Jupiter, because the length of its years is measured by the passage of that planet, by its mean motion, through one sign of the zodiac. According to the data given in the text of this Siddhânta, the length of Jupiter's year is $361^d\ 0^h\ 38^m$; the correction of the *bîja* makes it about 12^m longer. It was doubtless on account of the near coincidence of this period with the true solar year that it was adopted as a measure of time; but it has not been satisfactorily ascertained, so far as we are aware, where the cycle originated, or what is its age, or why it was made to consist of sixty years, including five whole revolutions of the planet. There was, indeed, also in use a cycle of twelve of Jupiter's years, or the time of one sidereal revolution : see below, xiv. 17. Davis (As. Res. iii. 209, etc.) and Warren (Kâla Sankalita, p. 197, etc.) have treated at some length of the greater cycle, and of the different modes of reckoning and naming its years usual in the different provinces of India.

In illustration of the rule, let us ascertain the year of the cycle corresponding to the present year, A. D. 1859. It is not necessary to make the calculation from the creation, as the rule contemplates; for, since the number of Jupiter's revolutions in the period of 1,080,000 years is divisible by five, a certain number of whole cycles, without a remainder, will have elapsed at the beginning of the Iron Age. The revolutions of the planet since that time, as stated in the table last given, are 418, and it is in the 3rd sign of the 419th revolution ; the reduction of the whole amount of movement to signs shows us that the current year is the 5019th since the epoch: divide this by 60, to cast out whole cycles, and the remainder, 39, is the number of the year in the current cycle. This treatise nowhere gives the names of the years of Jupiter, but, as in the case of the months, the signs of the zodiac, and other similar matters, assumes them to be already familiarly known in their succession: we accordingly present them below. We take them from Mr. Davis's paper, alluded to above, not having access at present to any original authority which contains them.

1. Vijaya.
2. Jaya.
3. Manmatha.
4. Durmukha.
5. Hemalamba.
6. Vilamba.
7. Vikârin.
8. Çarvarî.
9. Plava.
10. Çubhakṛt.
11. Çobhana.
12. Krodhin.
13. Viçvâvasu.
14. Parâbhava.
15. Plavanga.
16. Kîlaka.
17. Sâumya.
18. Sâdhâraṇa.
19. Virodhakṛt.
20. Paridhâvin.
21. Pramâdin.
22. Ânanda.
23. Râkshasa.
24. Anala.
25. Pingala.
26. Kâlayukta.
27. Siddhârthin.
28. Râudra.
29. Durmati.
30. Dundubhi.
31. Rudhirodgârin.
32. Raktâksha.
33. Krodhana.
34. Kshaya.
35. Prabhava.
36. Vibhava.
37. Çukla.
38. Pramoda.
39. Prajâpati.
40. Angiras.
41. Çrîmukha.
42. Bhâva.
43. Yuvan.
44. Dhâtar.
45. Îçvara.
46. Bahudhanya.
47. Pramâthin.
48. Vikrama.
49. Bhṛçya.
50. Citrabhânu.
51. Subhânu.
52. Târaṇa.
53. Pârthiva.
54. Vyaya.
55. Sarvajit.
56. Sarvadhârin.
57. Virodhin.
58. Vikṛta.
59. Khara.
60. Nandana.

It appears, then, that the current year of Jupiter's cycle is named Prajâpati: upon dividing by the planet's mean daily motion the part of the current sign already passed over, it will be found that, according to the text, that year commenced on the twenty-third of February, 1859; or, if the correction of the *bîja* be admitted, on the third of April.

Although it is thus evident that the Sûrya-Siddhânta regards both the existing order of things and the Iron Age as having begun with Vijaya, that year is not generally accounted as the first, but as the twenty-seventh, of the cycle, which is thus made to commence with Prabhava. An explanation of this discrepancy might perhaps throw important light upon the origin or history of the cycle.

This method of reckoning time is called (see below, xiv. 1, 2) the *bârhaspatya mâna,* "measure of Jupiter."

56. The processes which have thus been stated in full detail, are practically applied in an abridged form. The calculation of the mean place of the planets may be made from any epoch (*yuga*) that may be fixed upon.

57. Now, at the end of the Golden Age (*kṛta yuga*), all the planets, by their mean motion—excepting, however, their nodes and apsides (*mandocca*)—are in conjunction in the first of Aries.

58. The moon's apsis (*ucca*) is in the first of Capricorn, and its node is in the first of Libra; and the rest, which have been

stated above to have a slow motion—their position cannot be expressed in whole signs.

It is curious to observe how the Sûrya-Siddhânta, lest it should seem to admit a later origin than that which it claims in the second verse of this chapter, is compelled to ignore the real astronomical epoch, the beginning of the Iron Age; and also how it avoids any open recognition of the lesser cycle of 1,080,000 years, by which its calculations are so evidently intended to be made.

The words at the end of verse 56 the commentator interprets to mean: "from the beginning of the current, i. e., the Silver, Age." In this he is only helping to keep up the pretence of the work to immemorial antiquity, even going therein beyond the text itself, which expressly says: "from any desired (*ishtatas*) *yuga*." Possibly, however, we have taken too great a liberty in rendering *yuga* by "epoch," and it should rather be "Age," i. e., "beginning of an Age." The word *yuga* comes from the root *yuj*, "to join" (Latin, *jungo*; Greek, *ζεύγνυμι*: the word itself is the same with *jugum*, *ζυγόν*), and seems to have been originally applied to indicate a cycle, or period, by means of which the conjunction or correspondence of discordant modes of reckoning time was kept up; thus it still signifies also the *lustrum*, or cycle of five years, which, with an intercalated month, anciently maintained the correspondence of the year of 360 days with the true solar year. From such uses it was transferred to designate the vaster periods of the Hindu chronology.

As half an Age, or two of the lesser periods, are accounted to have elapsed between the end of the Golden and the beginning of the Iron Age, the planets, at the latter epoch, have again returned to a position of mean conjunction: the moon's node, also, is still in the first of Libra, but her apsis has changed its place half a revolution, to the first of Cancer (see above, under vv. 29–34). The positions of the apsides and nodes of the other planets at the same time have been given already, under verses 41–44.

The Hindu names of the signs correspond in signification with our own, having been brought into India from the West. There is nowhere in this work any allusion to them as constellations, or as having any fixed position of their own in the heavens: they are simply the names of the successive signs (*râçi*, *bha*) into which any circle is divided, and it is left to be determined by the connection, in any case, from what point they shall be counted. Here, of course, it is the initial point of the fixed Hindu sphere (see above, under v. 27). As the signs are, in the sequel, frequently cited by name, we present annexed, for the convenience of reference of those to whose memory they are not familiar in the order of their succession, their names, Latin and Sanskrit, their numbers, and the figures generally used to represent them. Those enclosed in brackets do not chance to occur in our text.

1.	Aries,	♈	*mesha*; *aja*.	7.	Libra,	♎	*tulâ*.
2.	Taurus,	♉	*vṛshan*.	8.	Scorpio,	♏	[*vṛçcika*,] *âli*.
3.	Gemini,	♊	*mithuna*.	9.	Sagittarius,	♐	*dhanus*.
4.	Cancer,	♋	*karka*, *karkaṭa*.	10.	Capricornus,	♑	*makara*, *mṛga*.
5.	Leo,	♌	[*sinha*].	11.	Aquarius,	♒	*kumbha*.
6.	Virgo,	♍	*kanyâ*.	12.	Pisces,	♓	[*mina*].

In the translation given above of the second half of verse 58, not a little violence is done to the natural construction. This would seem to require that it be rendered: "and the rest are in whole signs (have come to a position which is without a remainder of degrees); they, being of slow motion, are not stated here." But the actual condition of things at the epoch renders necessary the former translation, which is that of the commentator also. We cannot avoid conjecturing that the natural rendering was perhaps the original one, and that a subsequent alteration of the elements of the treatise compelled the other and forced interpretation to be put upon the passage.

The commentary gives the positions of the apsides and nodes (those of the nodes, however, in reverse) for the epoch of the end of the Golden Age, but, strangely enough, both in the printed edition and in our manuscript, commits the blunder of giving the position of Saturn's node a second time, for that of his apsis, and also of making the seconds of the position of the node of Mars 12, instead of 24. We therefore add them below, in their correct form.

Motion of the Apsides and Nodes of the Planets, to the End of the last Golden Age.

Planet.	Apsis.					Node.				
	(rev.)	s	°	′	″	(rev.)	s	°	′	″
Sun,	(175)	0	7	28	12					
Mercury,	(166)	5	4	4	48	(220)	8	11	16	48
Venus,	(241)	11	13	21	0	(408)	4	17	25	48*
Mars,	(92)	3	3	14	24	(96)	9	11	20	24
Jupiter,	(407)	0	9	0	0	(78)	8	8	56	24
Saturn,	(17)	7	19	35	24	(299)	4	20	13	12

The method of finding the mean places of the planets for midnight on the prime meridian having been now fully explained, the treatise proceeds to show how they may be found for other places, and for other times of the day. To this the first requisite is to know the dimensions of the earth.

59. Twice eight hundred *yojanas* are the diameter of the earth: the square root of ten times the square of that is the earth's circumference.

60. This, multiplied by the sine of the co-latitude (*lambajyâ*) of any place, and divided by radius (*trijîvâ*), is the corrected (*sphuṭa*) circumference of the earth at that place. . . .

There is the same difficulty in the way of ascertaining the exactness of the Hindu measurement of the earth as of the Greek; the uncertain value, namely, of the unit of measure employed. The *yojana* is ordinarily divided into *kroça*, "cries" (i. e., distances to which a certain cry may be heard); the *kroça* into *dhanus*, "bow-lengths," or *daṇḍa*, "poles;" and these again into *hasta*, "cubits." By its origin, the latter

* The printed edition, by an error of the press, gives 4.

ought not to vary far from eighteen inches; but the higher measures differ greatly in their relation to it. The usual reckoning makes the yojana equal 32,000 cubits, but it is also sometimes regarded as composed of 16,000 cubits; and it is accordingly estimated by different authorities at from four and a half to rather more than ten miles English. This uncertainty is no merely modern condition of things: Hiuen-Thsang, the Chinese monk who visited India in the middle of the seventh century, reports (see Stanislas Julien's Mémoires de Hiouen-Thsang, i. 59, etc.) that in India "according to ancient tradition a yojana equals forty *li;* according to the customary use of the Indian kingdoms, it is thirty *li;* but the yojana mentioned in the sacred books contains only sixteen *li:*" this smallest yojana, according to the value of the *li* given by Williams (Middle Kingdom, ii. 154), being equal to from five to six English miles. At the same time, Hiuen-Thsang states the subdivisions of the yojana in a manner to make it consist of only 16,000 cubits. Such being the condition of things, it is clearly impossible to appreciate the value of the Hindu estimate of the earth's dimensions, or to determine how far the disagreement of the different astronomers on this point may be owing to the difference of their standards of measurement. Âryabhaṭṭa (see Colebrooke's Hind. Alg. p. xxxviii; Essays, ii. 468) states the earth's diameter to be 1050 yojanas; Bhâskara (Siddh.-Çir. vii. 1) gives it as 1581: the latter author, in his Lîlâvatî (i. 5, 6), makes the yojana consist of 32,000 cubits.

The ratio of the diameter to the circumference of a circle is here made to be 1 : √10, or 1 : 3.1623, which is no very near approximation. It is not a little surprising to find this determination in the same treatise with the much more accurate one afforded by the table of sines given in the next chapter (vv. 17–21), of 3438 : 10,800, or 1 : 3.14136; and then farther, to find the former, and not the latter, made use of in calculating the dimensions of the planetary orbits (see below, xii. 83). But the same inconsistency is found also in other astronomical and mathematical authorities. Thus Âryabhaṭṭa (see Colebrooke, as above) calculates the earth's circumference from its diameter by the ratio 7 : 22, or 1 : 3.14286, but makes the ratio 1 : √10 the basis of his table of sines, and Brahmagupta and Çrîdhara also adopt the latter. Bhâskara, in stating the earth's circumference at 4967 yojanas, is very near the truth, since 1581 : 4967 : : 1 : 3.14168: his Lîlâvatî (v. 201) gives 7 : 22, and also, as more exact, 1250 : 3927, or 1 : 3.1416. This subject will be reverted to in connection with the table of sines.

The greatest circumference of the earth, as calculated according to the data and method of the text, is 5059.556 yojanas. The astronomical yojana must be regarded as an independent standard of measurement, by which to estimate the value of the other dimensions of the solar system stated in this treatise. To make the earth's mean diameter correct as determined by the Sûrya-Siddhânta, the yojana should equal 4.94 English miles; to make the circumference correct, it should equal 4.91 miles.

The rule for finding the circumference of the earth upon a parallel of latitude is founded upon a simple proportion, viz., rad. : cos. latitude : : circ. of earth at equator : do. at the given parallel; the cosine of the latitude being, in effect, the radius of the circle of latitude. Radius and cosine of latitude are tabular numbers, derived from the table to be given afterward (see below, ii. 17–21). This treatise is not accustomed to employ cosines directly in its calculations, but has special names for the complements of the different arcs which it has occasion to use. Terrestrial latitude is styled *aksha*, "axle," which term, as appears from xii. 42, is employed elliptically for *akshonnati*, "elevation of the axle," i. e., "of the pole:" *lamba*, co-latitude, which properly signifies "lagging, dependence, falling off," is accordingly the depression of the pole, or its distance from the zenith. Directions for finding the co-latitude are given below (iii. 13, 14).

The latitude of Washington being 38° 54′, the sine of its co-latitude is 2675′; the proportion 3438 : 2675 : : 5059.64 : 3936.75 gives us, then, the earth's circumference at Washington as 3936.75 yojanas.

60. . . . Multiply the daily motion of a planet by the distance in longitude (*deçântara*) of any place, and divide by its corrected circumference;

61. The quotient, in minutes, subtract from the mean position of the planet as found, if the place be east of the prime meridian (*rekhâ*); add, if it be west; the result is the planet's mean position at the given place.

The rules previously stated have ascertained the mean places of the planets at a given midnight upon the prime meridian; this teaches us how to find them for the same midnight upon any other meridian, or, how to correct for difference of longitude the mean places already found. The proportion is: as the circumference of the earth at the latitude of the point of observation is to the part of it intercepted between that point and the prime meridian, so is the whole daily motion of each planet to the amount of its motion during the time between midnight on the one meridian and on the other. The distance in longitude (*deçântara*, literally "difference of region") is estimated, it will be observed, neither in time nor in arc, but in yojanas. How it is ascertained is taught below, in verses 63–65.

The geographical position of the prime meridian (*rekhâ*, literally "line") is next stated.

62. Situated upon the line which passes through the haunt of the demons (*râkshasa*) and the mountain which is the seat of the gods, are Rohîtaka and Avantî, as also the adjacent lake.

The "haunt of the demons" is Lankâ, the fabled seat of Râvaṇa, the chief of the Râkshasas, the abduction by whom of Râma's wife, with the expedition to Lankâ of her heroic husband for her rescue, its accomplishment, and the destruction of Râvaṇa and his people, form the subject of the epic poem called the Râmâyana. In that poem, and to the general apprehension of the Hindus, Lankâ is the island Ceylon; in the astronomical geography, however (see below, xii. 39), it is a city, situated upon the equator. How far those who established the meridian may have regarded the actual position of Ceylon as identical with that assigned to Lankâ might not be easy to determine. The "seat of the gods" is Mount Meru, situated at the north pole (see below, xii. 34, etc.). The meridian is usually styled that of Lankâ, and "at Lankâ" is the ordinary phrase made use of in this treatise (as, for instance, above, v. 50; below, iii. 43) to designate a situation either of no longitude or of no latitude.

But the circumstance which actually fixes the position of the prime meridian is the situation of the city of Ujjayinî, the *Οζηνη* of the Greeks, the modern Ojein. It is called in the text by one of its ancient names, Avantî. It is the capital of the rich and populous province of Mâlava, occupying the plateau of the Vindhya mountains just north of the principal ridge and of the river Narmadâ (Nerbudda), and from old time a chief seat of Hindu literature, science, and arts. Of all the centres of Hindu culture, it lay nearest to the great ocean-route by which, during the first three centuries of our era, so important a commerce was carried on between Alexandria, as the mart of Rome, and India and the countries lying still farther east. That the prime meridian was made to pass through this city proves it to have been the cradle of the Hindu science of astronomy, or its principal seat during its early history. Its actual situation is stated by Warren (Kâla Sankalita, p. 9) as lat. 23° 11′ 30″ N., long. 75° 53′ E. from Greenwich: a later authority, Thornton's Gazetteer of India (London: 1857), makes it to be in lat. 23° 10′ N., long. 75° 47′ E.; in our farther calculations, we shall assume the latter position to be the correct one.

The situation of Rohîtaka is not so clear; we have not succeeded in finding such a place mentioned in any work on the ancient geography of India to which we have access, nor is it to be traced upon Lassen's map of ancient India. A city called Rohtuk, however, is mentioned by Thornton (Gazetteer, p. 836), as the chief place of a modern British district of the same name, and its situation, a little to the north-west of Delhi, in the midst of the ancient Kurukshetra, leads us to regard it as identical with the Rohîtaka of the text. That the meridian of Lankâ was expressly recognized as passing over the Kurukshetra, the memorable site of the great battle described by the Mahâbhârata, seems clear. Bhâskara (Siddh.-Çir., Gaṇ., vii. 2) describes it as follows: "the line which, passing above Lankâ and Ujjayinî, and touching the region of the Kurukshetra, etc., goes through Meru—that line is by the wise regarded as the central meridian (*madhyarekhâ*) of the earth." Our own commentary also explains *sannihitaṁ sarah*, which we have translated "adjacent lake," as signifying Kurukshetra. Warren (as above) takes the same expression to be the name of a city, which seems to us highly improbable; nor do we see that the word *saras* can properly be applied to a tract of country: we have therefore thought it safest to translate literally the words of the text, confessing that we do not know to what they refer.

If Rohîtaka and Rohtuk signify the same place, we have here a measure of the accuracy of the Hindu determinations of longitude; Thornton gives its longitude as 76° 38′, or 51′ to the east of Ujjayinî.

The method by which an observer is to determine his distance from the prime meridian is next explained.

63. When, in a total eclipse of the moon, the emergence (*unmîlana*) takes place after the calculated time for its occurrence, then the place of the observer is to the east of the central meridian;

64. When it takes place before the calculated time, his place is to the west: the same thing may be ascertained likewise from the immersion (*nimîlana*). Multiply by the difference of the two times in nâḍîs the corrected circumference of the earth at the place of observation,

65. And divide by sixty: the result, in yojanas, indicates the distance of the observer from the meridian, to the east or to the west, upon his own parallel; and by means of that is made the correction for difference of longitude.

Choice is made, of course, of a lunar eclipse, and not of a solar, for the purpose of the determination of longitude, because its phenomena, being unaffected by parallax, are seen everywhere at the same instant of absolute time; and the moments of total disappearance and first reappearance of the moon in a total eclipse are farther selected, because the precise instant of their occurrence is observable with more accuracy than that of the first and last contact of the moon with the shadow. For the explanation of the terms here used see the chapters upon eclipses (below, iv–vi).

The interval between the computed and observed time being ascertained, the distance in longitude (*deçântara*) is found by the simple proportion: as the whole number of nâḍîs in a day (sixty) is to the interval of time in nâḍîs, so is the circumference of the earth at the latitude of the point of observation to the distance of that point from the prime meridian, measured on the parallel. Thus, for instance, the distance of Ujjayinî from Greenwich, in time, being $5^h\ 3^m\ 8^s$, and that of Washington from Greenwich $5^h\ 8^m\ 11^s$ (Am. Naut. Almanac), that of Ujjayinî from Washington is $10^h\ 11^m\ 19^s$, or, in Hindu time, $25^n\ 28^v\ 1^p.8$, or $25^n.4718$: and by the proportion 60 : 25.4718 : : 3936.75 : 1671.28, we obtain 1671.28 yojanas as the distance in longitude (*deçântara*) of Washington from the Hindu meridian, the constant quantity to be employed in finding the mean places of the planets at Washington.

We might have expected that calculators so expert as the Hindus would employ the interval of time directly in making the correction for difference of longitude, instead of reducing it first to its value in yojanas. That they did not measure longitude in our manner, in degrees, etc., is owing to the fact that they seem never to have thought of applying to the globe of the earth the system of measurement by circles and divisions of circles which they used for the sphere of the heavens, but, even when dividing the earth into zones (see below, xii. 59–66) reduced all their distances laboriously to yojanas.

66. The succession of the week-day (*vâra*) takes place, to the east of the meridian, at a time after midnight equal to the difference of longitude in nâḍîs; to the west of the meridian, at a corresponding time before midnight.

This verse appears to us to be an astrological precept, asserting the regency of the sun and the other planets, in their order, over the successive portions of time assigned to each, to begin everywhere at the same instant of absolute time, that of their true commencement upon the prime meridian; so that, for instance, at Washington, Sunday, as the day placed under the guardianship of the sun, would really begin at eleven minutes before two on Saturday afternoon, by local time. The commentator, however, sees in it merely an intimation of what moment of local time, in places east and west of the meridian, corresponds to the true beginning of the day upon the prime meridian, and he is at much pains to defend the verse from the charge of being superfluous and unnecessary, to which it is indeed liable, if that be its only meaning.

The rules thus far given have directed us only how to find the mean places of the planets at a given midnight. The following verse teaches the method of ascertaining their position at any required hour of the day.

67. Multiply the mean daily motion of a planet by the number of nâḍîs of the time fixed upon, and divide by sixty: subtract the quotient from the place of the planet, if the time be before midnight; add, if it be after: the result is its place at the given time.

The proportion is as follows: as the number of nâḍîs in a day (sixty) is to those in the interval between midnight and the time for which the mean place of the planet is sought, so is the whole daily motion of the planet to its motion during the interval; and the result is additive or subtractive, of course, according as the time fixed upon is after or before midnight.

In order to furnish a practical test of the accuracy of this text-book of astronomy, and of its ability to yield correct results at the present time, we have calculated, by the rule given in this verse, the mean longitudes of the planets for a time after midnight of the first of January, 1860, on the meridian of Ujjayinî, which is equal to the distance in time of the meridian of Washington, viz. $25^{n}\ 28^{v}\ 1^{p}.8$, or $0^{d}.42453$; and we present the results in the annexed table. The longitudes are given as reckoned from the vernal equinox of that date, which we make to be distant 18° 5′ 8″.25 from the point established by the Sûrya-Siddhânta as the beginning of the Hindu sidereal sphere; this is (see below, chap. viii) 10′ east of ζ Piscium. We have ascertained the mean places both as determined by the text of our Siddhânta, and by the same with the correction of the *bîja*. Added are the actual mean places at the time designated: those of the primary planets have been found from Le Verrier's elements, presented in Biot's treatise, as cited above;* those of the moon, and of her apsis and node, were kindly furnished us from the office of the American Nautical Almanac, at Cambridge.

* We would warn our readers, however, of a serious error of the press in the table as given by Biot; as the yearly motion of the earth, read 1,295,977.38, instead of . . . 972.88.

Mean Longitudes of the Planets, Jan. 1st, 1860, midnight, at Washington.

Planet.	According to Sûrya-Siddhânta: text.			According to Sûrya-Siddhânta: with *bîja*.			According to moderns.		
	°	′	″	°	′	″	°	′	″
Sun,	96	18	21	96	18	21	100	5	6
Mercury,	155	2	30	148	25	39	151	28	20
Venus,	339	54	55	334	57	18	336	13	36
Mars,	192	36	5	192	36	5	197	26	32
Jupiter,	104	7	22	100	48	56	103	35	17
Saturn,	128	17	11	133	14	49	137	10	10
Moon,	9	4	9	9	4	9	12	41	23
" apsis,	327	50	24	326	11	11	326	47	35
" node,	312	29	51	310	50	38	312	48	10

In the next following table is farther given a view of the errors of the Hindu determinations—both the absolute errors, as compared with the actual mean place of each planet, and the relative, as compared with the place of the sun, to which it is the aim of the Hindu astronomical systems to adapt the elements of the other planets. Annexed to each error is the approximate date at which it was nothing, or at which it will hereafter disappear, ascertained by dividing the amount of present error by the present yearly loss or gain, absolute or relative, of each planet; excepting in the case of the moon, where we have made allowance, according to the formula used by the American Nautical Almanac, for the acceleration of her motion.

Errors of the Mean Longitudes of the Planets, as calculated according to the Sûrya-Siddhânta.

Planet.	Errors according to text: absolute.			when correct.	rel. to sun.			when correct.	The same, with *bîja*: absolute.			when correct.	rel. to sun.			when correct.
	°	′	″	A. D.	°	′	″	A. D.	°	′	″	A. D.	°	′	″	A. D.
Sun,	−3	46	45	250	0	0	0	....	−3	46	45	250	0	0	0	....
Mercury,	+3	34	10	2332	+7	20	55	3271	−3	2	41	1517	+0	44	5	1970
Venus,	+3	41	19	1222	+7	28	4	941	−1	16	18	2126	+2	30	27	1509
Mars,	−4	50	27	886	−1	3	42	1455	−4	50	27	886	−1	3	42	1455
Jupiter,	+0	32	5	1571	+4	18	50	832	−2	46	21	4203	+1	0	24	1575
Saturn,	−8	52	59	666	−5	6	14	857	−3	55	21	1250	−0	8	36	1825
Moon,	−3	37	14	115	+0	9	31	1067	−3	37	14	115	+0	9	31	1067
" apsis,	+1	2	49	1679	+4	49	34	1252	−0	36	24	1969	+3	10	21	1459
" node,	−0	18	19	1976	+3	28	26	1162	−1	57	32	2714	+1	49	13	1468

To complete the view of the planetary motions, and the statement of the elements requisite for ascertaining their position in the sky, it only remains to give the movement in latitude of each, its deviation from the general planetary path of the ecliptic. This is done in the concluding verses of the chapter.

68. The moon is, by its node, caused to deviate from the limit of its declination (*krânti*), northward and southward, to a distance, when greatest, of an eightieth part of the minutes of a circle;

69. Jupiter, to the ninth part of that multiplied by two; Mars, to the same amount multiplied by three; Mercury, Venus, and Saturn are by their nodes caused to deviate to the same amount multiplied by four.

70. So also, twenty-seven, nine, twelve, six, twelve, and twelve, multiplied respectively by ten, give the number of minutes of mean latitude (*vikshepa*) of the moon and the rest, in their order.

The deviation of the planets from the plane of the ecliptic is here stated in two different ways, which give, however, the same results; thus:

Moon,	$\frac{21600'}{80} = 270'$	or	$27' \times 10 = 270' = 4^\circ\ 30'$
Mars,	$\frac{270'}{9} \times 3 = 90'$	or	$9' \times 10 = 90' = 1^\circ\ 30'$
Mercury,	$\frac{270'}{9} \times 4 = 120'$	or	$12' \times 10 = 120' = 2^\circ$
Jupiter,	$\frac{270'}{9} \times 2 = 60'$	or	$6' \times 10 = 60' = 1^\circ$
Venus,	$\frac{270'}{9} \times 4 = 120'$	or	$12' \times 10 = 120' = 2^\circ$
Saturn,	$\frac{270'}{9} \times 4 = 120'$	or	$12' \times 10 = 120' = 2^\circ$

The subject of the latitude of the planets is completed in verses 6–8, and verse 57, of the following chapter; the former passage describes the manner, and indicates the direction, in which the node produces its disturbing effect; the latter gives the rule for calculating the apparent latitude of a planet at any point in its revolution.

There is a little discrepancy between the two specifications presented in these verses, as regards the description of the quantities specified: the one states them to be the amounts of greatest (*parama*) deviation from the ecliptic; the other, of mean (*madhya*) deviation. Both descriptions are also somewhat inaccurate. The first is correct only with reference to the moon, and the two terms require to be combined, in order to be made applicable to the other planets. The moon has its greatest latitude at 90° from its node, and this latitude is obviously equal to the inclination of its orbit to the ecliptic; for although its absolute distance from the ecliptic at this point of its course varies, as does its distance from the earth, on account of the eccentricity of its orbit, and the varying relation of the line of its apsides to that of its nodes, its angular distance remains unchanged. So, to an observer stationed at the sun, the greatest latitude of any one of the primary planets would be the same in its successive revolutions from node to node, and equal to the inclination of its orbit. But its greatest latitude as seen from the earth is very different in different revolutions, both on account of the difference of its absolute distance from the ecliptic when at the point of greatest removal from it in the two halves of its orbit, and, much more, on account of its varying distance from the earth. The former of these two causes of variation was not recognized by the

Hindus: in this treatise, at least, the distance of the node from the apsis (*mandocca*) is not introduced as an element into the process for determining a planet's latitude. The other cause of variation is duly allowed for (see below, ii. 57). Its effect, in the case of the three superior planets, is to make their greatest latitude sometimes greater, and sometimes less, than the inclination of their orbits, according as the planet is nearer to us than to the sun, or the contrary; hence the values given in the text for Mars, Jupiter, and Saturn, as they represent the mean apparent values, as latitude, of the greatest distance of each planet from the ecliptic, should nearly equal the inclination. In the case of Mercury and Venus, also, the quantities stated are the mean of the different apparent values of the greatest heliocentric latitude, but this mean is of course less, and for Mercury very much less, than the inclination. Ptolemy, in the elaborate discussion of the theory of the latitude contained in the thirteenth book of his Syntaxis, has deduced the actual inclination of the orbits of the two inferior planets: this the Hindus do not seem to have attempted.

We present below a comparative table of the inclinations of the orbits of the planets as determined by Ptolemy and by modern astronomers, with those of the Hindus, so far as given directly by the Sûrya-Siddhânta.

Inclination of the Orbits of the Planets, according to Different Authorities.

Planet.	Sûrya-Siddhânta.		Ptolemy.		Moderns.		
	°	′	°	′	°	′	″
Mercury,	. . . . .		7		7	0	8
Venus,	. . . . .		3	30	3	23	31
Mars,	1	30	1		1	51	5
Jupiter,	1		1	30	1	18	40
Saturn,	2		2	30	2	29	28
Moon,	4	30	5		5	8	40

The verb in verses 68 and 69, which we have translated "caused to deviate," is *vi kshipyate*, literally "is hurled away," *disjicitur;* from it is derived the term used in this treatise to signify celestial latitude, *vikshepa*, "disjection." The Hindus measure the latitude, however, as we shall have occasion to notice more particularly hereafter, upon a circle of declination, and not upon a secondary to the ecliptic. In the words chosen to designate it is seen the influence of the theory of the node's action, as stated in the first verses of the next chapter. The forcible removal is from the point of declination (*krânti*, "gait," or *apakrama*, "withdrawal," i. e., from the celestial equator) which the planet ought at the time to occupy.

The title given to this first chapter (*adhikâra*, "subject, heading") is *madhyamâdhikâra*, which we have represented in the title by "mean motions of the planets," although it would be more accurately rendered by "mean places of the planets;" that is to say, the data and methods requisite for ascertaining their mean places. Now follows the *spashṭâdhikâra*, "chapter of the true, or corrected, places of the planets."

CHAPTER II.

OF THE TRUE PLACES OF THE PLANETS.

CONTENTS:—1–3, causes of the irregularities of the planetary motions; 4–5, disturbing influence of the apsis and conjunction; 6–8, of the node; 9–11, different degree of irregularity of the motion of the different planets; 12–13, different kinds of planetary motion; 14, purpose of this chapter; 15–16, rule for constructing the table of sines; 17–22, table of sines; 22–27, table of versed sines; 28, inclination of the ecliptic, and rule for finding the declination of any point in it; 29–30, to find the sine and cosine of the anomaly; 31–32, to find, by interpolation, the sine or versed sine corresponding to any given arc; 33, to find, in like manner, the arc corresponding to a given sine or versed sine; 34–37, dimensions of the epicycles of the planets; 38, to find the true dimensions of the epicycle at any point in the orbit; 39, to find the equation of the apsis, or of the centre; 40–42, to find the equation of the conjunction, or the annual equation; 43–45, application of these equations in finding the true places of the different planets; 46, correction of the place of a planet for difference between mean and apparent solar time; 47–49, how to correct the daily motion of the planets for the effect of the apsis; 50–51, the same for that of the conjunction; 51–55, retrogradation of the lesser planets; 56, correction of the place of the node; 57–58, to find the celestial latitude of a planet, and its declination as affected by latitude; 59, to find the length of the day of any planet; 60, to find the radius of the diurnal circle; 61–63, to find the day-sine, and the respective length of the day and night; 64, to find the number of asterisms traversed by a planet, and of days elapsed, since the commencement of the current revolution; 65, to find the *yoga;* 66, to find the current lunar day, and the time in it of a given instant; 67–69, of the divisions of the lunar month called *karaṇa.*

1. Forms of Time, of invisible shape, stationed in the zodiac (*bhagaṇa*), called the conjunction (*çîghrocca*), apsis (*mandocca*), and node (*pâta*), are causes of the motion of the planets.

2. The planets, attached to these beings by cords of air, are drawn away by them, with the right and left hand, forward or backward, according to nearness, toward their own place.

3. A wind, moreover, called *provector* (*pravaha*) impels them toward their own apices (*ucca*); being drawn away forward and backward, they proceed by a varying motion.

4. The so-called apex (*ucca*), when in the half-orbit in front of the planet, draws the planet forward; in like manner, when in the half-orbit behind the planet, it draws it backward.

5. When the planets, drawn away by their apices (*ucca*), move forward in their orbits, the amount of the motion so caused is called their excess (*dhana*); when they move backward, it is called their deficiency (*ṛṇa*).

In these verses is laid before us the Hindu theory of the general nature of the forces which produce the irregularities of the apparent motions, regarded as being the real motions, of the planets. The world-wide difference between the spirit of the Hindu astronomy and that of the Greek is not less apparent here than in the manner of presentation of the elements in the last chapter: the one is purely scientific, devising methods for representing and calculating the observed motions, and attempting nothing farther; the other is not content without fabricating a fantastic and absurd theory respecting the superhuman powers which occasion the movements with which it is dealing. The Hindu method has this convenient peculiarity, that it absolves from all necessity of adapting the disturbing forces to one another, and making them form one consistent system, capable of geometrical representation and mathematical demonstration; it regards the planets as actually moving in circular orbits, and the whole apparatus of epicycles, given later in the chapter, as only a device for estimating the amount of the force, and of its resulting motion, exerted at any given point by the disturbing cause.

The commentator gives two different explanations of the provector wind, spoken of in the third verse: one, that it is the general current, mentioned below, in xii. 73, as impelling the whole firmament of stars, and which, though itself moving westward, drives the planets, in some unexplained way, towards its own apex of motion, in the east; the other, that a separate vortex for each planet, called provector on account of its analogy with that general current, although not moving in the same direction, carries them around in their orbits from west to east, leaving only the irregularities of their motion to be produced by the disturbing forces. This latter we regard as the proper meaning of the text: neither is very consistent with the theory of the lagging behind of the planets, given above, in i. 25, 26, as the explanation of their apparent eastward motion. The commentary also states more explicitly the method of production of the disturbance: a cord of air, equal in length to the orbit of each planet less the disk of the latter itself, is attached to the extremities of its diameter, and passes through the two hands of the being stationed at the point of disturbance; and he always draws it toward himself by the shorter of the two parts of the cord. The term *ucca*, which we have translated "apex," applies both to the apsis (*manda*, *mandocca*, "apex of slowest motion"—the apogee in the case of the sun and moon, the aphelion, though not recognized as such, in the case of the other planets), and to the conjunction (*çîghra*, *çîghrocca*, "apex of swiftest motion"). The statement made of the like effect of the two upon the motion of the planet is liable to cause difficulty, if it be not distinctly kept in mind that the Hindus understand by the influence of the disturbing cause, not its acceleration and retardation of the rate of the planet's motion, but its effect in giving to the planet a position in advance of, or behind, its mean place. It may be well, for the sake of aiding some of our readers to form a clearer apprehension of the Hindu view of the planetary motions, to expand and illustrate a little this statement of the effect upon them of the two principal disturbing forces.

First, as regards the apsis. This is the remoter extremity of the major axis of the planet's proper orbit, and the point of its slowest motion.

Upon passing this point, the planet begins to fall behind its mean place, but at the same time to gain velocity, so that at the quadrature it is farthest behind, but is moving at its mean rate; during the next quadrant it gains both in rate of motion and in place, until at the perigee, or perihelion, it is moving most rapidly, and has made up what it before lost, so that the mean and true places coincide. Upon passing that point again, it gains upon its mean place during the first quadrant, and loses what it thus gained during the second, until mean and true place again coincide at the apsis. Thus the equation of motion is greatest at the apsides, and nothing at the quadratures, while the equation of place is greatest at the quadratures, and nothing at the apsides; and thus the planet is always behind its mean place while passing from the higher to the lower apsis, and always in advance of it while passing from the lower to the higher; that is, it is constantly drawn away from its mean place toward the higher apsis, *mandocca.*

In treating of the effect of the conjunction, the *çīghrocca*, we have to distinguish two kinds of cases. With Mercury and Venus (see above, i. 29, 31, 32), the revolution of the conjunction takes the place, in the Hindu system as in the Greek, of that of the planet itself, the conjunction being regarded as making the circuit of the zodiac in the same time, and in the same direction, as the planet really revolves about the sun; while the mean place of these planets is always that of the sun itself. While, therefore, the conjunction is making the half-tour of the heavens eastward from the sun, the planet is making its eastward elongation and returning to the sun again, being all the time in advance of its mean place, the sun; when the conjunction reaches a point in the heavens opposite to the sun, the planet is in its inferior conjunction, or at its mean place; during the other half of the revolution of the conjunction, when it is nearest the planet upon the western side, the latter is making and losing its western elongation, or is behind its mean place. Accordingly, as stated in the text, the planet is constantly drawn away from its mean place, the sun, toward that side of the heavens in which the conjunction is.

Once more, as concerns the superior planets. The revolutions assigned to these by the Hindus are their true revolutions; their mean places are their mean heliocentric longitudes; and the place of the conjunction (*çīghrocca*) of each is the mean place of the sun. Since they move but slowly, as compared with the sun, it is their conjunction which approaches, overtakes, and passes them, and not they the conjunction. Their time of slowest motion is when in opposition with the sun; of swiftest, when in conjunction with him: from opposition on to conjunction, therefore, or while the sun is approaching them from behind, they are, with constantly increasing velocity of motion, all the while behind their mean places, or drawn away from them in the direction of the sun; but no sooner has the sun overtaken and passed them, than they, leaving with their most rapid motion the point of coincidence between mean and true place, are at once in advance, and continue to be so until opposition is reached again; that is to say, they are still drawn away from their mean place in the direction of the conjunction.

The words used in verse 5 for "excess" and "deficiency," or for additive and subtractive equation, mean literally "wealth" (*dhana*) and "debt" (*ṛṇa*).

6. In like manner, also, the node, Râhu, by its proper force, causes the deviation in latitude (*vikshepa*) of the moon and the other planets, northward and southward, from their point of declination (*apakrama*).

7. When in the half-orbit behind the planet, the node causes it to deviate northward; when in the half-orbit in front, it draws it away southward.

8. In the case of Mercury and Venus, however, when the node is thus situated with regard to the conjunction (*çîghra*), these two planets are caused to deviate in latitude, in the manner stated, by the attraction exercised by the node upon the conjunction.

The name Râhu, by which the ascending node is here designated, is properly mythological, and belongs to the monster in the heavens, which, by the ancient Hindus, as by more than one other people, was believed to occasion the eclipses of the sun and moon by attempting to devour them. The word which we have translated "force" is *ranhas*, more properly "rapidity, violent motion:" in employing it here, the text evidently intends to suggest an etymology for *râhu*, as coming from the root *rah* or *ranh*, "to rush on": with this same root Weber (Ind. Stud. i. 272) has connected the group of words in which *râhu* seems to belong. For the Hindu fable respecting Râhu, see Wilson's Vishṇu Purâṇa, p. 78. The moon's descending node was also personified in a similar way, under the name of Ketu, but to this no reference is made in the present treatise.

The description of the effect of the node upon the movement of the planet is to be understood, in a manner analogous with that of the effect of the apices in the next preceding passage, as referring to the direction in which the planet is made to deviate from the ecliptic, and not to that in which it is moving with reference to the ecliptic. From the ascending node around to the descending, of course, or while the node is nearest to the planet from behind, the latitude is northern; in the other half of the revolution it is southern.

For an explanation of some of the terms used here, see the note to the last passage of the preceding chapter.

As, in the case of Mercury and Venus, the revolution of the conjunction takes the place of that of the planet itself in its orbit, it is necessary, in order to give the node its proper effect, that it be made to exercise its influence upon the planet through the conjunction. The commentator gives himself here not a little trouble, in the attempt to show why Mercury and Venus should in this respect constitute an exception to the general rule, but without being able to make out a very plausible case.

9. Owing to the greatness of its orb, the sun is drawn away only a very little; the moon, by reason of the smallness of its orb, is drawn away much more;

10. Mars and the rest, on account of their small size, are, by the supernatural beings (*dâivata*) called conjunction (*çîghrocca*) and apsis (*mandocca*), drawn away very far, being caused to vacillate exceedingly.

11. Hence the excess (*dhana*) and deficiency (*ṛṇa*) of these latter is very great, according to their rate of motion. Thus do the planets, attracted by those beings, move in the firmament, carried on by the wind.

The dimensions of the sun and moon are stated below, in iv. 1; those of the other planets, in vii. 13.

We have ventured to translate *ativegita*, at the end of the tenth verse, as it is given above, because that translation seemed so much better to suit the requirements of the sense than the better-supported rendering "caused to move with exceeding velocity." In so doing, we have assumed that the noun *vega*, of which the word in question is a denominative, retains something of the proper meaning of the root *vij*, "to tremble," from which it comes.

12. The motion of the planets is of eight kinds: retrograde (*vakra*), somewhat retrograde (*anuvakra*), transverse (*kuṭila*), slow (*manda*), very slow (*mandatara*), even (*sama*); also, very swift (*çîghratara*), and swift (*çîghra*).

13. Of these, the very swift (*atiçîghra*), that called swift, the slow, the very slow, the even—all these five are forms of the motion called direct (*ṛju*); the somewhat retrograde is retrograde.

This minute classification of the phases of a planet's motion is quite gratuitous, so far as this Siddhânta is concerned, for the terms here given do not once occur afterward in the text, with the single exception of *vakra*, which, with its derivatives, is in not infrequent use to designate retrogradation. Nor does the commentary take the trouble to explain the precise differences of the kinds of motion specified. According to Mr. Hoisington (Oriental Astronomer [Tamil and English], Jaffna: 1848, p. 133), *anuvakra* is applied to the motion of a planet, when, in retrograding, it passes into a preceding sign. From the classification given in the second of the two verses it will be noticed that *kuṭila* is omitted: according to the commentator, it is meant to be included among the forms of retrograde motion; we have conjectured, however, that it might possibly be used to designate the motion of a planet when, being for the moment stationary in respect to longitude, and accordingly neither advancing nor retrograding, it is changing its latitude; and we have translated the word accordingly.

14. By reason of this and that rate of motion, from day to day, the planets thus come to an accordance with their observed places (*dṛç*)—this, their correction (*sphuṭîkaraṇa*), I shall carefully explain.

Having now disposed of matters of general theory and preliminary explanation, the proper subject of this chapter, the calculation of the true (*sphuṭa*) from the mean places of the different planets, is ready to be taken up. And the first thing in order is the table of sines, by means of which all the after calculations are performed.

15. The eighth part of the minutes of a sign is called the first sine (*jyârdha*); that, increased by the remainder left after subtracting from it the quotient arising from dividing it by itself, is the second sine.

16. Thus, dividing the tabular sines in succession by the first, and adding to them, in each case, what is left after subtracting the quotients from the first, the result is twenty-four tabular sines (*jyârdhapiṇḍa*), in order, as follows:

17. Two hundred and twenty-five; four hundred and forty-nine; six hundred and seventy-one; eight hundred and ninety; eleven hundred and five; thirteen hundred and fifteen;

18. Fifteen hundred and twenty; seventeen hundred and nineteen; nineteen hundred and ten; two thousand and ninety-three;

19. Two thousand two hundred and sixty-seven; two thousand four hundred and thirty-one; two thousand five hundred and eighty-five; two thousand seven hundred and twenty-eight;

20. Two thousand eight hundred and fifty-nine; two thousand nine hundred and seventy-eight; three thousand and eighty-four; three thousand one hundred and seventy-seven;

21. Three thousand two hundred and fifty-six; three thousand three hundred and twenty-one; three thousand three hundred and seventy-two; three thousand four hundred and nine;

22. Three thousand four hundred and thirty-one; three thousand four hundred and thirty-eight. Subtracting these, in reversed order, from the half-diameter, gives the tabular versed-sines (*utkramajyârdhapiṇḍaka*):

23. Seven; twenty-nine; sixty-six; one hundred and seventeen; one hundred and eighty-two; two hundred and sixty-one; three hundred and fifty-four;

24. Four hundred and sixty; five hundred and seventy-nine; seven hundred and ten; eight hundred and fifty-three; one thousand and seven; eleven hundred and seventy-one;

25. Thirteen hundred and forty-five; fifteen hundred and twenty-eight; seventeen hundred and nineteen; nineteen hundred and eighteen;

26. Two thousand one hundred and twenty-three; two thousand three hundred and thirty-three; two thousand five hundred and forty-eight; two thousand seven hundred and sixty-seven;

27. Two thousand nine hundred and eighty-nine; three thousand two hundred and thirteen; three thousand four hundred and thirty-eight: these are the versed sines.

We first present, in the following table, in a form convenient for reference and use, the Hindu sines and versed sines, with the arcs to which they belong, the latter expressed both in minutes and in degrees and minutes. To facilitate the practical use of the table in making calculations after the Hindu method, we have added a column of the differences of the sines, and have farther turned the sines themselves into decimal parts of the radius. For the purpose of illustrating the accuracy of the table, we have also annexed the true values of the sines, in minutes, as found by our modern tables. Comparison may also be made of the decimal column with the corresponding values given in our ordinary tables of natural sines.

Table of Sines and Versed Sines.

No.	Arcs, in ° ′	in ′	Hindu Sines, in ′	Diff.	in parts of rad.	True Sines, in ′	Versed Sines, in ′
1	3° 45′	225′	225′	224′	.065445	224′.84	7′
2	7° 30′	450′	449′	222′	.130599	448′.72	29′
3	11° 15′	675′	671′	219′	.195172	670′.67	66′
4	15°	900′	890′	215′	.258871	889′.76	117′
5	18° 45′	1125′	1105′	210′	.321408	1105′.03	182′
6	22° 30′	1350′	1315′	205′	.382489	1315′.57+	261′
7	26° 15′	1575′	1520′	199′	.442117	1520′.48	354′
8	30°	1800′	1719′	191′	.500000	1718′.88	460′
9	33° 45′	2025′	1910′	183′	.555555	1909′.91	579′
10	37° 30′	2250′	2093′	174′	.608784	2092′.77	710′
11	41° 15′	2475′	2267′	164′	.659395	2266′.67	853′
12	45°	2700′	2431′	154′	.707097	2430′.86	1007′
13	48° 45′	2925′	2585′	143′	.751894	2584′.64	1171′
14	52° 30′	3150′	2728′	131′	.793484	2727′.35–	1345′
15	56° 15′	3375′	2859′	119′	.831588	2858′.38–	1528′
16	60°	3600′	2978′	106′	.866201	2977′.18–	1719′
17	63° 45′	3825′	3084′	93′	.897033	3083′.22–	1918′
18	67° 30′	4050′	3177′	79′	.924084	3176′.07–	2123′
19	71° 15′	4275′	3256′	65′	.947062	3255′.31–	2333′
20	75°	4500′	3321′	51′	.965969	3320′.61	2548′
21	78° 45′	4725′	3372′	37′	.980803	3371′.70	2767′
22	82° 30′	4950′	3409′	22′	.991565	3408′.34–	2989′
23	86° 15′	5175′	3431′	7′	.997964	3430′.39–	3213′
24	90°	5400′	3438′		1.000000	3437′.75	3438′

The rule by which the sines are, in the text, directed to be found, may be illustrated as follows. Let s, s', s'', s''', s'''', etc., represent the successive sines. The first of the series, s, is assumed to be equal to its arc, or 225′, from which quantity, as is shown in the table above, it differs only by an amount much smaller than the table takes any account of. Then

$$s' = s + s - \frac{s}{s}$$

$$s'' = s' + s - \frac{s}{s} - \frac{s'}{s}$$

$$s''' = s'' + s - \frac{s}{s} - \frac{s'}{s} - \frac{s''}{s}$$

$$s'''' = s''' + s - \frac{s}{s} - \frac{s'}{s} - \frac{s''}{s} - \frac{s'''}{s}$$

and so on, through the whole series, any fraction larger than a half being counted as one, and a smaller fraction being rejected. In the majority of cases, as is made evident by the table, this process yields correct results: we have marked in the column of "true sines" with a plus or minus sign such modern values of the sines as differ by more than half a minute from those assigned by the Hindu table.

It is not to be supposed, however, that the Hindu sines were originally obtained by the process described in the text. That process was, in all probability, suggested by observing the successive differences in the values of the sines as already determined by other methods. Nor is it difficult to discover what were those methods; they are indicated by the limitation of the table to arcs differing from one another by 3° 45′, and by what we know in general of the trigonometrical methods of the Hindus. The two main principles, by the aid of which the greater portion of all the Hindu calculations are made, are, on the one hand, the equality of the square of the hypothenuse in a right-angled triangle to the sum of the squares of the other two sides, and, on the other hand, the proportional relation of the corresponding parts of similar triangles. The first of these principles gave the Hindus the sine of the complement of any arc of which the sine was already known, it being equal to the square root of the difference between the squares of radius and of the given sine. This led farther to the rule for finding the versed sine, which is given above in the text: it was plainly equal to the difference between the sine complement and radius. Again, the comparison of similar triangles showed that the chord of an arc was a mean proportional between its versed sine and the diameter; and this led to a method of finding the sine of half any arc of which the sine was known: it was equal to half the square root of the product of the diameter into the versed sine. That the Hindus had deduced this last rule does not directly appear from the text of this Siddhânta, nor from the commentary of Ranganâtha, which is the one given by our manuscript and by the published edition; but it is distinctly stated in the commentary which Davis had in his hands (As. Res. ii. 247); and it might be confidently assumed to be known upon the evidence of the table itself; for the principles and rules which we have here stated would give a table just such as the one here constructed. The sine of 90° was obviously equal to radius, and the sine of 30° to half radius: from the first could be found the sines of 45°, 22° 30′, and 11° 15′; from the latter, those of 15°, 7° 30′, and 3° 45′. The sines thus obtained would give those of the complementary arcs, or of 86° 15′, 82° 30′, 78° 45′, 75°, etc.; and the sine of 75°, again, would give those of 37° 30′ and 18° 45′. By continuing the same processes, the table of sines would soon be made complete for the twenty-four divisions of the quadrant; but these processes could yield nothing farther, unless by introducing fractions of minutes; which was undesirable, because the symmetry of the table would thus be destroyed, and no corresponding advantage gained; the table was already sufficiently extended to furnish, by interpolation, the sines intermediate between those given, with all the accuracy which the Hindu calculations required.

If, now, an attempt were made to ascertain a law of progression for the series, and to devise an empirical rule by which its members might be developed, the one from the other, in order, nothing could be more natural than to take the differences of the successive sines, and the differences of those differences, as we have given them under the headings Δ' and Δ'' in the annexed table.

Hindu Sines, with their First and Second Differences.

No.	Sine.	Δ'	Δ''	No.	Sine.	Δ'	Δ''
0	000	225		12	2431	154	10
1	225	224	1	13	2585	143	11
2	449	222	2	14	2728	131	12
3	671	219	3	15	2859	119	12
4	890	215	4	16	2978	106	13
5	1105	210	5	17	3084	93	13
6	1315	205	5	18	3177	79	14
7	1520	199	6	19	3256	65	14
8	1719	191	8	20	3321	51	14
9	1910	183	8	21	3372	37	14
10	2093	174	9	22	3409	22	15
11	2267	164	10	23	3431	7	15
12	2431		10	24	3438		

With these differences before him, an acute observer could hardly fail to notice the remarkable fact that the differences of the second order increase as the sines; and that each, in fact, is about the $\frac{1}{225}$th part of the corresponding sine. Now let the successive sines be represented by 0, s, s', s'', s''', s'''', and so on; and let q equal $\frac{1}{225}$, or $\frac{1}{s}$; let the first differences be $d = s - 0$, $d' = s' - s$, $d'' = s'' - s'$, $d''' = s''' - s''$, etc. The second differences will be: $-sq = d' - d$, $-s'q = d'' - d'$, $-s''q = d''' - d''$, etc. These last expressions give

$$\begin{aligned} d' &= d - sq = s - sq \\ d'' &= d' - s'q = s - sq - s'q \\ d''' &= d'' - s''q = s - sq - s'q - s''q, \text{ etc.} \end{aligned}$$

Hence, also,

$$\begin{aligned} s' &= s + d' = s + s - sq \\ s'' &= s' + d'' = s' + s - sq - s'q \\ s''' &= s'' + d''' = s'' + s - sq - s'q - s''q, \end{aligned}$$

and so on, according to the rule given in the text.

That the second differences in the values of the sines were proportional to the sines themselves, was probably known to the Hindus only by observation. Had their trigonometry sufficed to demonstrate it, they might easily have constructed a much more complete and accurate table of sines. We add the demonstration given by Delambre (Histoire de l'Astronomie Ancienne, i. 458), from whom the views here expressed have been substantially taken.

Let a be any arc in the series, and put $3° 45' = n$. Then $\sin(a-n)$, $\sin a$, $\sin(a+n)$, will be three successive terms in the series: $\sin a - \sin(a-n)$, and $\sin(a+n) - \sin a$, will be differences of the first order; and their difference, $\sin(a+n) + \sin(a-n) - 2\sin a$, will be a difference of the second order. But this last expression, by virtue of the formula $R \sin(a \pm n) = \sin a \cos n \pm \cos a \sin n$, reduces to $2 \sin a \cos n \div R - 2 \sin a$, or $2\left(\frac{\cos n}{R} - 1\right) \sin a$. That is to say, the second difference is equal to the product of the sine of the arc a into a certain constant quantity, or it varies as the sine. When n equals 3° 45′, as in the Hindu table, it is easy to show, upon working out the last expression by means of the tables, that the constant factor is, as stated by Delambre, $\frac{1}{233.5}$, instead of being $\frac{1}{225}$, as empirically determined by the Hindus.

It deserves to be noticed, that the commentary of Ranganâtha recognizes the dependence of the rule given in the text upon the value of the second differences. According to him, however, it is by describing a circle upon the ground, laying off the arcs, drawing the sines, and determining their relations by inspection, that the method is obtained. The differences of the sines, he says, will be observed to decrease, while the differences of those differences increase; and it will be noticed that the last second difference is 15′ 16″ 48‴. A proportion is then made: if at the radius the second difference is of this value, what will it be at any sine? or, taking the first sine as an example, 3438′ : 15′ 16″ 48‴ : : 225 : 1. Nothing can be clearer, however, than that this pretended result of inspection is one of calculation merely. It would be utterly impossible to estimate by the eye the value of a difference with such accuracy, and, were it possible, that difference would be found very considerably removed from the one here given, being actually only about 14′ 45″. The value 15′ 16″ 48‴ is assumed only in order to make its ratio to the radius exactly $\frac{1}{225}$.

The earliest substitution of the sines, in calculation, for the chords, which were employed by the Greeks, is generally attributed (see Whewell's History of the Inductive Sciences, B. III. ch. iv. 8) to the Arab astronomer Albategnius (al-Battânî), who flourished in the latter part of the ninth century of our era. It can hardly admit of question, however, that sines had already at that time been long employed by the Hindus. And considering the derivation by the Arabs from India of their system of notation, and of so many of the elements of their mathematical science, it would seem not unlikely that the first hint of this so convenient and practical improvement of the methods of calculation may also have come to them from that country. This cannot be asserted, however, with much confidence, because the substitution of the sines for the chords seems so natural and easy, that it may well enough have been hit upon independently by the Arabs; it is a matter for astonishment, as remarked by Delambre (Histoire de l'Astronomie du Moyen Age, p. 12), that Ptolemy himself, who came so near it, should have failed of it. If Albategnius got the suggestion from India, he, at any rate, got no more than that. His table of sines, much more complete than that of the Hindus, was made from Ptolemy's table of chords, by simply halving them. The method, too, which in India remained comparatively barren, led to valuable developments in the hands of the Arab mathematicians, who went on by degrees to form also tables of tangents and co-tangents, secants and co-secants; while the Hindus do not seem to have distinctly appreciated the significance even of the cosine.

In this passage, the sine is called *jyârdha*, "half-chord;" hereafter, however, that term does not once occur, but *jyâ* "chord" (literally "bow-string") is itself employed, as are also its synonyms *jîvâ*, *mâurvikâ*, to denote the sine. The usage of Albategnius is the same. The sines of the table are called *piṇḍa*, or *jyâpiṇḍa*, "the quantity corresponding to the sine." The term used for versed sine, *utkramajyâ*, means "inverse-order sine," the column of versed sines being found by subtracting that of sines in inverse order from radius.

The ratio of the diameter to the circumference involved in the expression of the value of radius by 3438′ is, as remarked above (under i. 59, 60), 1 : 3.14136. The commentator asserts that value to come from the ratio 1250 : 3927, or 1 : 3.1416, and it is, in fact, the nearest whole number to the result given by that ratio. If the ratio were adopted which has been stated above (in i. 59), of $1 : \sqrt{10}$, the value of radius would be only 3415′. It is to be observed with regard to this latter ratio, that it could not possibly be the direct result of any actual process adopted for ascertaining the value of the diameter from that of the circumference, or the contrary. It was probably fixed upon by the Hindus because it looked and sounded well, and was at the same time a sufficiently near approximation to the truth to be applied in cases where exactness was neither attainable by their methods, nor of much practical consequence; as in fixing the dimensions of the earth, and of the planetary orbits. The nature of the system of notation of the Hindus, and their constantly recurring extraction of square roots in their trigonometrical processes, would cause the suggestion to them, much more naturally than to the Greeks, of this artificial ratio, as not far from the truth; and their science was just of that character to choose for some uses a relation expressed in a manner so simple, and of an aspect so systematical, even though known to be inaccurate. We do not regard the ratio in question, although so generally adopted among the Hindu astronomers, as having any higher value and significance than this.

28. The sine of greatest declination is thirteen hundred and ninety-seven; by this multiply any sine, and divide by radius; the arc corresponding to the result is said to be the declination.

The greatest declination, that is to say, the inclination of the plane of the ecliptic, is here stated to be 24°, 1397′ being the sine of that angle. The true inclination in the year 300 of our era, which we may assume to have been not far from the time when the Hindu astronomy was established, was a little less than 23° 40′, so that the error of the Hindu determination was then more than 20′: at present, it is 32′ 34″. The value assigned by Ptolemy (Syntaxis, i) to the inclination was between 23° 50′ and 23° 52′ 30″; an error, as compared with its true value in the time of Hipparchus, of only about 7′.

The second half of the verse gives, in the usual vague and elliptical language of the treatise, the rule for finding the declination of any given point in the ecliptic. We have not in this case supplied the ellipses in our translation, because it could not be done succinctly, or without introducing an element, that of the precession, which possibly was not taken into account when the rule was made. See what is said upon this subject under verses 9 and 10 of the next chapter. The "sine" employed is, of course, the sine of the distance from the vernal equinox, or of the longitude as corrected by the precession.

The annexed figure will explain the rule, and the method of its demonstration.

Let A C E represent a quadrant of the plane of the equatorial, and A C G a quadrant of that of the ecliptic, A C being the line of their intersection: then A P is the equinoctial colure, P E the solstitial, G E, or the angle G C E, the inclination of the ecliptic, or the greatest declination (*paramâpakrama*, or *paramakrânti*), and G D its sine (*paramakrântijyâ*). Let S be the position of the sun, and draw the circle of declination P H; S H, or the angle S C H, is the declination of the sun at that point, and S F the sine of declination (*krântijyâ*). From S and F draw S B and F B at right angles to A C; then S B is the sine of the arc A S, or of the sun's longitude. But G C D and S B F are similar right-angled triangles, having their angles at C and B each equal to the inclination. Therefore C G : G D : : S B : S F; and $SF = \frac{GD \times SB}{CG}$;

that is, $\text{sin decl.} = \frac{\text{sin incl.} \times \text{sin long.}}{R}$.

Fig. 1.

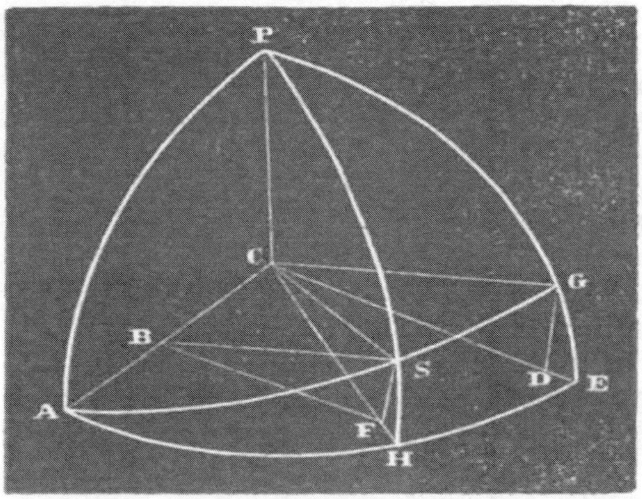

The same result is, by our modern methods, obtained directly from the formula in right-angled spherical trigonometry: $\sin c = \sin a \sin C$; or, in the triangle A S H, right-angled at H, sin S H = sin S A sin S A H.

29. Subtract the longitude of a planet from that of its apsis (*mandocca*); so also, subtract it from that of its conjunction (*çîghra*); the remainder is its anomaly (*kendra*); from that is found the quadrant (*pada*); from this, the base-sine (*bhujajyâ*), and likewise that of the perpendicular (*koṭi*).

30. In an odd (*vishama*) quadrant, the base-sine is taken from the part past, the perpendicular from that to come; but in an even (*yugma*) quadrant, the base-sine (*bâhujyâ*) is taken from the part to come, and the perpendicular-sine from that past.

The distance of a planet from either of its two apices of motion, or centres of disturbance, is called its *kendra ;* according to the commentary, its distance from the apsis (*mandocca*) is called *mandakendra*, and that from the conjunction (*çîghrocca*) is called *çîghrakendra :* the Sûrya-Siddhânta, however, nowhere has occasion to employ these terms. The former of the two corresponds to what in modern astronomy is called the anomaly, the latter to what is known as the commutation. The word *kendra* is not of Sanskrit origin, but is the Greek *κέντρον ;* it is a circumstance no less significant to meet with a Greek word thus at the very foundation of the method of calculating the true place of a planet by means of a system of epicycles, than to find one, as noticed above (under i. 52), at the base of the theory of planetary regency upon which depend the names and succession of the days of the week. Both anomaly and commutation, it will be noticed, are, according to this treatise, to be reckoned always forward from the planet to its apsis and conjunction respectively; excepting that, in the case of Mercury and Venus, owing to the exchange with regard to those planets of the place of the planet itself with that of its conjunction, the commutation is really reckoned the other way. The functions of any arc being the same with those of its negative, it makes no difference, of course, whether the distance is measured from the planet to the apex (*ucca*), or from the apex to the planet.

The quantities actually made use of in the calculations which are to follow are the sine and cosine of the anomaly, or of the commutation. The terms employed in the text require a little explanation. *Bhuja* means "arm;" it is constantly applied, as are its synonyms *bâhu* and *dos*, to designate the base of a right-angled triangle; *koṭi* is properly "a recurved extremity," and, as used to signify the perpendicular in such a triangle, is conceived of as being the end of the *bhuja*, or base, bent up to an upright position: *bhujajyâ* and *koṭijyâ*, then, are literally the values, as sines, of the base and perpendicular of a right-angled triangle of which the hypothenuse is made radius: owing to the relation to one another of the oblique angles of such a triangle, they are respectively as sine and cosine. We have not been willing to employ these latter terms in translating them, because, as before remarked, the Hindus do not seem to have conceived of the cosine, the sine of the complement, of an arc, as being a function of the arc itself.

To find the sine and cosine of the planet's distance from either of its apices (*ucca*) is accordingly the object of the directions given in verse 30 and the latter part of the preceding verse. The rule itself is only the awkward Hindu method of stating the familiar truth that the sine and cosine of an arc and of its supplement are equal. The accompanying figure will, it is believed, illustrate the Hindu manner of looking at the subject. Let P be the place of a planet, and divide its orbit into the four quadrants P Q, Q R, R S, and S P; the first and third of these are called the odd (*vishama*) quadrants; the second and fourth, the even (*yugma*) quadrants. Let A, B, C, and D, be four positions of the apsis (or of the conjunction); then the arcs P A, P Q B, P Q R C, P Q R S D will be the values of the anomaly in each case. A M, the base-sine, or sine of anomaly, when the apsis is in the first quadrant, is

Fig. 2.

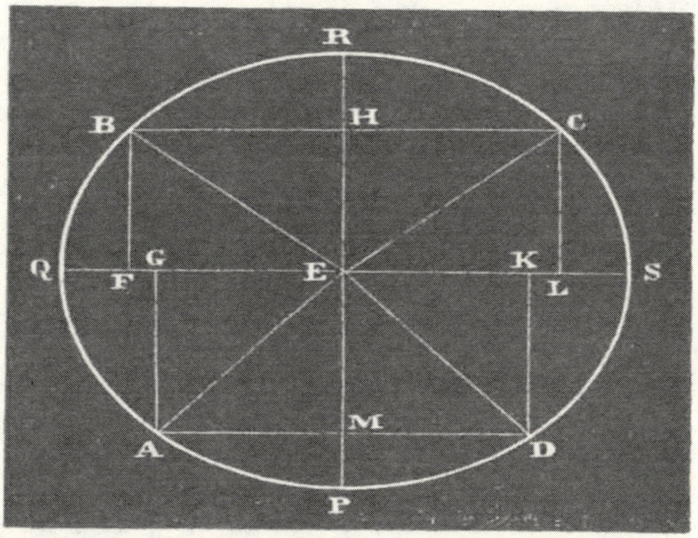

determined by the arc A P, the arc passed over in reckoning the anomaly, while A G or E M, the perpendicular-sine, or cosine, is taken from the arc A Q, the remaining part of the quadrant. The same is true in the other odd quadrant, R S; the sine C H, or E L, comes from R C, the part of the quadrant between the planet and the apsis; the cosine C L is from its complement. But in the even quadrants, Q R and S P, the case is reversed; the sines, B H, or E F, and D M, are determined by the arcs B R and D P, the parts of the quadrant not included in the anomaly, and the cosines, B F and K D, or E M, correspond to the other portions of each quadrant respectively.

This process, of finding what portion of any arc greater than a quadrant is to be employed in determining its sine, is ordinarily called in Hindu calculations "taking the *bhuja* of an arc."

31. Divide the minutes contained in any arc by two hundred and twenty-five; the quotient is the number of the preceding tabular sine (*jyâpiṇḍaka*). Multiply the remainder by the difference of the preceding and following tabular sines, and divide by two hundred and twenty-five;

32. The quotient thus obtained add to the tabular sine called the preceding; the result is the required sine. The same method is prescribed also with respect to the versed sines.

33. Subtract from any given sine the next less tabular sine; multiply the remainder by two hundred and twenty-five, and divide by the difference between the next less and next greater tabular sines; add the quotient to the product of the serial number of the next less sine into two hundred and twenty-five: the result is the required arc.

The table of sines and versed sines gives only those belonging to arcs which are multiples of 3° 45′; the first two verses of this passage state the method of finding, by simple interpolation, the sine or versed sine of any intermediate arc; while the third verse gives the rule for the contrary process, for converting any given sine or versed sine in the same manner into the corresponding arc.

In illustration of the first rule, let us ascertain the sine corresponding to an arc of 24°, or 1440′. Upon dividing the latter number by 225, we obtain the quotient 6, and the remainder 90′. This preliminary step is necessary, because the Hindu table is not regarded as containing any designation of the arcs to which the sines belong, but as composed simply of the sines themselves in their order. The sine corresponding to the quotient obtained, or the sixth, is 1315′: the difference between it and the next following sine is 205′. Now a proportion is made: if, at this point in the quadrant, an addition of 225′ to the arc causes an increase in the sine of 205′, what increase will be caused by an addition to the arc of 90′: that is to say, 225 : 205 : : 90 : 82. Upon adding the result, 82′, to the sixth sine, the amount, 1397′, is the sine of the given arc, as stated in verse 28. The actual value, it may be remarked, of the sine of 24°, is 1398′.26.

The other rule is the reverse of this, and does not require illustration.

The extreme conciseness aimed at in the phraseology of the text, and not unfrequently carried by it beyond the limit of distinctness, or even of intelligibility, is well illustrated by verse 33, which, literally translated, reads thus: "having subtracted the sine, the remainder, multiplied by 225, divided by its difference, having added to the product of the number and 225, it is called the arc." In verse 31, also, the important word "remainder" is not found in the text.

The proper place for this passage would seem to be immediately after the table of sines and versed sines: it is not easy to see why verses 28–30 should have been inserted between, or indeed, why the subject of the inclination of the ecliptic is introduced at all in this part of the chapter, as no use is made of it for a long time to come.

34. The degrees of the sun's epicycle of the apsis (*manda-paridhi*) are fourteen, of that of the moon, thirty-two, at the end of the even quadrants; and at the end of the odd quadrants, they are twenty minutes less for both.

35. At the end of the even quadrants, they are seventy-five, thirty, thirty-three, twelve, forty-nine; at the odd (*oja*) they are seventy-two, twenty-eight, thirty-two, eleven, forty-eight,

36. For Mars and the rest; farther, the degrees of the epicycle of the conjunction (*çîghra*) are, at the end of the even quadrants, two hundred and thirty-five, one hundred and thirty-three, seventy, two hundred and sixty-two, thirty-nine;

37. At the end of the odd quadrants, they are stated to be two hundred and thirty-two, one hundred and thirty-two, seventy-two, two hundred and sixty, and forty, as made use of in the calculation for the conjunction (*çîghrakarman*).

38. Multiply the base-sine (*bhujajyâ*) by the difference of the epicycles at the odd and even quadrants, and divide by radius (*trijyâ*); the result, applied to the even epicycle (*vṛta*), and additive (*dhana*) or subtractive (*ṛṇa*), according as this is less or greater than the odd, gives the corrected (*sphuṭa*) epicycle.

The corrections of the mean longitudes of the planets for the disturbing effect of the apsis (*mandocca*) and conjunction (*çîghrocca*) of each—that is to say, for the effect of the ellipticity of their orbits, and for that of the annual parallax, or of the motion of the earth in its orbit—are made in Hindu astronomy by the Ptolemaic method of epicycles, or secondary circles, upon the circumference of which the planet is regarded as moving, while the centre of the epicycle revolves about the general centre of motion. The details of the method, as applied by the Hindus, will be made clear by the figures and processes to be presented a little later; in this passage we have only the dimensions of the epicycles assumed for each planet. For convenience of calculation, they are measured in degrees of the orbits of the planets to which they severally belong; hence only their relative dimensions, as compared with the orbits, are given us. The data of the text belong to the planets in the order in which these succeed one another as regents of the days of the week, viz., Mars, Mercury, Jupiter, Venus, and Saturn (see above, under i. 51, 52). The annexed table gives the dimensions of the epicycles, both their circumferences, which are presented directly by the text, and their radii, which we have calculated after the method of this Siddhânta, assuming the radius of the orbit to be 3438′.

Dimensions of the Epicycles of the Planets.

Planet.	Epicycle of the apsis: at even quadrant, circ.	rad.	at odd quadrant, circ.	rad.	Epicycle of the conjunction: at even quadrant, circ.	rad.	at odd quadrant, circ.	rad.
Sun,	14°	133′.70	13° 40′	130′.52	....	....	....	....
Moon,	32°	305′.60	31° 40′	302′.42	....	....	....	....
Mercury,	30°	286′.50	28°	267′.40	133°	1270′.15	132°	1260′.60
Venus,	12°	114′.60	11°	105′.05	262°	2502′.10	260°	2483′.00
Mars,	75°	716′.25	72°	687′.60	235°	2244′.25	232°	2215′.60
Jupiter,	33°	315′.15	32°	305′.60	70°	668′ 50	72°	687′.60
Saturn,	49°	467′.95	48°	458′.40	39°	372′.45	40°	382′.00

A remarkable peculiarity of the Hindu system is that the epicycles are supposed to contract their dimensions as they leave the apsis or the conjunction respectively (excepting in the case of the epicycles of the conjunction of Jupiter and Saturn, which expand instead of contracting), becoming smallest at the quadrature, then again expanding till the lower apsis, or opposition, is reached, and decreasing and increasing in like manner in the other half of the orbit; the rate of increase and diminution being as the sine of the distance from the apsis, or conjunction. Hence the rule in verse 38, for finding the true dimensions of the epicycle at any point in the orbit. It is founded upon the simple proportion: as radius, the sine of the distance at which the diminution (or increase) is greatest, is to the amount of diminution (or of increase) at that point, so is the sine of the given distance to the corresponding diminution (or increase); the application of the correction thus obtained to the dimensions of the epicycle at the apsis, or conjunction, gives the true epicycle.

We shall revert farther on to the subject of this change in the dimensions of the epicycle.

The term employed to denote the epicycle, *paridhi*, means simply "circumference," or "circle;" it is the same which is used elsewhere in this treatise for the circumference of the earth, etc. In a single instance, in verse 38, we have *vṛtta* instead of *paridhi;* its signification is the same, and its other uses are closely analogous to those of the more usual term.

39. By the corrected epicycle multiply the base-sine (*bhujajyâ*) and perpendicular-sine (*koṭijyâ*) respectively, and divide by the number of degrees in a circle: then, the arc corresponding to the result from the base-sine (*bhujajyâphala*) is the equation of the apsis (*mânda phala*), in minutes, etc.

All the preliminary operations having been already performed, this is the final process by which is ascertained the equation of the apsis, or the amount by which a planet is, at any point in its revolution, drawn away from its mean place by the disturbing influence of the apsis. In modern phraseology, it is called the first inequality, due to the ellipticity of the orbit; or, the equation of the centre.

Figure 3, upon the next page, will serve to illustrate the method of the process.

Let A M M′ P represent a part of the orbit of any planet, which is supposed to be a true circle, having E, the earth, for its centre. Along this orbit the planet would move, in the direction indicated by the arrow, from A through M and M′ to P, and so on, with an equable motion, were it not for the attraction of the beings situated at the apsis (*mandocca*) and conjunction (*çighrocca*) respectively. The general mode of action of these beings has been explained above, under verses 1–5 of this chapter: we have now to ascertain the amount of the disturbance produced by them at any given point in the planet's revolution. The method devised is that of an epicycle, upon the circumference of which the planet revolves with an equable motion, while the centre of the epicycle traverses the orbit with a velocity equal to that of the planet's mean motion, having always a position coincident with the mean place of the planet. At present, we have to do only with the epicycle which represents the disturbing effect of the apsis (*mandocca*). The period of the planet's revolution about the centre of the epicycle is the time which it takes the latter to make the circuit of the orbit from the apsis around to the apsis again, or the period of its anomalistic revolution. This is almost precisely equal to the period of sidereal revolution in the case of all the planets excepting the moon, since their apsides are regarded by the Hindus as stationary (see above, under i. 41–44): the moon's apsis, however, has a forward motion of more than 40° in a year; hence the moon's anomalistic revolution is very perceptibly longer than its sidereal, being $27^d\ 13^h\ 18^m$. The arc of the epicycle traversed by the planet at any mean point in its revolution is accordingly always equal to the arc of the orbit intercepted between that point and the apsis, or to the mean anomaly, when the latter is reckoned, in the usual manner, from the apsis forward to the planet. Thus, in the figure, suppose A to be the place of the apsis (*mandocca*, the apogee of the sun and moon, the aphelion of the other planets), and P that of the opposite point (perigee, or perihelion; it has in this treatise no distinctive name); and let M and M′ be two mean positions of the planet, or actual positions of the centre of the epicycle; the lesser circles drawn about these four points represent the epicycle: this is made, in the figure, of twice the size of that assumed for the moon, or a little smaller than that of Mars. Then, when the centre of the epicycle is at A, the planet's place in the epicycle is at *a*; as the centre advances to M, M′, and P, the planet moves in the opposite direction, to *m*, *m′*, and *p*, the arc *a′ m* being equal to A M, *a″ m′* to A M′, and *a‴ p* to A P. It is as if, while the axis E *a* revolves about E, the part of it A *a* remained constant in direction, parallel to E A, assuming the positions M *m*, M′ *m′*, and P *p* successively. The effect of this combination of motions is to make the planet virtually traverse the orbit indicated in the figure by the broken line, which is a circle of equal radius with the true orbit, but having its centre removed from E, toward A, by a distance equal to

Fig. 8.

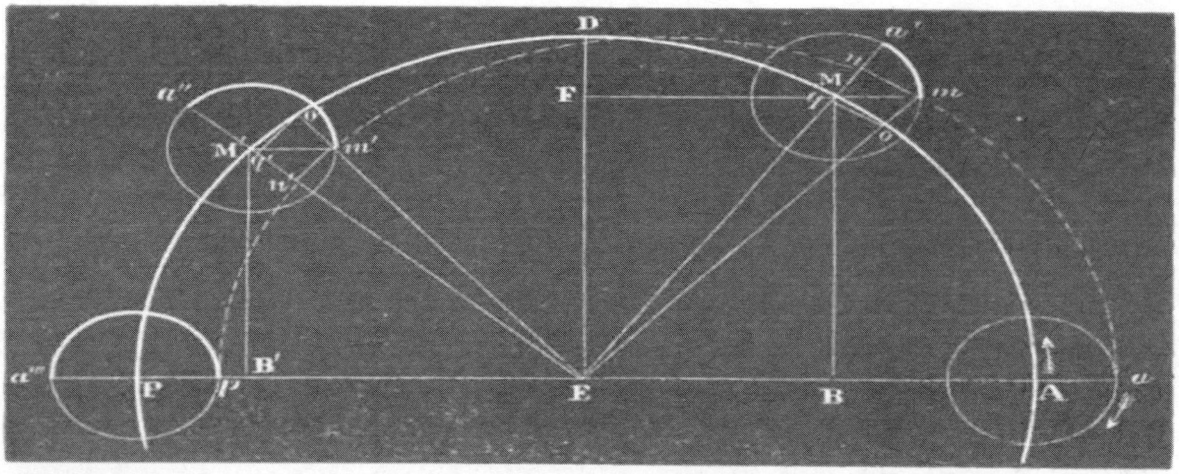

A *a*, the radius of the epicycle. This identity of the virtual orbit with an eccentric circle, of which the eccentricity is equal to the radius of the epicycle, was doubtless known to the Hindus, as to Ptolemy: the latter, in the third book of his Syntaxis, demonstrates the equivalence of the suppositions of an epicycle and an eccentric, and chooses the latter to represent the first inequality: the Hindus have preferred the other supposition, as better suited to their methods of calculation, and as admitting a general similarity in the processes for the apsis and the conjunction. The Hindu theory, however, as remarked above (under vv. 1–5 of this chapter), rejects the idea of the actual motion of the planet in the epicycle, or on the eccentric circle: the method is but a device for ascertaining the effect of the attractive force of the being at the apsis. Thus the planet really moves in the circle A M M′ P, and if the lines E *m*, E *m′* be drawn, meeting the orbit in *o* and *o′*, its actual place is at *o* and *o′*, when its mean place is at M and M′ respectively. To ascertain the value of the arcs *o* M and *o′* M′, which are the amount of removal from the mean place, or the equation, is the object of the process prescribed by the text.

Suppose the planet's mean place to be M, its mean distance from the apsis being A M: it has traversed, as above explained, an equal arc, *a′m*, in the epicycle. From M draw M B and M F, and from *m* draw *m n*, at right angles to the lines upon which they respectively fall: then M B is the base-sine (*bhujajyâ*), or the sine of mean anomaly, and M F, or its equal E B, is the perpendicular-sine (*koṭijyâ*), or cosine, and *m n* and *n* M are corresponding sine and cosine in the epicycle. But as the relation of the circumference of the orbit to that of the epicycle is known, and as all corresponding parts of two circles are to one another as their respective circumferences, the values of *m n* and *n* M are found by a proportion, as follows: as 360° is to the number of degrees in the circumference of the epicycle at M, so is M B to *m n*, and E B to *n* M. Hence *m n* is called the "result from the base-sine" (*bhujajyâphala*, or, more briefly, *bhujaphala*, or *bâhuphala*), and *n* M the "result from the perpendicular-sine" (*koṭijyâphala*, or *koṭiphala*): the latter of the two, however, is not employed in the process for calculating the equation of the apsis. Now, as the dimensions of the epicycle of the apsis are in all cases small, *m n* may without any considerable error be assumed to be equal to *o q*, which is the sine of the arc *o* M, the equation: this assumption is accordingly made, and the conversion of *m n*, as sine, into its corresponding arc, gives the equation required.

The same explanation applies to the position of the planet at M′: *a″ m′*, the equivalent of A M M′, is here the arc of the epicycle traversed; *m′ n′*, its sine, is calculated from M′ B′, as before, and is assumed to equal *o′ q′*, the sine of the equation *o′* M′.

To give a farther and practical illustration of the process, we will proceed to calculate the equation of the apsis for the moon, at the time for which her mean place has been found in the notes to the last chapter, viz., the 1st of January, 1860, midnight, at Washington.

Moon's mean longitude, midnight, at Ujjayinî (i. 53),	11ˢ	15°	23′	24″
add the equation for difference of meridian (*deçântaraphala*), or for her motion between midnight at Ujj. and Wash. (i. 60, 61),		5	35	37
Moon's mean longitude at required time,	11	20	59	1
Longitude of moon's apsis, midnight, at Ujjayinî (i. 53),	10	9	42	26
add for difference of meridian, as above,			2	50
Longitude of moon's apsis at required time,	10	9	45	16
deduct moon's mean longitude (ii. 29),	11	20	59	1
Moon's mean anomaly (*mandakendra*),	10	18	46	15

The anomaly being reckoned forward on the orbit from the planet, the position thus found for the moon relative to the apsis is, nearly enough for purposes of illustration, represented by M in the figure. By the rule given above, in verse 30, the base-sine (*bhujajyâ*)—since the anomaly is in the fourth, an even, quadrant—is to be taken from the part of the quadrant not included in the anomaly, or A M; the perpendicular-sine (*koṭijyâ*) is that corresponding to its complement, or M D. That is to say:

From the anomaly,	10ˢ	18°	46′	15″
deduct three quadrants,	9			
remains the arc M D,	1	18	46	15
take this from a quadrant,	3			
remains the arc A M,	1	11	13	45

And by the method already illustrated under verses 31, 32, the sine corresponding to the latter arc, which is the base-sine (*bhujajyâ*), or the sine of mean anomaly, M B, is found to be 2266′; that from M D, which is M F, or E B, the perpendicular-sine (*koṭijyâ*), or cosine of mean anomaly, is 2585′.

The next point is to find the true size of the epicycle at M. By verse 34, the contraction of its circumference amounts at D to 20′; hence, according to the rule in verse 38, we make the proportion, sin A D : 20′ : : sin A M : diminution at M; or,

$$3438 : 20 :: 2266 : 13$$

Deducting from 32°, the circumference of the epicycle at A, the amount of diminution thus ascertained, we have 31° 47′ as its dimensions at M.

Once more, by verse 39, we make the proportion, circ. of orbit : circ. of epicycle : : M B : *m n* ; or,

$$360° : 31° 47' :: 2266 : 200$$

The value, then, of *m n*, the result from the base-sine (*bhujajyâphala*), is 200′; which, as *m n* is assumed to equal *o q*, is the sine of the equation. Being less than 225′, its arc (see the table of sines, above) is of the same value: 3° 20′, accordingly, is the moon's equation of the apsis (*mânda phala*) at the given time: the figure shows it to be subtractive (*ṛṇa*), as the rule in verse 45 also declares it. Hence, from the

Moon's mean longitude,	11ˢ	20°	59′
deduct the equation,		3	20
Moon's true longitude,	11	17	39

We present below, in a briefer form, the results of a similar calculation made for the sun, at the same time.

Sun's mean longitude, midnight, at Ujjayinî (i. 53),	8ˢ	17°	48′	7″
add for difference of meridian (i. 60, 61),			25	6
Sun's mean longitude at required time,	8	18	13	13
Longitude of sun's apsis (i. 41),	2	17	17	24
Sun's mean anomaly (ii. 29),	5	29	4	11
subtract from two quadrants (ii. 30),	6			
Arc determining base-sine,			55′	49″
Base-sine (*bhujajyâ*),			56′	
Dimensions of epicycle (ii. 38),		14°		
Result from base-sine (*bhujajyâphala*), or sine of equation (ii. 39),			2′	
Equation (*mânda phala*, ii. 45),			+2′	
Sun's true longitude,	8ˢ	18°	15′	

In making these calculations, we have neglected the seconds, rejecting the fraction of a minute, or counting it as a minute, according as it was less or greater than a half. For, considering that this method is followed in the table of sines, which lies at the foundation of the whole process, and considering that the sine of the arc in the epicycle is assumed to be equal to that of the equation, it would evidently be a waste of labor, and an affectation of an exactness greater than the process contemplates, or than its general method renders practicable, to carry into seconds the data employed.

As stated below, in verse 43, the equation thus found is the only one required in determining the true longitude of the sun and of the moon: in the case of the other planets, however, of which the apparent place is affected by the motion of the earth, a much longer and more complicated process is necessary, of which the explanation commences with the next following passage.

The Ptolemaic method of making the calculation of the equation of the centre for the sun and moon is illustrated by the annexed figure (Fig. 4). The points E, A, M, *a*, *m*, and *o*, correspond with those similarly marked in the last figure (Fig. 3). The centre of the eccentric circle is at *e*, and E*e*, which equals A*a*, is the eccentricity, which is given. Join *em*; the angle *mea* equals MEA, the mean anomaly, and E*me* equals ME*o*, the equation. Extend *me* to *d*, where it meets E*d*, a perpendicular let fall upon it from E. Then, in the right-angled triangle E*ed*, the side E*e* and the angles—since E*ed* equals *mea*—are given, to find the other sides, *ed* and *d*E. Add *ed* to *em*, the radius; add the square of the sum to that of E*d*; the square root of their sum is E*m*: then, in the right-angled triangle *m*E*d*, all the sides and the right angle are given, to find the angle E*me*, the equation.

Fig. 4.

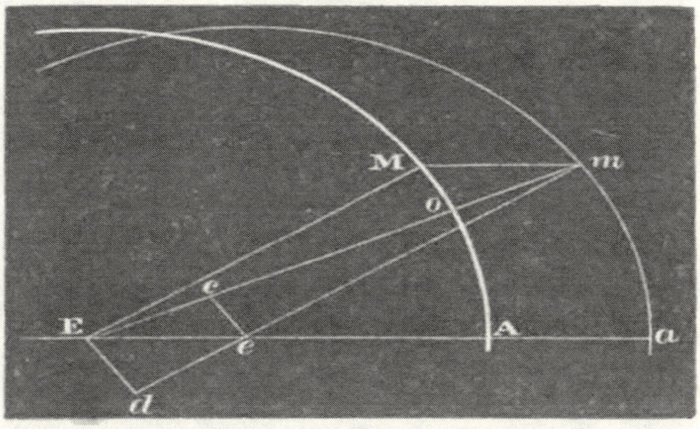

This process is equivalent to a transfer of the epicycle from M to E; E*d* becomes the result from the base-sine (*bhujajyâphala*), and *de* that from the perpendicular-sine (*koṭijyâphala*), and the angle of the equation is found in the same manner as its sine, *ec*, is found in the Hindu process next to be explained; while, in that which we have been considering, E*d* is assumed to be equal to *ec*.

Ptolemy also adds to the moon's orbit an epicycle, to account for her second inequality, the evection, the discovery of which does him so much honor. Of this inequality the Hindus take no notice.

40. The result from the perpendicular-sine (*koṭiphala*) of the distance from the conjunction is to be added to radius, when the distance (*kendra*) is in the half-orbit beginning with Capricorn; but when in that beginning with Cancer, the result from the perpendicular-sine is to be subtracted.

41. To the square of this sum or difference add the square of the result from the base-sine (*bâhuphala*); the square root of their sum is the hypothenuse (*karṇa*) called variable (*cala*). Multiply the result from the base-sine by radius, and divide by the variable hypothenuse:

42. The arc corresponding to the quotient is, in minutes, etc., the equation of the conjunction (*çâighrya phala*); it is employed in the first and in the fourth process of correction (*karman*) for Mars and the other planets.

The process prescribed by this passage is essentially the same with that explained and illustrated under the preceding verse, the only difference being that here the sine of the required equation, instead of being assumed equal to that of the arc traversed by the planet in the epicycle, is obtained by calculation from it. The annexed figure (Fig. 5) will exhibit the method pursued.

The larger circle, CMM′O, represents, as before, the orbit in which any one of the planets, as also the being at its conjunction (*çîghrocca*) are

Fig. 5.

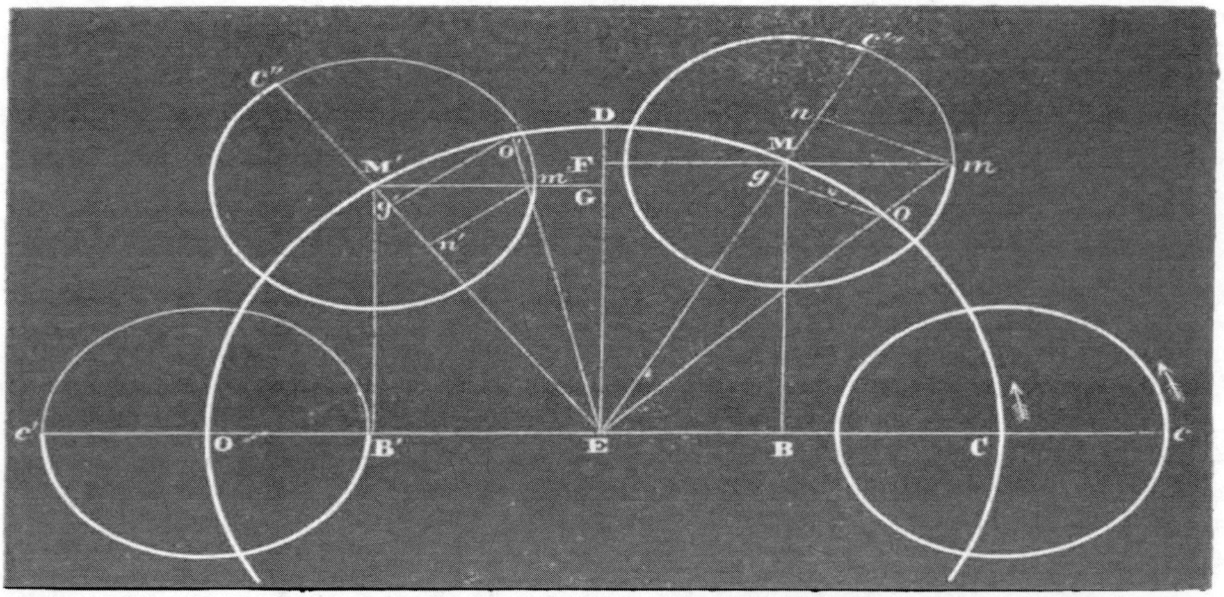

making the circuit of the heavens about E, the earth, as a centre, in the direction indicated by the arrow, from C through M and M′ to O, and so on. But since, in every case, the conjunction moves more rapidly eastward than the planet, overtaking and passing it, if we suppose the conjunction stationary at C, the virtual motion of the planet relative to that point is backward, or from O through M′ and M to C, its mean rate of approach toward C being the difference between the mean motion of the planet and that of the sun. As before, the amount to which the planet is drawn away from its mean place toward the conjunction is calculated by means of an epicycle. The circles drawn in the figure to represent the epicycle are of the relative dimensions of that assigned to Mercury, or a little more than half that of Mars. The direction of the planet's motion in the epicycle is the reverse of that in the epicycle of the apsis, as regards the actual motion of the planet in its orbit, being eastward at the conjunction; as regards the motion of the planet relative to the conjunction, it is the same as in the former case, being in the contrary direction at the conjunction; its effect, of course, is to increase the rate of the eastward movement at that point. The time of the planet's revolution about the centre of the epicycle is the interval between two successive passages through the point C, the conjunction: that is to say, it is equal to the period of synodical revolution of each planet. These periods are, according to the elements presented in the text of this Siddhânta, as follows;

Mercury,	115d	21h	42m
Venus,	583	21	37
Mars,	779	22	11
Jupiter,	398	21	20
Saturn,	378	2	4

The arc of the epicycle traversed by the planet, at any point in its revo-tion, is equal to its distance from the conjunction, when reckoned for-d from the planet, according to the method prescribed in verse 29.

Suppose, now, the mean place of the planet, relative to its conjunction (*çîghrocca*) at C, to be at M: its place in the epicycle is at *m*, as far from *c'''*, in either direction, as M from C. The arc of the epicycle already traversed is indicated in this figure, as in Fig. 3, by the heavier line. Draw E*m*, cutting the orbit in *o*; then *o* is the planet's true place, and *o*M is the equation, or the amount of removal from the mean place by the attraction of the being at C.

The sine and cosine of the distance from the conjunction, the dimensions of the epicycle, and the value of the correspondents in the epicycle to the sine and cosine, are found as in the preceding process. Add *n*M, the result from the cosine (*koṭijyâphala*), to ME, the radius: the result is the perpendicular, E*n*, of the triangle E*nm*. To the square of E*n* add that of the base *nm*, the result from the sine (*bhujajyâphala*); the square root of the sum is the line E*m*, the hypothenuse: it is termed the variable hypothenuse (*cala karṇa*) from its constantly changing its length. We have now the two similar triangles E*mn* and E*og*, a comparison of the corresponding parts of which gives us the proportion E*m* : *mn* :: E*o* : *og*; that is to say, *og*, which is the sine of the equation *o*M, equals the product of E*o*, the radius, into *mn*, the result from the base-sine, divided by the variable hypothenuse, E*m*.

When the planet's mean place is in the quadrant D O, as at M′, the result from the perpendicular-sine (*koṭijyâphala*), or M′*n*′, is subtracted from radius, and the remainder, E*n*′, is employed as before to find the value of E*m*′, the variable hypothenuse: and the comparison of the similar triangles E*m*′*n*′ and E*o*′*g*′ gives *o*′*g*′, the sine of the equation, *o*′M′.

It is obvious that when the mean distance of a planet from its conjunction is less than a quadrant in either direction, as at M, the base E*n* is greater than radius; when that distance is more than a quadrant, as at M′, the base E*n*′ is less than radius: the cosine is to be added to radius in the one case, and subtracted from it in the other. This is the meaning of the rule in verse 40: compare the notes to i. 58 and ii. 30.

In illustration of the process, we will calculate the equation of the conjunction of Mercury for the given time, or for midnight preceding January 1st, 1860, at Washington.

Since the Hindu system, like the Greek, interchanges in the case of the two inferior planets the motion and place of the planet itself and of the sun, giving to the former as its mean motion that which is the mean apparent motion of the sun, and assigning to the conjunction (*çîghrocca*) a revolution which is actually that of the planet in its orbit, the mean position of Mercury at the given time is that found above (under v. 39) to be that of the sun at the same time, while to find that of its conjunction we have to add the equation for difference of meridian (*deçântaraphala*, i. 60, 61), to the longitude given under i. 53 as that of the planet.

	s	°	′	″
Longitude of Mercury's conjunction (*çîghrocca*), midnight, at Ujjayinî,	4	15	13	8
add for difference of meridian,		1	44	14
Longitude of conjunction at required time,	4	16	57	22
Mean longitude of Mercury,	8	18	13	13
Mean commutation (*çîghrakendra*,),	7	28	44	9

The position of Mercury with reference to the conjunction is accordingly very nearly that of M′, in Fig. 5. The arc which determines the base-sine (*bhujajyâ*), or O M′, is 58° 44′, while M′ D, its complement, from which the perpendicular-sine (*koṭijyâ*) is taken, is 31° 16′. The corresponding sines, M′ B′ and M′ G, are 2938′ and 1784′ respectively.

The epicycle of Mercury is one degree less at D than at O. Hence the proportion

$$3438 : 60 :: 2938 : 51$$

gives 51′ as the diminution at M′: the circumference of the epicyle at M, then, is 132° 9′. The two proportions

$$360° : 132° \; 9' :: 2938 : 1078, \quad \text{and} \quad 360° : 132° \; 9' :: 1784 : 655,$$

give us the value of $m'n'$ as 1078′, and that of n'M′ as 655′. The commutation being more than three and less than nine signs, or in the half-orbit beginning with Cancer, the fourth sign, n'M′ is to be subtracted from EM′, or radius, 3438′; the remainder, 2783′, is the perpendicular En'.

To the square of En',	7,745,089
add the square of $n'm'$,	1,162,084
of their sum,	8,907,173
the square root,	2984

is the variable hypothenuse (*cala karṇa*), Em'. The comparison of the triangles E$m'n'$ and E$o'g'$ gives the proportion Em' : $m'n'$: : Eo' : $o'g'$, or

$$2984 : 1078 :: 3438 : 1242$$

The value of $o'q'$, the sine of the equation, is accordingly 1242′: the corresponding arc, o'M, is found by the process prescribed in verse 33 to be 21° 12′. The figure shows the equation to be subtractive.

The annexed table presents the results of the calculation of the equation of the conjunction (*çîghrakarman*) for the five planets.

Results of the First Process for finding the True Places of the Planets.

Planet.	Mean Longitude.				Longitude of Conjunction.				Mean Commutation.				Base-sine.	Corr. Epicycle.		Result from B.-sine.	Result from P.-sine.	Variable Hyp.	Equat'n of Conj.	
	s	°	′	″	s	°	′	″	s	°	′	″	′	°	′	′	′	′	°	′
Mercury,	8	18	13	13	4	16	57	22	7	28	44	9	2938	132	9	1078	655	2984	−21	12
Venus,	8	18	13	13	10	21	49	47	2	3	36	34	3080	260	13	2226	1104	5058	+26	7
Mars,	5	24	30	57	8	18	13	13	2	23	42	16	3416	232	1	2202	225	4274	+31	1
Jupiter,	2	26	2	14	8	18	13	13	5	22	10	59	468	70	16	91	665	2774	+ 1	53
Saturn,	3	20	12	3	8	18	13	13	4	28	1	10	1820	39	32	200	320	3124	+ 3	40

This is, however, only a first step in the whole operation for finding the true longitudes of these five planets, as is laid down in the next passage.

43. The process of correction for the apsis (*mânda karman*) is the only one required for the sun and moon: for Mars and the other planets are prescribed that for the conjunction (*çâighrya*), that for the apsis (*mânda*), again that for the apsis, and that for the conjunction—four, in succession.

44. To the mean place of the planet apply half the equation of the conjunction (*çighraphala*), likewise half the equation of the apsis; to the mean place of the planet apply the whole equation of the apsis (*mandaphala*), and also that of the conjunction.

45. In the case of all the planets, and both in the process of correction for the conjunction and in that for the apsis, the equation is additive (*dhana*) when the distance (*kendra*) is in the half-orbit beginning with Aries; subtractive (*ṛṇa*), when in the half-orbit beginning with Libra.

The rule contained in the last verse is a general one, applying to all the processes of calculation of the equations of place, and has already been anticipated by us above. Its meaning is, that when the anomaly, (*mandakendra*), or commutation (*çighrakendra*), reckoned always forward from the planet to the apsis or conjunction, is less than six signs, the equation of place is additive; when the former is more than six signs, the equation is subtractive. The reason is made clear by the figures given above, and by the explanations under verses 1–5 of this chapter.

It should have been mentioned above, under verse 29, where the word *kendra* was first introduced, that, as employed in this sense by the Hindus, it properly signifies the position (see note to i. 53) of the "centre" of the epicycle—which coincides with the mean place of the planet itself —relative to the apsis or conjunction respectively. In the text of the Sûrya-Siddhânta it is used only with this signification: the commentary employs it also to designate the centre of any circle.

Since the sun and moon have but a single inequality, according to the Hindu system, the calculation of their true places is simple and easy. With the other planets the case is different, on account of the existence of two causes of disturbance in their orbits, and the consequent necessity both of applying two equations, and also of allowing for the effect of each cause in determining the equation due to the other. For, to the apprehension of the Hindu astronomer, it would not be proper to calculate the two equations from the mean place of the planet; nor, again, to calculate either of the two from the mean place, and, having applied it, to take the new position thus found as a basis from which to calculate the other; since the planet is virtually drawn away from its mean place by the divinity at either apex (*ucca*) before it is submitted to the action of the other. The method adopted in this Siddhânta of balancing the two influences, and arriving at their joint effect upon the planet, is stated in verses 43 and 44. The phraseology of the text is not entirely explicit, and would bear, if taken alone, a different interpretation from that which the commentary puts upon it, and which the rules to be given later show to be its true meaning; this is as follows: first calculate from the mean place of the planet the equation of the conjunction, and apply the half of it to the mean place; from the position thus obtained calculate the equation of the apsis, and apply half of it to the longitude as already once equated; from this result find once more the equation of the apsis, and apply it to the original mean place of the planet; and finally, calculate from, and apply to, this last place the whole equation of the conjunction.

We have calculated by this method the true places of the five planets, and present the results of the processes in the following tables. Those of the first process have been already given under the preceding passage: the application of half the equations there found to the mean longitude gives us the longitude once equated as a basis for the next process.

Results of the Second Process for finding the True Places of the Planets.

Planet.	Equated Longitude.			Longitude of Apsis.				Equated Anomaly.			Base-sine.	Corrected Epicycle.		Equation of Apsis.	
	s	°	′	s	°	′	″	s	°	′	′	°	′	°	′
Mercury,	8	7	37	7	10	28	20	11	2	51	1568	29	5	− 2	7
Venus,	9	1	17	2	19	52	17	5	18	35	681	11	48	+ 0	22
Mars,	6	10	1	4	10	2	40	10	0	2	2977	72	24	−10	2
Jupiter,	2	26	59	5	21	22	19	2	24	23	3420	32	0	+ 5	5
Saturn,	3	22	1	7	26	37	34	4	4	37	2829	48	11	+ 6	20

Again, the application of half these equations to the longitudes as once equated furnishes the data for the third process. The longitudes of the apsides, being the same as in the second operation, are not repeated in this table.

Results of the Third Process for finding the True Places of the Planets.

Planet.	Equated Longitude.			Equated Anomaly.			Base-sine.	Corrected Epicycle.		Equation of Apsis.	
	s	°	′	s	°	′	′	°	′	°	′
Mercury,	8	6	34	11	3	54	1512	29	7	−2	2
Venus,	9	1	28	5	18	24	691	11	48	+0	23
Mars,	6	5	0	10	5	3	2814	72	33	−9	30
Jupiter,	2	29	30	2	21	52	3403	32	1	+5	4
Saturn,	3	25	11	4	1	27	2932	48	9	+6	33

The original mean longitudes are now corrected by the results of the third process, to obtain a position from which shall be once more calculated the equation of the conjunction; and the application of this to the position which furnished it yields, as a final result, the true place of each planet.

Results of the Fourth Process for finding the True Places of the Planets.

Planet.	Equated Longitude.			Equated Commutation.			Base-sine.	Corr. Epicycle.		Result from B.-sine.	Result from P.-sine.	Variable Hypoth	Equation of Conj.		True Longitude.		
	s	°	′	s	°	′	′	°	′	′	′	′	°	′	s	°	′
Mercury,	8	16	11	8	0	46	3000	132	8	1101	616	3029	−21	20	7	24	51
Venus,	8	18	36	2	3	14	3069	260	13	2218	1118	5067	+25	59	9	14	35
Mars,	5	15	1	3	3	12	3432	232	0	2212	124	3984	+33	44	6	18	45
Jupiter,	3	1	6	5	17	7	766	70	27	150	656	2786	+ 3	5	3	4	11
Saturn,	3	26	45	4	21	28	2141	39	37	236	296	3151	+ 4	17	4	1	2

We cannot furnish a comparison of the Hindu determinations of the true places of the planets with their actual positions as ascertained by our modern methods, until after the subject of the latitude has been dealt with: see below, under verses 56–58.

The Hindu method of finding the true longitudes of the five planets whose apparent position is affected by the parallax of the earth's motion having thus been fully explained, we will proceed to indicate, as succinctly as possible, the way in which the same problem is solved by the great Greek astronomer. The annexed figure (Fig. 6) will illustrate his method: it is taken from those presented in the Syntaxis, but with such modifications of form as to make it correspond with the figures previously given here: the conditions which it represents are only hypothetical, not according with the actual elements of any of the planetary orbits.

Let E be the earth's place, and let the circle A *p* C, described about E as a centre, represent the mean orbit of any planet, E A being the direction of its line of apsides, and E C that of its conjunction (*çîghra*), called by Ptolemy the apogee of its epicycle. Let E X be the double eccentricity, or the equivalent to the radius of the Hindu epicycle of the apsis; and let E X be bisected in Q. Then, as regards the influence of the eccentricity of the orbit upon the place of the planet, the centre of equable angular motion is at X, but the centre of equal distance is at Q: the planet virtually describes the circle A′ P, of which Q is the centre, but at the same rate as if it were moving equably upon the dotted circle, of which the centre is at X. The angle of mean anomaly, accordingly, which increases proportionally to the time, is *x* X A″, but P is the planet's place, P E A the true anomaly, and E P X the equation of place. The value of E P X is obtained by a process analogous to that described above, under verse 39 (pp. 66, 67); E B and B X, and Q D and D X, are first found; then D P, which, by subtracting D X, gives X P; X P added to B X gives B P; and from B P and B E is derived E P B, the equation required; subtract this from P X A, and the remainder is R E A, the planet's true distance from the apsis. About P describe the epicycle of the conjunction, and draw the radius P T parallel to E C: then T is the planet's place in the epicycle, *p* its apparent position in the mean orbit, and T E P the equation of the epicycle, or of the conjunction. In order to arrive at the value of this equation, Ptolemy first finds that of S E R, the corresponding angle when the centre of the epicycle is placed at R, at the mean distance E R, or radius, from E: he then diminishes it by a complicated process, into the details of which it is not necessary here to enter, and which, as he himself acknowledges, is not strictly accurate, but yields results sufficiently near to the truth. The application of the equation thus obtained

Fig. 6.

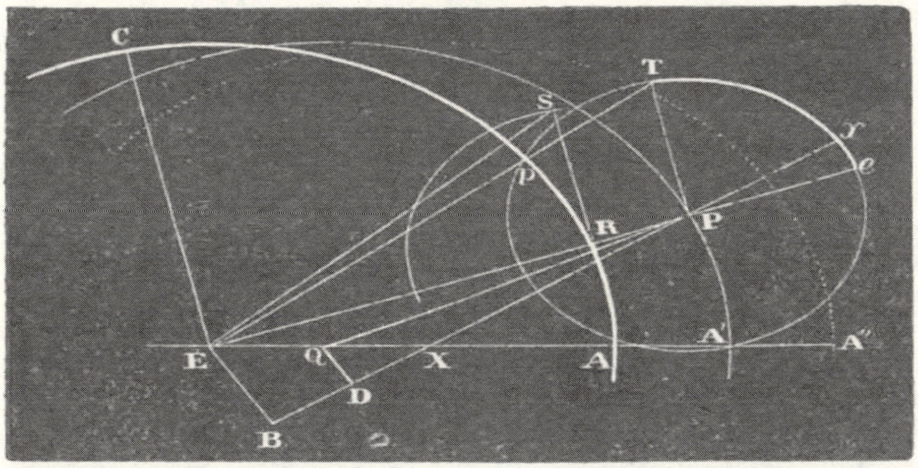

to the place of the planet as already once equated gives the final result sought for, its geocentric place.

In the case of Mercury, Ptolemy introduces the additional supposition that the centre of equal distances, instead of being fixed at Q, revolves in a retrograde direction upon the circumference of a circle of which X is the centre, and X Q the radius.

After a thorough discussion of the observations upon which his data and his methods are founded, and a full exposition of the latter, Ptolemy proceeds himself to construct tables, which are included in the body of his work, from which the true places of the planets at any given time may be found by a brief and simple process. The Hindus are also accustomed to employ such tables, although their construction and use are nowhere alluded to in this treatise. Hindu tables, in part professing to be calculated according to the Sûrya-Siddhânta, have been published by Bailly (Traité de l'Astr. Ind. et Or., p. 335, etc.), by Bentley (Hind. Ast., p. 219, etc.), by Warren (Kâla Sankalita, Tables), by Mr. Hoisington (Oriental Astronomer, p. 61, etc.), and, for the sun and moon, by Davis (As. Res., ii. 255, 256).

We are now in a condition to compare the planetary system of the Hindus with that of the Greeks, and to take note of the principal resemblances and differences between them. And it is evident, in the first place, that in all their grand features the two are essentially the same. Both alike analyze, with remarkable success, the irregularities of the apparent motions of the planets into the two main elements of which they are made up, and both adopt the same method of representing and calculating those irregularities. Both alike substitute eccentric circles for the true elliptic orbits of the planets. Both agree in assigning to Mercury and Venus the same mean orbit and motion as to the sun, and in giving them epicycles which in fact correspond to their heliocentric orbits, making the centre of those epicycles, however, not the true, but the mean place of the sun, and also applying to the latter the correction due to the eccentricity of the orbit. Both transfer the centre of the orbits of the superior planets from the sun to the earth, and then assign to each, as an epicycle, the earth's orbit; not, however, in the form of an ellipse, nor even of an eccentric, but in that of a true circle; and here, too, both make the place of the centre of the epicycle to depend upon the mean, instead of the true, place of the sun. The key to the whole system of the Greeks, and the determining cause both of its numerous accordances with the actual conditions of things in nature, and of its inaccuracies, is the principle, distinctly laid down and strictly adhered to by them, that the planetary movements are to be represented by a combination of equable circular motions alone, none other being deemed suited to the dignity and perfection of the heavenly bodies. By the Hindus, this principle is nowhere expressly recognized, so far as we are aware, as one of binding influence, and although their whole system, no less than that of the Greeks, seems in other respects inspired by it, it is in one point, as we shall note more particularly hereafter, distinctly abandoned and violated by them (see below, under vv. 50, 51). We cannot but regard with the highest admiration the acuteness and industry, the power of observation, analysis, and deduction of the Greeks, that, hampered by false assumptions, and imperfectly provided with instruments, they were able to construct a science containing so much of truth, and serving as a secure basis for the improvements of after time: whether we pay the same tribute to the genius of the Hindu will depend upon whether we consider him also, like all the rest of the world, to have been the pupil of the Greek in astronomical science, or whether we shall believe him to have arrived independently at a system so closely the counterpart of that of the West.

The differences between the two systems are much less fundamental and important. The assumption of a centre of equal distance different from that of equal angular motion—and, in the case of Mercury, itself also movable—is unknown to the Hindus: this, however, appears to be an innovation introduced into the Greek system by Ptolemy, and unknown before his time; it was adopted by him, in spite of its seeming arbitrariness, because it gave him results according more nearly with his observations. The moon's evection, the discovery of Ptolemy, is equally wanting in the Hindu astronomy. As regards the combined application of the equations of the apsis and the conjunction, the two systems are likewise at variance. Ptolemy follows the truer, as well as the simpler, method: he applies first the whole correction for the eccentricity of the orbit, obtaining as a result, in the case of the superior planets, the planet's true heliocentric place; and this he then corrects for the parallax of the earth's position. Here, too, ignorant as he was of the actual relation between the two equations, we may suppose him to have been guided by the better coincidence with observation of the results of his processes when thus conducted. The Hindus, on the other hand, not knowing to which of the two supernatural beings at the apsis and conjunction should be attributed the priority of influence, conceived them to act simultaneously, and adopted the method stated above, in verse 44, of obtaining an average place whence their joint effect should be calculated. This is the only point where they forsook the geometrical method, and suffered their theory respecting the character of the forces producing the inequalities of motion to modify their processes and results. The change of dimensions of the epicycles is also a striking peculiarity of the Hindu system, and to us, thus far, its most enigmatical feature. The virtual effect of the alteration upon the epicycles themselves is to give them a form approximating to the elliptical. But, although the epicycles of the conjunction of the inferior planets represent the proper orbits of those planets, and those of the superior the orbit of the earth, it is not possible to see in this alteration an unconscious recognition of the principle of ellipticity, because the major axis of the quasi-ellipse—or, in the case of Jupiter and Saturn, the minor axis—is constantly pointed toward the earth. Its effect upon the orbit described by the planet is, as concerns the epicycle of the apsis, to give to the eccentric circle an ovoid shape, flattened in the first and fourth quadrants, bulging in the second and third: this is, so far as it goes, an approximation toward Ptolemy's virtual orbit, a circle described about a centre distant from the earth's place by only half the equivalent of the radius of the Hindu epicycle (the circle A′ P in figure 6): but the approximation seems too distant to furnish any hint of an explanation. A diminution of the epicycle also effects a corresponding diminution of the equation, carrying the planet forward where the equation is subtractive, and backward where it is additive: but we hardly feel justified in assuming that it is to be regarded as an empirical correction, applied to make the results of calculation agree more nearly with those of observation, because its amount and place stand in no relation which we have been able to trace to the true elements of the planetary orbits, nor is the accuracy of either the Hindu calculations or observations so great as to make such slight corrections of appreciable importance. We are compelled to leave the solution of this difficulty, if it shall prove soluble, to later investigation, and a more extended comparison of the different text-books of Hindu astronomical science.

As regards the numerical value of the elements adopted by the two systems—their mutual relation, and their respective relations to the true elements established by modern science, are exhibited in the annexed table. The first part of it presents the comparative dimensions of the planetary orbits, or the value of the radius of each in terms of that of the earth's orbit. In the case of Mercury and Venus, this is represented by the relation of the radius of the epicycle (of the conjunction) to that of the orbit; in the case of the superior planets, by that of the radius of the orbit to the radius of the epicycle. For the Hindu system it was necessary to give two values in every case, derived respectively from the greatest and least dimensions of the epicycles. Such a relative determination of the moon's orbit, of course, could not be obtained: its absolute dimensions will be found stated later (see under iv. 3 and xii. 84). The second part of the table gives, as the fairest practicable comparison of the values assigned by each system to the eccentricities, the greatest equations of the centre. For Mercury and Venus, however, the ancient and modern determinations of these equations are not at all comparable, the latter giving their actual heliocentric amount, the former their apparent value, as seen from the earth.

Relative Dimensions and Eccentricities of the Planetary Orbits, according to Different Authorities.

Planet.	Radius of the Orbit.				Greatest Equation of the Centre.		
	Sûrya-Siddhânta.		Ptolemy.	Moderns	Sûrya-Sidddhânta.	Ptolemy.	Moderns.
	even quad.	odd quad			° ′ ″	° ′	° ′ ″
Sun,	1.0000	1.0000	1.0000	1.0000	2 10 31	2 23	1 55 27
Moon,	.....	.....	.....	.....	5 2 46	5 1	6 17 13
Mercury,	.3694	.3667	.3750	.3871	4 27 35	2 52	23 40 43
Venus,	.7278	.7222	.7194	.7233	1 45 3	2 23	0 47 11
Mars,	1.5139	1.5513	1.5190	1.5237	11 32 3	11 32	10 41 33
Jupiter,	5.1429	5.0000	5.2174	5.2028	5 5 58	5 16	5 31 14
Saturn,	9.2308	9.0000	9.2308	9.5389	7 39 32	6 32	6 26 12

46. Multiply the daily motion (*bhukti*) of a planet by the sun's result from the base-sine (*bâhuphala*), and divide by the number of minutes in a circle (*bhacakra*); the result, in minutes, apply to the planet's true place, in the same direction as the equation was applied to the sun.

By this rule, allowance is made for that part of the equation of time, or of the difference between mean and apparent solar time, which is due to the difference between the sun's mean and true places. The instruments employed by the Hindus in measuring time are described, very briefly and insufficiently, in the thirteenth chapter of this work: in all probability the gnomon and shadow was that most relied upon; at any rate, they can have had no means of keeping mean time with any accuracy, and it appears from this passage that apparent time alone is regarded as ascertainable directly. Now if the sun moved in the equinoctial instead of in the ecliptic, the interval between the passage of his mean and his true place across the meridian would be the same part of a day, as the difference of the two places is of a circle: hence the proportion upon which the rule in the text is founded: as the number of minutes in a circle is to that in the sun's equation (which is the same with his "result from the base-sine:" see above, v. 39), so is the whole daily motion of any planet to its motion during the interval. And since, when the sun is in advance of his true place, he comes later to the meridian, the planet moving on during the interval, and the reverse, the result is additive to the planet's place, or subtractive from it, according as the sun's equation is additive or subtractive.

The other source of difference between true and apparent time, the difference in the daily increment of the arcs of the ecliptic, in which the sun moves, and of those of the equinoctial, which are the measures of time, is not taken account of in this treatise. This is the more strange, as that difference is, for some other purposes, calculated and allowed for.

At the time for which we have ascertained above the true places of the planets, the sun is so near the perigee, and his equation of place is so small, that it renders necessary no modification of the places as given: even the moon moves but a small fraction of a second during the interval between mean and apparent midnight.

By *bhukti*, as used in this verse, we are to understand, of course, not the mean, but the actual, daily motion of the planet: the commentary also gives the word this interpretation. How the actual rate of motion is found at any given time, is taught in the next passage.

47. From the mean daily motion of the moon subtract the daily motion of its apsis (*manda*), and, having treated the difference in the manner prescribed by the next rule, apply the result, as an additive or subtractive equation, to the daily motion.

48. The equation of a planet's daily motion is to be calculated like the place of the planet in the process for the apsis: multiply the daily motion by the difference of tabular sines corresponding to the base-sine (*dorjyâ*) of anomaly, and then divide by two hundred and twenty-five;

49. Multiply the result by the corresponding epicycle of the apsis (*mandaparidhi*), and divide by the number of degrees in a circle (*bhagaṇa*); the result, in minutes, is additive when in the half-orbit beginning with Cancer, and subtractive when in that beginning with Capricorn.

Only the effect of the apsis upon the daily rate of motion is treated of in these verses; the farther modification of it by the conjunction is the subject of those which succeed.

Verse 47 is a separate specification under the general rule given in the following verse, applying to the moon alone. The rate of a planet's motion in its epicycle being equal to its mean motion from the apsis, or its anomalistic motion, it is necessary in the case of the moon, whose apsis has a perceptible forward movement, to subtract the daily amount of this movement from that of the planet in order to obtain the daily rate of removal from the apsis.

In the first half of verse 48 the commentary sees only an intimation that, as regards the apsis, the equation of motion is found in the same general method as the equations of place, a certain factor being multiplied by the circumference of the epicycle and divided by that of the orbit. Such a direction, however, would be altogether trifling and superfluous, and not at all in accordance with the usual compressed style of the treatise; and moreover, were it to be so understood, we should lack any direction as to which of the several places found for a planet in the process for ascertaining its true place should be assumed as that for which this first equation of motion is to be calculated. The true meaning of the line, beyond all reasonable question, is, that the equation is to be derived from the same data from which the equation of place for the apsis was finally obtained, to be applied to the planet's mean position, as this is applied to its mean motion; from the data, namely, of the third process, as given above.

The principle upon which the rule is founded may be explained as follows. The equation of motion for any given time is evidently equal to the amount of acceleration or of retardation effected during that time by the influence of the apsis. Thus, in Fig. 3 (p. 64), $m\,n$, the sine of $a'm$, is the equation of motion for the whole time during which the centre of the epicycle has been traversing the arc A M. If that arc, and the arc $a'm$, be supposed to be divided into any number of equal portions, each equal to a day's motion, the equation of motion for each successive day will be equal to the successive increments of the sines of the increasing arcs in the epicycle; and these will be equal to the successive increments from day to day of the sines of mean anomaly, reduced to the dimensions of the epicycle. But the rate at which the sine is increasing or decreasing at any point in the quadrant is approximately measured by the difference of the tabular sines at that point: and as the arcs of mean daily motion are generally quite small—being, except in the case of the moon, much less than 3° 45′, the unit of the table—we may form this proportion: if, at the point in the orbit occupied by the planet, a difference of 3° 45′ in arc produces an increase or decrease of a given amount in sine, what increase or decrease of sine will be produced by a difference of arc equal to the planet's daily motion? or, 225 : diff. of tab. sines : : planet's daily motion : corresponding diff. of sine. The reduction of the result of this proportion to the dimensions of the epicycle gives the equation sought.

We will calculate by this method the true daily motion of the moon at the time for which her true longitude has been found above.

Moon's mean daily motion (i. 30),	790′ 35″
deduct daily motion of apsis (i. 33),	6 41
Moon's mean anomalistic motion,	783 54

From the process of calculation of the moon's true place, given above, we take

Moon's mean anomaly,	10ˢ 18° 46′ 15″
Sine of anomaly (*bhujajyâ*),	2266′

From the table of sines (ii. 15–27), we find

Corresponding difference of tabular sines,	174′

Hence the proportion

$$225' : 174' :: 783' \; 54'' : 606' \; 13''$$

shows the increase of the sine of anomaly in a day at this point to be 606′ 13″. The dimensions of the epicycle were found to be 31° 47′. Hence the proportion

$$360° : 31° \; 47' :: 606' \; 13'' : 53' \; 31''$$

give us the desired equation of motion, as 53′ 31″. By verse 49 it is subtractive, the planet being less than a quadrant from the apsis, or its anomaly being more than nine and less than three signs. Therefore, from the

Moon's mean daily motion,	790′ 35″
subtract the equation,	53 31
Moon's true daily motion at given time,	737 4

The roughness of the process is well illustrated by this example. Had the sine of anomaly been but 2′ greater, the difference of sines would have been 10′ less, and the equation only about 50′.

The equation of the sun's motion, calculated in a similar manner, is found to be +2′ 18″, and his true motion 61′ 26″.

The corrected rate of motion of the other planets will be given under the next following passage.

50. Subtract the daily motion of a planet, thus corrected for the apsis (*manda*), from the daily motion of its conjunction (*çîghra*); then multiply the remainder by the difference between the last hypothenuse and radius,

51. And divide by the variable hypothenuse (*cala karṇa*): the result is additive to the daily motion when the hypothenuse is greater than radius, and subtractive when this is less; if, when subtractive, the equation is greater than the daily motion, deduct the latter from it, and the remainder is the daily motion in a retrograde (*vakra*) direction.

The commentary gives no demonstration of the rule by which we are here taught to calculate the variation of the rate of motion of a planet occasioned by the action of its conjunction: the following figure, however (Fig. 7), will illustrate the principle upon which it is founded.

As in a previous figure (Fig. 5, p. 68), C M M′ represents the mean orbit of a planet, E the earth, and M the planet's mean position, at a given time, relative to its conjunction, C: the circle described about M is the epicycle of the conjunction: it is drawn, in the figure, of the relative dimensions of that assumed for Mars. Suppose M′ M to be the amount of motion of the centre of the epicycle, or the (equated) mean synodical motion of the planet, during one day; *m′ m* is the arc of the epicycle traversed by the planet in the same time. As the amount of daily synodical motion is in every case small, these arcs are necessarily greatly exaggerated in the figure, being made about twenty-four times too great for Mars. Had the planet remained stationary in the epicycle at *m′* while the centre of the epicycle moved from M′ to M, its place at the given time would be at *s*; having moved to *m*, it is seen at *t*: hence *s t* is the equation of daily motion, of which it is required to ascertain the value. Produce E *m′* to *n*, making E *n* equal to E *m*, and join *m n*; from M draw M *o* at right angles to E *m*. Then, since the arc *m m′* is very small, the angles E *m n* and E *n m*, as also M *m m′* and M *m′ m*, may be regarded as right angles; M *m o* and *n m m′* are therefore equal, each being the complement of E *m m′*, and the triangles *m n m′* and M *m o* are similar. Hence

Fig. 7.

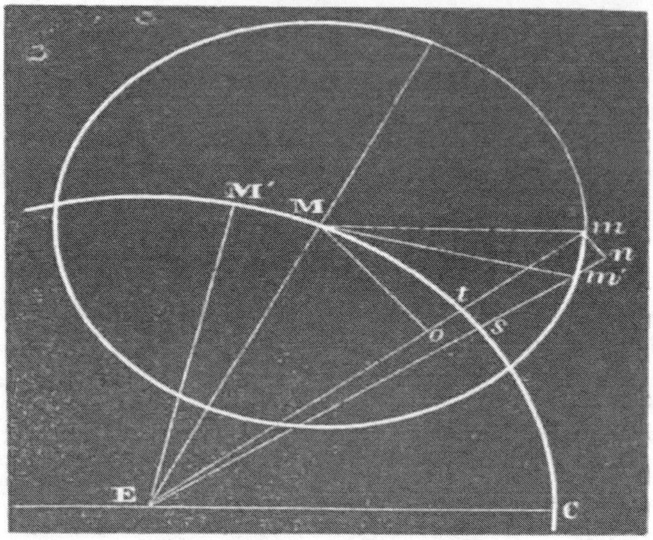

	$M m : m o :: m m' : m n$
But	$E M : M m :: M M' : m m'$
Hence, by combining terms,	$E M : m o :: M M' : m n$
But	$t s : E t :: m n : E m$
therefore, since E M equals E *t*, by again combining,	$t s : m o :: M M' : E m$

and, reducing the proportion to an equation, *t s*, the required equation of motion, equals M M′, the equated mean synodical motion in a day, multiplied by *m o*, and divided by E *m*, the variable hypothenuse. This, however, is not precisely the rule given above; for in the text of this Siddhânta, *m t*, the difference between the variable hypothenuse and radius, is substituted for *m o*, as if the two were virtually equivalent: a highly inaccurate assumption, since they differ from one another by the versed sine, *o t*, of the equation of the conjunction, M *t*, which equation is sometimes as much as 40°: and indeed, the commentary, contrary to its usual habit of obsequiousness to the inspired text with which it has to deal, rejects this assumption, and says, without even an apology for the liberty it is taking, that by the word "radius" in verse 50 is to be understood the cosine (*koṭijyâ*) of the second equation of the conjunction.

In illustration of the rule, we will calculate the true rate of daily motion of the planet Mars, at the same time for which the previous calculations have been made.

By the process already illustrated under the preceding passage, the equation of Mars's daily motion for the effect of the apsis, as derived from the data of the third process for ascertaining his true place, is found to be −3′ 41″, the difference of tabular sines being 131′. Accordingly,

from the mean daily motion of Mars (i. 34),	31′ 26″
deduct the equation for the apsis,	3 41
Mars's equated daily motion,	27 45

Now, to find the equated daily synodical motion,

from the daily motion of Mars's conjunction (the sun),	59′ 8″
deduct his equated daily motion,	27 45
Mars's equated daily synodical motion,	31 23

The variable hypothenuse used in the last process for finding the true place was 3984′; its excess above radius is 546′. The proportion

3984′ : 546′ :: 31′ 23″ : 4′ 18″

shows, then, that the equation of motion due to the conjunction at the given time is 4′ 18″. Since the hypothenuse is greater than radius—that is to say, since the planet is in the half-orbit in which the influence of the conjunction is accelerative—the equation is additive. Therefore,

to Mars's equated daily motion,	27′ 45″
add the equation for the conjunction,	4 18
Mars's true daily motion at the given time,	32 3

In this calculation we have followed the rule stated in the text: had we accepted the amendment of the commentary, and, in finding the second term of our proportion, substituted for radius the cosine of 33° 44′, the resulting equation would have been more than doubled, becoming 8′ 51″, instead of 4′ 18″; this happening to be a case where the difference is nearly as great as possible. We have deemed it best, however, in making out the corresponding results for all the five planets, as presented in the annexed table, to adhere to the directions of the text itself. The inaccuracy, it may be observed, is greatest when the equation of motion is least, and the contrary; so that, although sometimes very large relatively to the equation, it never comes to be of any great importance absolutely.

Results of the Processes for finding the True Daily Motion of the Planets.

Planet.	Diff. of sines.	Equation of Apsis.		Equated Motion.		Equated Synod. Motion.		Equation of Conjunction.		True Motion.	
	′	′	″	′	″	′	″	′	″	′	″
Mercury,	205	−4	21	54	47	190	45	−25	45	+29	2
Venus,	219	+1	53	61	1	35	7	+11	17	+72	18
Mars,	131	−3	41	27	45	31	23	+ 4	18	+32	3
Jupiter,	37	−0	4	4	55	54	13	−12	41	− 7	46
Saturn,	119	+0	8	2	8	57	0	− 5	11	− 3	3

The final abandonment by the Hindus of the principle of equable circular motion, which lies at the foundation of the whole system of eccentrics and epicycles, is, as already pointed out above (under vv. 43–45), distinctly exhibited in this process: *m′ m* (Fig. 7), the arc in the epicycle traversed by the planet during a given interval of time, is no fixed and equal quantity, but is dependent upon the arc M′ M, the value of which, having suffered correction by the result of a triply complicated process, is altogether irregular and variable. This necessarily follows from the assumption of simultaneous and mutual action on the part of the beings at the apsis and conjunction, and the consequent impossibility of constructing a single connected geometrical figure which shall represent the joint effect of the two disturbing influences. By the Ptolemaic method the principle is consistently preserved: the fixed axis of the epicycle (see Fig. 6, p. 217), to the revolution of which that of the epicycle itself is bound, is *x* P X; and as the angle *x* P T, like *x* X A″, increases equably, the planet traverses the circumference of the epicycle with an unvarying motion relative to the fixed point *x*; although the equation is derived, not from the arc *x* T, but from *e* T, the equivalent of C R, its part *e x* varying with the varying angle E P X.

In case the reverse motion of the planet upon the half-circumference of the epicycle within the mean orbit is, when projected upon the orbit, greater than the direct motion of the centre of the epicycle, the planet will appear to move backward in its orbit, at a rate equal to the excess of the former over the latter motion. This is, as the last table shows, the case with Jupiter and Saturn at the given time. The subject of the retrogradation of the planets is continued and completed in the next following passage.

52. When at a great distance from its conjunction (*çîghrocca*), a planet, having its substance drawn to the left and right by slack cords, comes then to have a retrograde motion.

53. Mars and the rest, when their degrees of commutation (*kendra*), in the fourth process, are, respectively, one hundred and sixty-four, one hundred and forty-four, one hundred and thirty, one hundred and sixty-three, one hundred and fifteen,

54. Become retrograde (*vakrin*): and when their respective commutations are equal to the number of degrees remaining after subtracting those numbers, in each several case, from a whole circle, they cease retrogradation.

55. In accordance with the greatness of their epicycles of the conjunction (*çîghraparidhi*), Venus and Mars cease retrograding in the seventh sign, Jupiter and Mercury in the eighth, Saturn in the ninth.

The subject of the stations and retrogradations of the planets is rather briefly and summarily disposed of in this passage, although treated with as much fullness, perhaps, as is consistent with the general method of the Siddhânta. Ptolemy devotes to it the greater part of the twelfth book of the Syntaxis.

The first verse gives the theory of the physical cause of the phenomenon: it is to be compared with the opening verses of the chapter, particularly verse 2. We note here, again, the entire disavowal of the system of epicycles as a representation of the actual movements of the planets. How the slackness of the cords by which each planet is attached to, and attracted by, the supernatural being at its conjunction, furnishes an explanation of its retrogradation which should commend itself as satisfactory to the mind even of one who believed in the supernatural being and the cords, we find it very hard to see, in spite of the explanation of the commentary: it might have been better to omit verse 52 altogether, and to suffer the phenomenon to rest upon the simple and intelligible explanation given at the end of the preceding verse, which is a true statement of its cause, expressed in terms of the Hindu system. The actual reason of the apparent retrogradation is, indeed, different in the case of the inferior and of the superior planets. As regards the former, when they are traversing the inferior portion of their orbits, or are nearly between the sun and the earth, their heliocentric eastward motion becomes, of course, as seen from the earth, westward, or retrograde; by the parallax of the earth's motion in the same direction this apparent retrogradation is diminished, both in rate and in continuance, but is not prevented, because the motion of the inferior planets is more rapid than that of the earth. The retrogradation of the superior planets, on the other hand, is due to the parallax of the earth's motion in the same direction when between them and the sun, and is lessened by their own motion in their orbits, although not done away with altogether, because their motion is less rapid than that of the earth. But, in the Hindu system, the revolution of the planet in the epicycle of the conjunction represents in the one case the proper motion of the planet, in the other, that of the earth, reversed; hence, whenever its apparent amount, in a contrary direction, exceeds that of the movement of the centre of the epicycle—which is, in the one case, that of the earth, in the other, that of the planet itself—retrogradation is the necessary consequence.

Verses 53–55 contain a statement of the limits within which retrogradation takes place. The data of verse 53 belong to the different planets in the order, Mars, Mercury, Jupiter, Venus, and Saturn (see above, under i. 51, 52). That is to say, Mercury retrogrades, when his equated commutation, as made use of in the fourth process for finding his true place (see above, under vv. 43–45), is more than 144° and less than 216°; Venus, when her commutation, in like manner, is between 163° and 197°; Mars, between 164° and 196°; Jupiter, between 130° and 230°; Saturn, between 115° and 245°. These limits ought not, however, even according to the theory of this Siddhânta, to be laid down with such exactness; for the precise point at which the subtractive equation of motion for the conjunction will exceed the proper motion of the planet must depend, in part, upon the varying rate of the latter as affected by its eccentricity, and must accordingly differ a little at different times. We have not thought it worth while to calculate the amount of this variation, nor to draw up a comparison of the Hindu with the Greek and the modern determinations of the limits of retrogradation, since these are dependent for their correctness upon the accuracy of the elements assumed, and the processes employed, both of which have been already sufficiently illustrated.

The last verse of the passage adds little to what had been already said, being merely a repetition, in other and less precise terms, of the specifications of the preceding verse, together with the assertion of a relation between the limits of retrogradation and the dimensions of the respective epicycles; a relation which is only empirical, and which, as regards Venus and Mars, does not quite hold good.

56. To the nodes of Mars, Saturn, and Jupiter, the equation of the conjunction is to be applied, as to the planets themselves respectively; to those of Mercury and Venus, the equation of the apsis, as found by the third process, in the contrary direction.

57. The sine of the arc found by subtracting the place of the node from that of the planet—or, in the case of Venus and Mercury, from that of the conjunction—being multiplied by the extreme latitude, and divided by the last hypothenuse—or, in the case of the moon, by radius—gives the latitude (*vikshepa*).

58. When latitude and declination (*apakrama*) are of like direction, the declination (*krânti*) is increased by the latitude; when of different direction, it is diminished by it, to find the true (*spashṭa*) declination: that of the sun remains as already determined.

How to find the declination of a planet at any given point in the ecliptic, or circle of declination (*krântivṛtta*), was taught us in verse 28 above, taken in connection with verses 9 and 10 of the next chapter: here we have stated the method of finding the actual declination of any planet, as modified by its deviation in latitude from the ecliptic.

The process by which the amount of a planet's deviation in latitude from the ecliptic is here directed to be found is more correct than might have been expected, considering how far the Hindus were from comprehending the true relations of the solar system. The three quantities employed as data in the process are, first, the angular distance of the planet from its node; second, the apparent value, as latitude, of its greatest removal from the ecliptic, when seen from the earth at a mean distance, equal to the radius of its mean orbit; and lastly, its actual distance from the earth. Of these quantities, the second is stated for each planet in the concluding verses of the first chapter; the third is correctly represented by the variable hypothenuse (*cala karṇa*) found in the fourth process for determining the planet's true place (see above, under vv. 43–45); the first is still to be obtained, and verse 56 with the first part of verse 57 teach the method of ascertaining it. The principle of this method is the same for all the planets, although the statement of it is so different; it is, in effect, to apply to the mean place of the planet, before taking its distance from the node, only the equation of the apsis, found as the result of the third process. In the case of the superior planets, this method has all the correctness which the Hindu system admits; for by the first three processes of correction is found, as nearly as the Hindus are able to find it, the true heliocentric place of the planet, the distance from which to the node determines, of course, the amount of removal from the ecliptic. Instead, however, of taking this distance directly, rejecting altogether the fourth equation, that for the parallax of the earth's place, the Hindus apply the latter both to the planet and to the node; their relative position thus remains the same as if the other method had been adopted.

Thus, for instance, the position of Jupiter's node upon the first of January, 1860, is found from the data already given above (see i. 41–44) to be 2^s 19° 40′; his true heliocentric longitude, employed as a datum in the fourth process (see p. 216), is 3^s 1° 6′; Jupiter's heliocentric distance from the node is, accordingly, 11° 26′. Or, by the Hindu method, the planet's true geocentric place is 3^s 4° 11′, and the corrected longitude of its node is 2^s 22° 45′; the distance remains, as before, 11° 26′.

In the case of the inferior planets, as the assumptions of the Hindus respecting them were farther removed from the truth of nature, so their method of finding the distance from the node is more arbitrary and less accurate. In their system the heliocentric position of the planet is represented by the place of its conjunction (*çighra*), and they had, as is shown above (see ii. 8), recognized the fact that it was the distance of the latter from the node which determined the amount of deviation from the ecliptic. Now, in ascertaining the heliocentric distance of an inferior planet from its node, allowance needs to be made, of course, for the effect upon its position of the eccentricity of its orbit. But the Hindu equation of the apsis is no true representative of this effect: it is calculated in order to be applied to the mean place of the sun, the assumed centre of the epicycle—that is, of the true orbit; its value, as found, is geocentric, and, as appears by the table on p. 220, is widely different from its heliocentric value; and its sign is plus or minus according as its influence is to carry the planet, as seen from the earth, eastward or westward; while, in either case, the true heliocentric effect may be at one time to bring the planet nearer to, at another time to carry it farther from, the node. The Hindus, however, overlooking these incongruities, and having, apparently, no distinct views of the subject to guide them to a correcter method, follow with regard to Venus and Mercury what seems to them the same rule as was employed in the case of the other planets—they apply the equation of the apsis, the result of the third process, to the mean place of the conjunction; only here, as before, by an indirect process: instead of applying it to the conjunction itself, they apply it with a contrary sign to the node, the effect upon the relative position of the two being the same.

Thus, for instance, the longitude of Mercury's conjunction at the given time is (see p. 214) 4^s 16° 57′; from this subtract 2° 2′, the equation of the apsis found by the third process, and its equated longitude is 4^s 14° 55′: now deducting the longitude of the node at the same time, which is 20° 41′, we ascertain the planet's distance from the node to be 3^s 24° 14′. Or, by the Hindu method, add the same equation to the mean position of the node, and its equated longitude is 22° 43′; subtract this from the mean longitude of the conjunction, and the distance is, as before, 3^s 24° 14′.

The planet's distance from the node being determined, its latitude would be found by a process similar to that prescribed in verse 28 of this chapter, if the earth were at the centre of motion; and that rule is accordingly applied in the case of the moon; the proportion being, as radius is to the sine of the distance from the node, so is the sine of extreme latitude (or the latitude itself, the difference between the sine and the arc being of little account when the arc is so small) to the latitude at the given point. In the case of the other planets, however, this proportion is modified by combination with another, namely: as the last variable hypothenuse (*cala karṇa*), which is the line drawn from the earth to the finally determined place of the planet, or its true distance, is to radius, its mean distance, so is its apparent latitude at the mean distance to its apparent latitude at its true distance. That is, with

R : sin nod. dist. : : extreme lat. : actual lat. at dist. R
combining var. hyp : R : : lat. at dist. R : lat. at true dist.
we have var. hyp : sin nod. dist. : : extreme lat. : actual lat. at true dist.

which, turned into an equation, is the rule in the latter half of v. 57.

The latitude, as thus found, is measured, of course, upon a secondary to the ecliptic. By the rule in verse 58, however, it is treated as if measured upon a circle of declination, and is, without modification, added to or subtracted from the declination, according as the direction of the two is the same or different. The commentary takes note of this error, but explains it, as in other similar cases, as being, "for fear of giving men trouble, and on account of the very slight inaccuracy, overlooked by the blessed Sun, moved with compassion."

We present in the annexed table the results of the processes for calculating the latitude, the declination, and the true declination as affected by latitude, of all the planets, at the time for which their longitude has already been found. The declination is calculated by the rule in verse 28 of this chapter, the precession at the given time being, as found under verses 9–12 of the next chapter, 20° 24′ 39″. Upon the line for the sun in the table are given the results of the process for calculating his declination, the equinox itself being accounted as a "node": it is, in fact, styled, in modern Hindu astronomy, *krântipâta*, "node of declination," although that term does not occur in this treatise.

Results of the Process for finding the Latitude and Declination of the Planets.

Planet.	Longitude of Node.	do. corrected.	Distance from Node.	Sine.	Latitude.	Declination.	Corrected Declination.
	s ° ′ ″	s ° ′	s ° ′	′	° ′	° ′	° ′
Sun,	0 20 24 38	....	9 8 40	3397	....	23 41 S.	....
Moon,	9 24 24 43	....	1 23 14	2754	3 36 N.	4 56 N.	8 32 N.
Mercury,	0 20 40 41	0 22 43	3 24 14	3134	2 4 N.	23 10 S.	21 6 S.
Venus,	1 29 39 22	1 29 16	8 22 34	3409	1 21 S.	20 27 S.	21 48 S.
Mars,	1 10 3 5	2 13 47	4 4 58	2816	1 4 N.	14 52 S.	13 48 S.
Jupiter,	2 19 40 5	2 22 45	0 11 26	682	0 15 N.	21 42 N.	21 57 N.
Saturn,	3 10 20 45	3 14 38	0 16 24	970	0 37 N.	14 40 N	15 17 N.

We are now able to compare the Hindu determinations of the true places and motions of the planets with their actual positions and motions, as obtained by modern science. The comparison is made in the annexed table. As the longitudes given by the Sûrya-Siddhânta contain a constant error of 2° 20′, owing to the incorrect rate of precession adopted by the treatise, and the false position thence assigned to the equinox, we give, under the head of longitude, the distance of each planet both from the Hindu equinox, and from the true vernal equinox of Jan. 1, 1860. The Hindu daily motions are reduced from longitude to right ascension by the rule given in the next following verse (v. 59). The modern data are taken from the American Nautical Almanac.

True Places and Motions of the Planets, Jan. 1*st*, 1860, *midnight, at Washington, according to the Sûrya-Siddhânta and to Modern Science.*

Planet.	True Longitude. Sûrya Siddhânta: from Hindu eq.	True Longitude. Sûrya Siddhânta: from true eq.	True Longitude. Moderns.	Declination. Sûrya-Siddhânta.	Declination. Moderns.	Daily Motion in Right Ascension. Sûrya-Siddhânta.	Daily Motion in Right Ascension. Moderns.
	° ′	° ′	° ′	° ′	° ′	′ ″	′ ″
Sun,	278 40	276 20	280 5	23 41 S.	23 5 S.	+ 66 2	+ 66 18
Moon,	8 4	5 44	7 27	8 32 N.	6 56 N.	+683 50	+655 14
Mercury,	255 16	252 56	257 25	21 6 S.	20 42 S.	+ 31 13	+ 52 39
Venus,	305 0	302 40	303 25	21 48 S.	20 58 S.	+ 72 59	+ 78 6
Mars,	219 10	216 50	221 33	13 48 S.	14 23 S.	+ 31 58	+ 36 19
Jupiter,	114 36	112 16	111 34	21 57 N.	22 1 N.	− 8 21	− 8 17
Saturn,	141 27	139 7	145 32	15 17 N.	14 15 N.	− 3 3	− 2 29

The proper subject of the second chapter, the determination of the true places of the planets, being thus brought to a close, we should expect to see the chapter concluded here, and the other matters which it contains put off to that which follows, in which they would seem more properly to belong. The treatise, however, is nowhere distinguished for its orderly and consistent arrangement.

59. Multiply the daily motion of a planet by the time of rising of the sign in which it is, and divide by eighteen hundred; the quotient add to, or subtract from, the number of respirations in a revolution: the result is the number of respirations in the day and night of that planet.

In the first half of this verse is taught the method of finding the increment or decrement of right ascension corresponding to the increment or decrement of longitude made by any planet during one day. For the "time of rising" (*udayaprâṇâs*, or, more commonly, *udayâsavas*, literally "respirations of rising") of the different signs, or the time in respirations (see i. 11), occupied by the successive signs of the ecliptic in passing the meridian—or, at the equator, in rising above the horizon—see verses 42–44 of the next chapter. The statement upon which the rule is founded is as follows: if the given sign, containing 1800′ of arc (each minute of arc corresponding, as remarked above, under i. 11–12, to a respiration of sidereal time), occupies the stated number of respirations in passing the meridian, what number of respirations will be occupied by the arc traversed by the planet on a given day? The result gives the amount by which the day of each planet, reckoned in the manner of this Siddhânta, or from transit to transit across the inferior meridian, differs from a sidereal day: the difference is additive when the motion of the planet is direct; subtractive, when this is retrograde.

Thus, to find the length of the sun's day, or the interval between two successive apparent transits, at the time for which his true longitude and rate of motion have already been ascertained. The sun's longitude, as corrected by the precession, is $9^s\ 8°\ 40'$; he is accordingly in the tenth sign, of which the time of rising (*udayâsavas*), or the equivalent in right ascension, is 1935^p. His rate of daily motion in longitude is $61'\ 26''$. Hence the proportion

$$1800' : 1935^p :: 61'\ 26'' : 66^p.04$$

shows that his day differs from the true sidereal day by $11^v\ 0^p.04$. As his motion is direct, the difference is additive: the length of the apparent day is therefore $60^n\ 11^v\ 0^p.04$, which is equivalent to $24^h\ 0^m\ 27^s.5$, mean solar time. According to the Nautical Almanac, it is $24^h\ 0^m\ 28^s.6$. By a similar process, the length of Jupiter's day at the same time is found to be $59^n\ 58^v\ 4^p$, or $23^h\ 55^m\ 30^s.8$; by the Nautical Almanac, it is $23^h\ 55^m\ 30^s$.

60. Calculate the sine and versed sine of declination: then radius, diminished by the versed-sine, is the day-radius: it is either south or north.

The quantities made use of, and the processes prescribed, in this and the following verses, may be explained and illustrated by means of the annexed figure (Fig. 8).

Let the circle Z S Z′ N represent the meridian of a given place, C being the centre, the place of the observer, S N the section of the plane of his horizon—S being the south, and N the north point—Z and Z′ the zenith and its opposite point, the nadir, P and P′ the north and south poles, and E and E′ the points on the meridian cut by the equator. Let E D be the declination of a planet at a given time; then D D′ will be the diameter of the circle of diurnal revolution described by the planet, and B D the radius of that circle: B D is the line which in verse 60 is called the "day-radius." Draw D F perpendicular to E C:

Fig. 8.

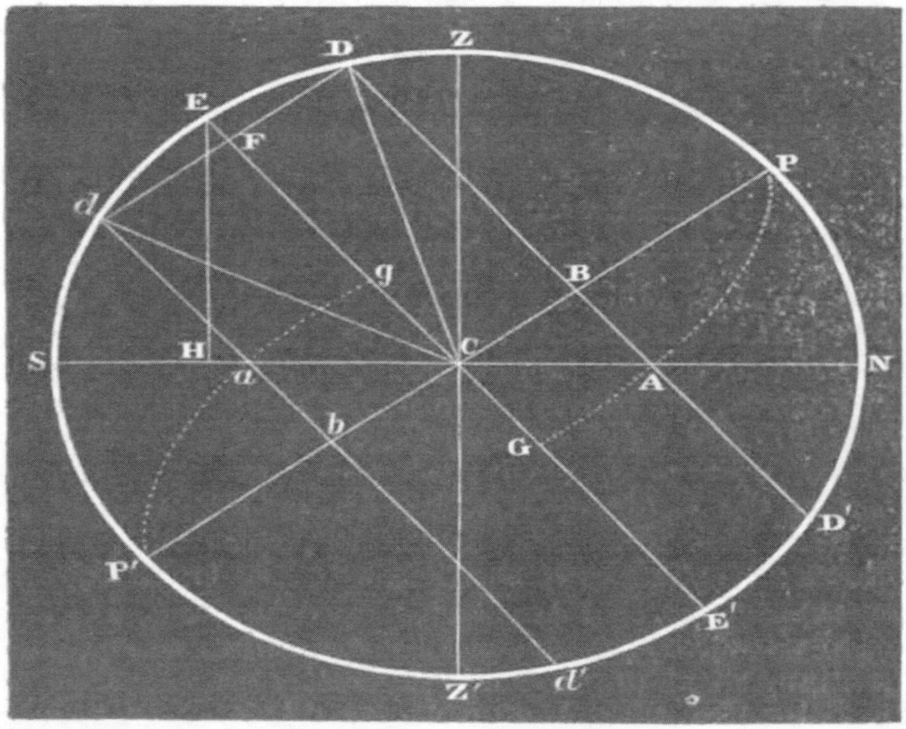

then it is evident that B D is equal to E C diminished by E F, which is the versed sine of E D, the declination.

For "radius" we have hitherto had only the term *trijyâ* (or its equivalents, *trijîvâ*, *tribhajîvâ*, *tribhajyâ*, *tribhamâurvikâ*), literally "the sine of three signs," that is, of 90°. That term, however, is applicable only to the radius of a great circle, or to tabular radius. In this verse, accordingly, we have for "day-radius" the word *dinavyâsadala*, "half-diameter of the day;" and other expressions synonymous with this are found used instead of it in other passages. A more frequent name for the same quantity in modern Hindu astronomy is *dyujyâ*, "day-sine:" this, although employed by the commentary, is not found anywhere in our text.

It is a matter for surprise that we do not find the day-radius declared equal simply to the cosine (*koṭijyâ*) of declination.

In illustration of the rule, it will be sufficient to find the radius of the diurnal circle described by the sun at the time for which his place has been determined. His declination, E *d* (Fig. 8) was found to be 23° 41′: of this the versed sine, E F, is, by the table given above (ii. 22–27), 290′: the difference between this and radius, E C, or 3438′, is 3148′, which is the value of C F or *b d*, the day-radius. The declination in this case being south, the day-radius is also south of the equator.

61. Multiply the sine of declination by the equinoctial shadow, and divide by twelve; the result is the earth-sine (*kshitijyâ*); this, multiplied by radius and divided by the day-radius, gives the sine of the ascensional difference (*cara*): the number of respirations due to the ascensional difference

62. Is shown by the corresponding arc. Add these to, and subtract them from, the fourth part of the corresponding day and night, and the sum and remainder are, when declination is north, the half-day and half-night;

63. When declination is south, the reverse; these, multiplied by two, are the day and the night. The day and the night of the asterisms (*bha*) may be found in like manner, by means of their declination, increased or diminished by their latitude.

We were taught in verse 59 how to find the length of the entire day of a planet at any given time; this passage gives us the method of ascertaining the length of its day and of its night, or of that part of the day during which the planet is above, and that during which it is below, the horizon.

In order to this, it is necessary to ascertain, for the planet in question, its ascensional difference (*cara*), or the difference between its right and oblique ascension, the amount of which varies with the declination of the planet and the latitude of the observer. The method of doing this is stated in verse 61: it may be explained by means of the last figure (Fig. 8). First, the value of the line A B, which is called the "earth-sine" (*kshitijyâ*), is found, by comparing the two triangles A B C and C H E, which are similar, since the angles A C B and C E H are each equal to the latitude of the observer. The triangle C H E is represented here by a triangle of which a gnomon of twelve digits is the perpendicular, and its equinoctial shadow, cast when the sun is in the equator and on the meridian (see the next chapter, verse 7, etc.), is the base. Hence the proportion EH : HC :: BC : AB is equivalent—since BC equals DF, the sine of declination—to gnom. : eq. shad. :: sin decl. : earth-sine. But the arc of which AB is sine is the same part of the circle of diurnal revolution as the ascensional difference is of the equator; hence the reduction of AB to the dimensions of a great circle, by the proportion BD : AB :: CE : CG, gives the value of CG, the sine of the ascensional difference. The corresponding arc is the measure in time of the amount by which the part of the diurnal circle intercepted between the meridian and the horizon differs from a quadrant, or by which the time between sun-rise or sun-set and noon or midnight differs from a quarter of the day.

In illustration of the process, we will calculate the respective length of the sun's day and night at Washington at the time for which our previous calculations have been made.

The latitude of Washington being 38° 54′, the length of the equinoctial shadow cast there by a gnomon twelve digits long is found, by the rule given below (iii. 17), to be $9^d.68$. The sine, *d* F or *b* C, of the sun's declination at the given time, 23° 41′ S, is 1380′. Hence the proportion

$$12 : 9.68 :: 1380 : 1113$$

gives us the value of the earth-sine, *a b*, as 1113′. This is reduced to the dimensions of a great circle by the proportion

$$3148 : 3438 :: 1113 : 1216$$

The value of C*g*, the sine of ascensional difference, is therefore 1216′: the corresponding arc is 20° 44′, or 1244′, which, as a minute of arc equals a respiration of time, is equivalent to $3^n\ 27^v\ 2^p$. The total length of the day was found above (under v. 59) to be $60^n\ 11^v$; increase and diminish the quarter of this by the ascensional difference, and double the sum and remainder, and the length of the night is found to be $37^n\ 0^v\ 1^p$, and that of the day $23^n\ 10^v\ 5^p$, which are equivalent respectively to $14^h\ 45^m\ 38^s.6$ and $9^h\ 14^m\ 48^s.9$, mean solar time.

Of course, the respective parts of a sidereal day during which each of the lunar mansions, as represented by its principal star, will remain above and below the horizon of a given latitude, may be found in the same manner, if the declination of the star is known; and this is stated in the chapter (ch. viii) which treats of the asterisms.

Why AB is called *kshitijyâ* is not easy to see. One is tempted to understand the term as meaning rather "sine of situation" than "earth-sine," the original signification of *kshiti* being "abode, residence": it might then indicate a sine which, for a given declination, varies with the situation of the observer. But that *kshiti* in this compound is to be taken in its other acceptation, of "earth," is at least strongly indicated by the other and more usual name of the sine in question, *kujyâ*, which is used by the commentary, although not in the text, and which can only mean "earth-sine." The word *cara*, used to denote the ascensional difference, means simply "variable"; we have elsewhere *carakhaṇḍa*, *caradala*, "variable portion"; that is to say, the constantly varying amount by which the apparent day and night differ from the equatorial day and night of one half the whole day each. The gnomon, the equinoctial shadow, etc., are treated of in the next chapter.

64. The portion (*bhoga*) of an asterism (*bha*) is eight hundred minutes; of a lunar day (*tithi*), in like manner, seven hundred and twenty. If the longitude of a planet, in minutes, be divided by the portion of an asterism, the result is its position in asterisms: by means of the daily motion are found the days, etc.

The ecliptic is divided (see ch. viii) into 27 lunar mansions or asterisms, of equal amount; hence the portion of the ecliptic occupied by each asterism is $13^\circ\ 20'$, or $800'$. In order to find, accordingly, in which asterism, at a given time, the moon or any other of the planets is, we have only to reduce its longitude, not corrected by the precession, to minutes, and divide by 800: the quotient is the number of asterisms traversed, and the remainder the part traversed of the asterism in which the planet is. The last clause of the verse is very elliptical and obscure; according to the commentary, it is to be understood thus: divide by the planet's true daily motion the part past and the part to come of the current asterism, and the quotients are the days and fractions of a day which the planet has passed, and is to pass, in that asterism. This interpretation is supported by the analogy of the following verses, and is doubtless correct.

The true longitude of the moon was found above (under v. 39) to be $11^s\ 17^\circ\ 39'$, or $20{,}859'$. Dividing by 800, we find that, at the given time, the moon is in the 27th, or last, asterism, named Revatî, of which it has traversed $59'$, and has $741'$ still to pass over. Its daily motion being $737'$, it has spent $28^v\ 4^p$, and has yet to continue $1^d\ 0^n\ 19^v\ 3^p$, in the asterism.

The latter part of this process proceeds upon the assumption that the planet's rate of motion remains the same during its whole continuance in the asterism. A similar assumption, it will be noticed, is made in all the processes from verse 59 onward; its inaccuracy is greatest, of course, where the moon's motion is concerned.

Respecting the lunar day (*tithi*) see below, under verse 66.

65. From the number of minutes in the sum of the longitudes of the sun and moon are found the *yogas*, by dividing that sum by the portion (*bhoga*) of an asterism. Multiply the minutes past and to come of the current yoga by sixty, and divide by the sum of the daily motions of the two planets: the result is the time in nâḍîs.

What the *yoga* is, is evident from this rule for finding it; it is the period, of variable length, during which the joint motion in longitude of the sun and moon amounts to $13^\circ\ 20'$, the portion of a lunar mansion. According to Colebrooke (As. Res., ix. 365; Essays, ii. 362, 363), the use of the yogas is chiefly astrological; the occurrence of certain movable festivals is, however, also regulated by them, and they are so frequently employed that every Hindu almanac contains a column specifying the yoga for each day, with the time of its termination. The names of the twenty-seven yogas are as follows:

1. Vishkambha.	10. Gaṇḍa.	19. Parigha.
2. Prîti.	11. Vṛddhi.	20. Çiva.
3. Âyushmant.	12. Dhruva.	21. Siddha.
4. Sâubhâgya.	13. Vyâghâta.	22. Sâdhya.
5. Çobhana.	14. Harshaṇa.	23. Çubha.
6. Atigaṇḍa.	15. Vajra.	24. Çukla.
7. Sukarman.	16. Siddhi.	25. Brahman.
8. Dhṛti.	17. Vyatîpâta.	26. Indra.
9. Çûla.	18. Varîyas.	27. Vâidhṛti.

There is also in use in India (see Colebrooke, as above) another system of yogas, twenty-eight in number, having for the most part different names from these, and governed by other rules in their succession. Of this system the Sûrya-Siddhânta presents no trace.

We will find the time in yogas corresponding to that for which the previous calculations have been made.

The longitude of the moon at that time is $11^s\ 17°\ 39'$, that of the sun is $8^s\ 18°\ 15'$; their sum is $8^s\ 5°\ 54'$, or 14,754′. Dividing by 800, we find that eighteen yogas of the series are past, and that the current one is the nineteenth, Parigha, of which 354′ are past, and 446′ to come. To ascertain the time at which the current yoga began and that at which it is to end, we divide these parts respectively by $798'\frac{1}{2}$, the sum of the daily motions of the sun and moon at the given time, and multiply by 60 to reduce the results to nâḍîs: and we find that Parigha began $26^n\ 36^v$ before, and will end $33^n\ 30^v\ 4^p$ after the given time.

The name *yoga*, by which this astrological period is called, is applied to it, apparently, as designating the period during which the "sum" (*yoga*) of the increments in longitude of the sun and moon amounts to a given quantity. It seems an entirely arbitrary device of the astrologers, being neither a natural period nor a subdivision of one, not being of any use that we can discover in determining the relative position, or aspect, of the two planets with which it deals, nor having any assignable relation to the asterisms, with which it is attempted to be brought into connection. Were there thirty yogas, instead of twenty-seven, the period would seem an artificial counterpart to the lunar day, which is the subject of the next verse; being derived from the sum, as the other from the difference, of the longitudes of the sun and moon.

66. From the number of minutes in the longitude of the moon diminished by that of the sun are found the lunar days (*tithi*), by dividing the difference by the portion (*bhoga*) of a lunar day. Multiply the minutes past and to come of the current lunar day by sixty, and divide by the difference of the daily motions of the two planets: the result is the time in nâḍîs.

The *tithi*, or lunar day, is (see i. 13) one thirtieth of a lunar month, or of the time during which the moon gains in longitude upon the sun a whole revolution, or 360°: it is, therefore, the period during which the difference of the increment of longitude of the two planets amounts to 12°, or 720′, which arc, as stated in verse 64, is its portion (*bhoga*). To find the current lunar day, we divide by this amount the whole excess of the longitude of the moon over that of the sun at the given time; and to find the part past and to come of the current day, we convert longitude into time in a manner analogous to that employed in the case of the yoga.

Thus, to find the date in lunar time of the midnight preceding the first of January, 1860, we first deduct the longitude of the sun from that of the moon; the remainder is 2^s 29° 24′, or 5364′: dividing by 720, it appears that the current lunar day is the eighth, and that 324′ of its portion are traversed, leaving 396′ to be traversed. Multiplying these numbers respectively by 60, and dividing by 675′ 38″, the difference of the daily motions at the time, we find that 28^n 46^v 2^p have passed since the beginning of the lunar day, and that it still has 35^n 10^v 8^p to run.

The lunar days have, for the most part, no distinctive names, but those of each half month (*paksha*—see above, under i. 48–51) are called first, second, third, fourth, etc., up to fourteenth. The last, or fifteenth, of each half has, however, a special appellation: that which concludes the first, the light half, ending at the moment of opposition, is called *pâurṇamâsî*, *pûrṇimâ*, *pûrṇamâ*, "day of full moon;" that which closes the month, and ends with the conjunction of the two planets, is styled *amâvâsyâ*, "the day of dwelling together."

Each lunar day is farther divided into two halves, called *karaṇa*, as appears from the next following passage.

67. The fixed (*dhruva*) *karaṇas*, namely *çakuni*, *nâga*, *catushpada* the third, and *kinstughna*, are counted from the latter half of the fourteenth day of the dark half-month.

68. After these, the karaṇas called movable (*cara*), namely *bava*, etc., seven of them: each of these karaṇas occurs eight times in a month.

69. Half the portion (*bhoga*) of a lunar day is established as that of the karaṇas . . . .

Of the eleven *karaṇas*, four occur only once in the lunar month, while the other seven are repeated each of them eight times to fill out the remainder of the month. Their names, and the numbers of the half lunar days to which each is applied, are presented below:

1. Kinstughna.	1st.
2. Bava.	2nd, 9th, 16th, 23rd, 30th, 37th, 44th, 51st.
3. Bâlava.	3rd, 10th, 17th, 24th, 31st, 38th, 45th, 52nd.
4. Kâulava.	4th, 11th, 18th, 25th, 32nd, 39th, 46th, 53rd.
5. Tâitila.	5th, 12th, 19th, 26th, 33rd, 40th, 47th, 54th.
6. Gara.	6th, 13th, 20th, 27th, 34th, 41st, 48th, 55th.
7. Baṇij.	7th, 14th, 21st, 28th, 35th, 42nd, 49th, 56th.
8. Vishṭi.	8th, 15th, 22nd, 29th, 36th, 43rd, 50th, 57th.
9. Çakuni.	58th.
10. Nâga.	59th.
11. Catushpada.	60th.

Most of these names are very obscure: the last three mean "hawk," "serpent," and "quadruped." *Karaṇa* itself is, by derivation, "factor, cause:" in what sense it is applied to denote these divisions of the month, we do not know. Nor have we found anywhere an explanation of the value and use of the karaṇas in Hindu astronomy or astrology.

The time which we have had in view in our other calculations being, as is shown under the preceding passage, in the first half of the eighth lunar day, is, of course, in the fifteenth karaṇa, which is named Vishṭi.

The remaining half-verse is simply a winding-up of the chapter.

69. . . . . Thus has been declared the corrected (*sphuṭa*) motion of the sun and the other planets.

The following chapter is styled the "chapter of the three inquiries" (*tripraçnâdhikâra*). According to the commentary, this means that it is intended by the teacher as a reply to his pupil's inquiries respecting the three subjects of direction (*diç*), place (*deça*), and time (*kâla*).

CHAPTER III.

OF DIRECTION, PLACE, AND TIME.

CONTENTS:—1–6, construction of the dial, and description of its parts; 7, the measure of amplitude; 8, of the gnomon, hypothenuse, and shadow, any two being given, to find the third; 9–12, precession of the equinoxes; 12–13, the equinoctial shadow; 13–14, to find, from the equinoctial shadow, the latitude and co-latitude; 14–17, the sun's declination being known, to find, from a given shadow at noon, his zenith-distance, the latitude, and its sine and cosine; 17, latitude being given, to find the equinoctial shadow; 17–20, to find, from the latitude and the sun's zenith-distance at noon, his declination and his true and mean longitude; 20–22, latitude and declination being given, to find the noon-shadow and hypothenuse; 22–23, from the sun's declination and the equinoctial shadow, to find the measure of amplitude; 23–25, to find, from the equinoctial shadow and the measure of amplitude at any given time, the base of the shadow; 25–27, to find the hypothenuse of the shadow when the sun is upon the prime vertical; 27–28, the sun's declination and the latitude being given, to find the sine and the measure of amplitude; 28–33, to find the sines of the altitude and zenith-distance of the sun, when upon the south-east and south-west vertical circles; 33–34, to find the corresponding shadow and hypothenuse; 34–36, the sun's ascensional difference and the hour-angle being given, to find the sines of his altitude and zenith-distance, and the corresponding shadow and hypothenuse; 37–39, to find, by a contrary process, from the shadow of a given time, the sun's altitude and zenith-distance, and the hour-angle; 40–41, the latitude and the sun's amplitude being known, to find his declination and true longitude; 41–42, to draw the path described by the extremity of the shadow; 42–45, to find the arcs of right and oblique ascension corresponding to the several signs of the ecliptic; 46–48, the sun's longitude and the time being known, to find the point of the ecliptic which is upon the horizon; 49, the sun's longitude and the hour-angle being known, to find the point of the ecliptic which is upon the meridian; 50–51, determination of time by means of these data.

1. On a stony surface, made water-level, or upon hard plaster, made level, there draw an even circle, of a radius equal to any required number of the digits (*angula*) of the gnomon (*çanku*).
2. At its centre set up the gnomon, of twelve digits of the measure fixed upon; and where the extremity of its shadow touches the circle in the former and after parts of the day,
3. There fixing two points upon the circle, and calling them the forenoon and afternoon points, draw midway between them, by means of a fish-figure (*timi*), a north and south line.
4. Midway between the north and south directions draw, by a fish-figure, an east and west line: and in like manner also, by fish-figures (*matsya*) between the four cardinal directions, draw the intermediate directions.
5. Draw a circumscribing square, by means of the lines going out from the centre; by the digits of its base-line (*bhujasûtra*) projected upon that is any given shadow reckoned.

In this passage is described the method of construction of the Hindu dial, if that can properly be called a dial which is without hour-lines, and does not give the time by simple inspection. It is, as will be at once remarked, a horizontal dial of the simplest character, with a vertical gnomon. This gnomon, whatever may be the length chosen for it, is regarded as divided into twelve equal parts called digits (*angula*, "finger"). The ordinary digit is one twelfth of a span (*vitasti*), or one twenty-fourth of a cubit (*hasta*): if made according to this measure, then, the gnomon would be about nine inches long. Doubtless the first gnomons were of such a length, and the rules of the gnomonic science were constructed accordingly, "twelve" and "the gnomon" being used, as they are used everywhere in this treatise, as convertible terms: thus twelve digits became the unvarying conventional length of the staff, and all measurements of the shadow and its hypothenuse were made to correspond. How the digit was subdivided, we have nowhere any hint. In determining the directions, the same method was employed which is still in use; namely, that of marking the points at which the extremity of the shadow, before and after noon, crosses a circle described about the base of the gnomon; these points being, if we suppose the sun's declination to have remained the same during the interval, at an equal distance upon either side from the meridian line. In order to bisect the line joining these points by another at right angles to it, which will be the meridian, the Hindus draw the figure which is called here the "fish" (*matsya* or *timi*); that is to say, from the two extremities of the line in question as centres, and with a radius equal to the line itself, arcs of circles are described, cutting one another in two points. The lenticular figure formed by the two arcs is the "fish;" through the points of intersection, which are called (in the commentary) the "mouth" and "tail," a line is drawn, which is the one required. The meridian being thus determined, the east and west line, and those for the intermediate points of directions, are laid down from it, by a repetition of the same process. A square (*caturasra*, "having four corners") is then farther described about the general centre, or about a circle drawn about that centre, the eastern and western sides of which are divided into digits; its use is, to aid in ascertaining the "base" (*bhuja*) of any given shadow, which is the value of the latter when projected upon a north and south line (see below, vv. 23–25); the square is drawn, as explained by the commentary, in order to insure the correctness of the projection, by a line strictly parallel to the east and west line.

The figure (Fig. 9) given below, under verse 7, will illustrate the form of the Hindu dial, as described in this passage.

The term used for "gnomon" is *çanku*, which means simply "staff." For the shadow, we have the common word *châyâ*, "shadow," and also, in many places, *prabhâ* and *bhâ*, which properly signify the very opposite of shadow, namely "light, radiance:" it is difficult to see how they should come to be used in this sense; so far as we are aware, they are applied to no other shadow than that of the gnomon.

6. The east and west line is called the prime vertical (*samamaṇḍala*); it is likewise denominated the east and west hour circle (*unmaṇḍala*) and the equinoctial circle (*vishuvanmaṇḍala*).

The line drawn east and west through the base of the gnomon may be regarded as the line of common intersection at that point of three great circles, as being a diameter to each of the three, and as thus entitled to represent them all. These circles are the ones which in the last figure (Fig. 8, p. 88) are shown projected in their diameters Z Z′, P P′, and E E′; the centre C, in which the diameters intersect, is itself the projection of the line in question here. Z Z′ represents the prime vertical, which is styled *samamaṇḍala*, literally "even circle:" P P′ is the hour circle, or circle of declination, which passes through the east and west points of the observer's horizon; it is called *unmaṇḍala*, "up-circle"—that is to say, the circle which in the oblique sphere is elevated; E E′ finally, the equator, has the name of *vishuvanmaṇḍala*, or *vishuvad-vṛtta*, "circle of the equinoxes;" the equinoctial points themselves being denominated *vishuvat*, or *vishuva*, which may be rendered "point of equal separation." The same line of the dial might be regarded as the representative in like manner of a fourth circle, that of the horizon (*kshitija*), projected, in the figure, in S N: hence the commentary adds it also to the other three; it is omitted in the text, perhaps, because it is represented by the whole circle drawn about the base of the gnomon, and not by this diameter alone.

The specifications of this verse, especially of the latter half of it, are of little practical importance in the treatise, for there hardly arises a case, in any of its calculations, in which the east and west axis of the dial comes to be taken as standing for these circles, or any one of them. In drawing the base (*bhuja*) of the shadow, indeed, it does represent the plane of the prime vertical (see below, under vv. 23–25); but this is not distinctly stated, and the name of the prime vertical (*samamaṇḍala*) occurs in only one other passage (below, v. 26): the east and west hour-circle (*unmaṇḍala*) is nowhere referred to again: and the equator, as will be seen under the next verse, is properly represented on the dial, not by its east and west axis, but by the line of the equinoctial shadow.

7. Draw likewise an east and west line through the extremity of the equinoctial shadow (*vishuvadbhâ*); the interval between any given shadow and the line of the equinoctial shadow is denominated the measure of amplitude (*agrâ*).

The equinoctial shadow is defined in a subsequent passage (vv. 12, 13); it is, as we have already had occasion to notice (under ii. 61–63), the shadow cast at mid-day when the sun is at either equinox—that is to say, when he is in the plane of the equator. Now as the equator is a circle of diurnal revolution, the line of intersection of its plane with that of the horizon will be an east and west line; and since it is also a great circle, that line will pass through the centre, the place of the observer: if, therefore, we draw through the extremity of the equinoctial shadow a line parallel to the east and west axis of the dial, it will represent the intersection with the dial of an equinoctial plane passing through the top of the gnomon, and in it will terminate the lines drawn through that point from any point in the plane of the equator; and hence, it will also coincide with the path of the extremity of the shadow on the day of the equinox. Thus, let the following figure (Fig. 9) represent the plane of the dial, N S and E W being its two axes, and *b* the base of the gnomon: and let the shadow cast at noon when the sun is upon the equator be, in a given latitude, *b e*: then *b e* is the equinoctial shadow, and Q Q′, drawn through *e* and parallel to E W, is the path of the equinoctial shadow, being the line in which a ray of the sun, from any point in the plane of the equator, passing through the top of the gnomon, will meet the face of the dial. In the figure as given, the circle is supposed to be described about the base of the gnomon with a radius of forty digits, and the graduation of the eastern and western sides of the circumscribing square, used in measuring the base (*bhuja*) of the shadow, is indicated: the length given to the equinoctial shadow corresponds to that which it has in the latitude of Washington.

Fig. 9.

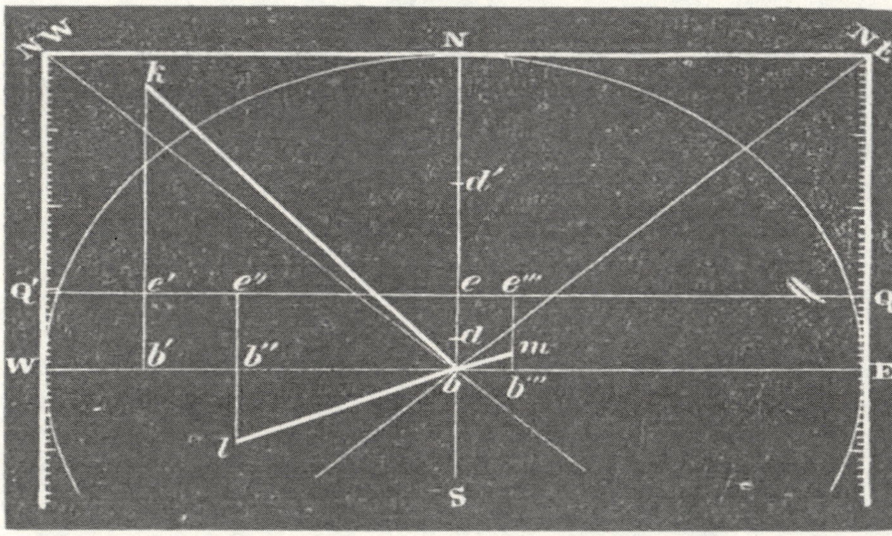

It is not, however, on account of the coincidence of Q Q′ with the path of the equinoctial shadow that it is directed to be permanently drawn upon the dial-face: its use is to determine for any given shadow its *agrâ*, or measure of amplitude. Thus, let *b d*, *b d′*, *b k*, *b l*, *b m*, be shadows cast by the gnomon, under various conditions of time and declination: then the distance from the extremity of each of them to the

line of the equinoctial shadow, or *d e, d′ e, k e′, l e″, m e‴* respectively, is denominated the *agrâ* of that shadow or of that time.

The term *agrâ* we have translated "measure of amplitude," because it does in fact represent the sine of the sun's amplitude—understanding by "amplitude" the distance of the sun at rising or setting from the east or west point of the horizon—varying with the hypothenuse of the shadow, and always maintaining to that hypothenuse the fixed ratio of the sine of amplitude to radius. That this is so, is assumed by the text in its treatment of the *agrâ*, but is nowhere distinctly stated, nor is the commentator at the pains of demonstrating the principle. Since, however, it is not an immediately obvious one, we will take the liberty of giving the proof of it.

In the annexed figure (Fig. 10) let C represent the top of the gnomon, and let K be any given position of the sun in the heavens. From K draw K B′ at right angles to the plane of the prime vertical, meeting that

Fig. 10.

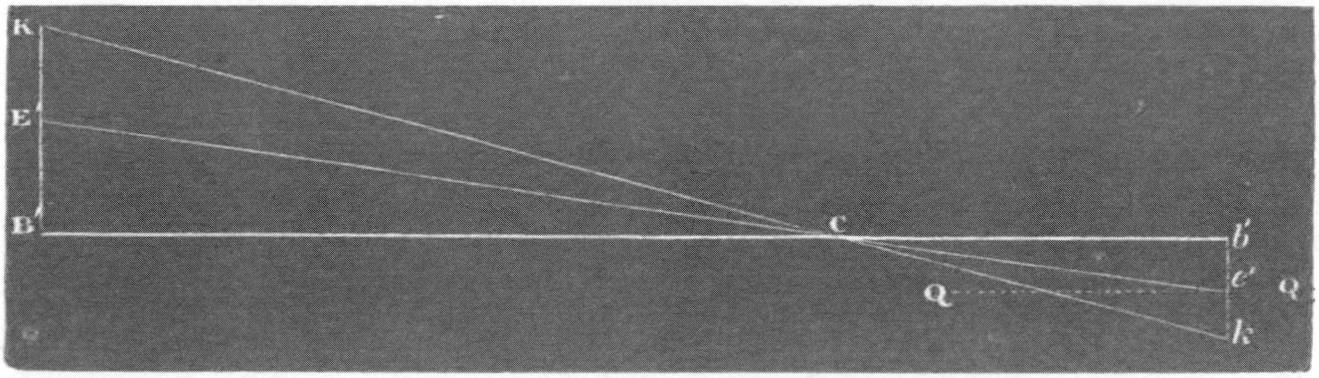

plane in B′, and let the point of its intersection with the plane of the equator be in E′. Join K C, E′ C, and B′ C. Then K C is radius, and E′ K is equal to the sine of the sun's amplitude: for if, in the sun's daily revolution, the point K is brought to the horizon, E′ B′ will disappear, K E′ C will become a right angle, K C E′ will be the amplitude, and E′ K its sine; but, with a given declination, the value of E′ K remains always the same, since it is a line drawn in a constant direction between two parallel planes, that of the circle of declination and that of the equator. Now conceive the three lines intersecting in C to be produced until they meet the plane of the dial in *b′*, *e′*, and *k* respectively; these three points will be in the same straight line, being in the line of intersection with the horizon of the plane K B′ C produced, and this line, *b′ k*, will be at right angles to B′ *b′*, since it is the line of intersection of two planes, each of which is at right angles to the plane of the prime vertical, in which B′ *b′* lies. K B′ and *k b′* are therefore parallel, and the triangles C E′ K and C *e′ k* are similar, and *e′ k* : C *k* : : E′ K : C K. But C *k* is the hypothenuse of the shadow at the given time, and *e′ k* is the *agrâ*, or measure of amplitude, since *e′*, by what was said above, is in the line of the equinoctial shadow; therefore meas. ampl. : hyp. shad. : : sin ampl. : R. Hence, if the declination and the latitude, which together determine the sine of amplitude, be given, the measure of amplitude will vary with the hypothenuse of the shadow, and the measure of amplitude of any given shadow will be to that of any other, as the hypothenuse of the former to that of the latter.

The lettering of the above figure is made to correspond, as nearly as may be, with that of the one preceding, and also with that of the one given later, under verses 13 and 14, in either of which the relations of the problem may be farther examined.

There are other methods of proving the constancy of the ratio borne by the measure of amplitude to the hypothenuse of the shadow, but we have chosen to give the one which seemed to us most likely to be that by which the Hindus themselves deduced it. Our demonstration is in one respect only liable to objection as representing a Hindu process—it is founded, namely, upon the comparison of oblique-angled triangles, which elsewhere in this treatise are hardly employed at all. Still, although the Hindus had no methods of solving problems excepting in right-angled trigonometry, it is hardly to be supposed that they refrained from deriving proportions from the similarity of oblique-angled triangles. The principle in question admits of being proved by means of right-angled triangles alone, but these would be situated in different planes.

Why the line on the dial which thus measures the sun's amplitude is called the *agrâ*, we have been unable thus far to discover. The word, a feminine adjective (belonging, probably, to *rekhâ*, "line," understood), literally means "extreme, first, chief." Possibly it may be in some way connected with the use of *antyâ*, "final, lowest," to designate the line E*g* or EG (Fig. 8, p. 88): see below, under v. 35. The sine of amplitude itself, *a*C or AC (Fig. 8), is called below (vv. 27–30) *agrajyâ*.

8. The square root of the sum of the squares of the gnomon and shadow is the hypothenuse: if from the square of the latter the square of the gnomon be subtracted, the square root of the remainder is the shadow: the gnomon is found by the converse process.

This is simply an application of the familiar rule, that in a right-angled triangle the square of the hypothenuse is equal to the sum of the squares of the other two sides, to the triangle produced by the gnomon as perpendicular, the shadow as base, and the hypothenuse of the shadow, the line drawn from the top of the gnomon to the extremity of the shadow, as hypothenuse.

The subject next considered is that of the precession of the equinoxes.

9. In an Age (*yuga*), the circle of the asterisms (*bha*) falls back eastward thirty score of revolutions. Of the result obtained after multiplying the sum of days (*dyugaṇa*) by this number, and dividing by the number of natural days in an Age,

10. Take the part which determines the sine, multiply it by three, and divide by ten; thus are found the degrees called those of the precession (*ayana*). From the longitude of a planet as corrected by these are to be calculated the declination, shadow, ascensional difference (*caradala*), etc.

11. The circle, as thus corrected, accords with its observed place at the solstice (*ayana*) and at either equinox; it has moved eastward, when the longitude of the sun, as obtained by calculation, is less than that derived from the shadow,

12. By the number of degrees of the difference; then, turning back, it has moved westward by the amount of difference, when the calculated longitude is greater. . . .

Nothing could well be more awkward and confused than this mode of stating the important fact of the precession of the equinoxes, of describing its method and rate, and of directing how its amount at any time is to be found. The theory which the passage, in its present form, is actually intended to put forth is as follows: the vernal equinox librates westward and eastward from the fixed point, near ζ Piscium, assumed as the commencement of the sidereal sphere—the limits of the libratory movement being 27° in either direction from that point, and the time of a complete revolution of libration being the six-hundredth part of the period called the Great Age (see above, under i. 15–17), or 7200 years; so that the annual rate of motion of the equinox is 54″. We will examine with some care the language in which this theory is conveyed, as important results are believed to be deducible from it.

The first half of verse 9 professes to teach the fundamental fact of the motion in precession. The words *bhânâm cakram*, which we have rendered "circle of the asterisms," i. e., the fixed zodiac, would admit of being translated "circle of the signs," i. e., the movable zodiac, as reckoned from the actual equinox, since *bha* is used in this treatise in either sense. But our interpretation is shown to be the correct one by the directions given in verses 11 and 12, which teach that when the sun's calculated longitude—which is his distance from the initial point of the fixed sphere—is less than that derived from the shadow by the process to be taught below (vv. 17–19)—which is his distance from the equinox—the circle has moved eastward, and the contrary: it is evident, then, that the initial point of the sphere is regarded as the movable point, and the equinox as the fixed one. Now this is no less strange than inconsistent with the usage of the rest of the treatise. Elsewhere ζ Piscium is treated as the one established limit, from which all motion commenced at the creation, and by reference to which all motion is reckoned, while here it is made secondary to a point of which the position among the stars is constantly shifting, and which hardly has higher value than a node, as which the Hindu astronomy in general treats it (see p. 86). The word used to express the motion (*lambate*) is the same with that employed in a former passage (i. 25) to describe the eastward motion of the planets, and derivatives of which (as *lamba*, *lambana*, etc.) are not infrequent in the astronomical language; it means literally to "lag, hang back, fall behind:" here we have it farther combined with the prefix *pari*, "about, round about," which seems plainly to add the idea of a complete revolution in the retrograde direction indicated by it, and we have translated the line accordingly. This verse, then, contains no hint of a libratory movement, but rather the distinct statement of a continuous eastward revolution. It should be noticed farther, although the circumstance is one of less significance, that the form in which the number of revolutions is stated, *trinçatkṛtyas*, "thirty twenties," has no parallel in the usage of this Siddhânta elsewhere.

We may also mention in this connection that Bhâskara, the great Hindu astronomer of the twelfth century, declares in his Siddhânta-Çiromaṇi (Golâdh., vi. 17) that the revolutions of the equinox are given by the Sûrya-Siddhânta as thirty in an Age (see Colebrooke, As. Res., xii. 209, etc.; Essays, ii. 374, etc., for a full discussion of this passage and its bearings); thus not only ignoring the theory of libration, but giving a very different number of revolutions from that presented by our text. As regards this latter point, however, the change of a single letter in the modern reading (substituting *trinçatkṛtvas*, "thirty times," for *trinçatkṛtyas*, "thirty twenties") would make it accord with Bhâskara's statement. We shall return again to this subject.

The number of revolutions, of whatever kind they may be, being 600 in an Age, the position at any given time of the initial point of the sphere with reference to the equinox is found by a proportion, as follows: as the number of days in an Age is to the number of revolutions in the same period, so is the given "sum of days" (see above, under i. 48–51) to the revolutions and parts of a revolution accomplished down to the given time. Thus, let us find, in illustration of the process, the amount of precession on the first of January, 1860. Since the number of years elapsed before the beginning of the present Iron Age (*kali yuga*) is divisible by 7200, it is unnecessary to make our calculation from the commencement of the present order of things: we may take the sum of days since the current Age began, which is (see above, under i. 53) 1,811,945. Hence the proportion

$$1,577,917,828^{d} : 600^{rev} :: 1,811,945^{d} : 0^{rev}\ 248^{\circ}\ 2'\ 8''.9$$

gives us the portion accomplished of the current revolution. Of this we are now directed (v. 10) to take the part which determines the sine (*dos*, or *bhuja*—for the origin and meaning of the phrase, see above, under ii. 29, 30). This direction determines the character of the motion as libratory. For a motion of 91°, 92°, 93°, etc., gives, by it, a precession of 89°, 88°, 87°, etc.; so that the movable point virtually returns upon its own track, and, after moving 180°, has reverted to its starting-place. So its farther motion, from 180° to 270°, gives a precession increasing from 0° to 90° in the opposite direction; and this, again, is reduced to 0° by the motion from 270° to 360°. It is as if the second and third quadrants were folded over upon the first and fourth, so that the movable point can never, in any quadrant of its motion, be more than 90° distant from the fixed equinox. Thus, in the instance taken, the *bhuja* of 248° 2′ 8″.9 is its supplement, or 68° 2′ 8″.9; the first 180° having only brought the movable point back to its original position, its present distance from that position is the excess over 180° of the arc obtained as the result of the first process. But this distance we are now farther directed to multiply by three and divide by ten: this is equivalent to reducing it to the measure of an arc of 27°, instead of 90°, as the quadrant of libration, since 3 : 10 :: 27 : 90. This being done, we find the actual distance of the initial point of the sphere from the equinox on the first of January, 1860, to be 20° 24′ 38″.67.

The question now arises, in which direction is the precession, thus ascertained, to be reckoned? And here especially is brought to light the awkwardness and insufficiency, and even the inconsistency, of the process as taught in the text. Not only have we no rule given which furnishes us the direction, along with the amount, of the precessional movement, but it would even be a fair and strictly legitimate deduction from verse 9, that that movement is taking place at the present time in an opposite direction from the actual one. We have already remarked above that the last complete period of libratory revolution closed with the close of the last Brazen Age, and the process of calculation has shown that we are now in the third quarter of a new period, and in the third quadrant of the current revolution. Therefore, if the revolution is an eastward one, as taught in the text, only taking place upon a folded circle, so as to be made libratory, the present position of the movable point, ζ Piscium, ought to be to the west of the equinox, instead of to the east, as it actually is. It was probably on account of this unfortunate flaw in the process, that no rule with regard to the direction was given, excepting the experimental one contained in verses 11 and 12, which, moreover, is not properly supplementary to the preceding rules, but rather an independent method of determining, from observation, both the direction and the amount of the precession. In verse 12, it may be remarked, in the word *âvṛtya*, "turning back," is found the only distinct intimation to be discovered in the passage of the character of the motion as libratory.

We have already above (under ii. 28) hinted our suspicions that the phenomenon of the precession was made no account of in the original composition of the Sûrya-Siddhânta, and that the notice taken of it by the treatise as it is at present is an afterthought: we will now proceed to expose the grounds of those suspicions.

It is, in the first place, upon record (see Colebrooke, As. Res., xii. 215; Essays, ii. 380, etc.) that some of the earliest Hindu astronomers were ignorant of, or ignored, the periodical motion of the equinoxes; Brahmagupta himself is mentioned among those whose systems took no account of it; it is, then, not at all impossible that the Sûrya-Siddhânta, if an ancient work, may originally have done the same. Among the positive evidences to that effect, we would first direct attention to the significant fact that, if the verses at the head of this note were expunged, there would not be found, in the whole body of the treatise besides, a single hint of the precession. Now it is not a little difficult to suppose that a phenomenon of so much consequence as this, and which enters as an element into so many astronomical processes, should, had it been borne distinctly in mind in the framing of the treatise, have been hidden away thus in a pair of verses, and unacknowledged elsewhere—no hint being given, in connection with any of the processes taught, as to whether the correction for precession is to be applied or not, and only the general directions contained in the latter half of verse 10, and ending with an "etc.," being even here presented. It has much more the aspect of an after-thought, a correction found necessary at a date subsequent to the original composition, and therefore inserted, with orders to "apply it wherever it is required." The place where the subject is introduced

looks the same way: as having to do with a revolution, as entering into the calculation of mean longitudes, it should have found a place where such matters are treated of, in the first chapter; and even in the second chapter, in connection with the rule for finding the declination, it would have been better introduced than it is here. Again, in the twelfth chapter, where the orbits of the heavenly bodies are given, in terms dependent upon their times of revolution, such an orbit is assigned to the asterisms (v. 88) as implies a revolution once in sixty years: it is, indeed, very difficult to see what can have been intended by such a revolution as this; but if the doctrine respecting the revolution of the asterisms given in verse 9 of this chapter had been in the mind of the author of the twelfth chapter, he would hardly have added another and a conflicting statement respecting the same or a kindred phenomenon. It appears to us even to admit of question whether the adoption by the Hindus of the sidereal year as the unit of time does not imply a failure to recognize the fact that the equinox was variable. We should expect, at any rate, that if, at the outset, the ever-increasing discordance between the solar and the sidereal year had been fully taken into account by them, they would have more thoroughly established and defined the relations of the two, and made the precession a more conspicuous feature of their general system than they appear to have done. In the construction of their cosmical periods they have reckoned by sidereal years only, at the same time assuming (as, for instance, above, i. 13, 14) that the sidereal year is composed of the two *ayanas*, "progresses" of the sun from solstice to solstice. The supposition of an after-correction likewise seems to furnish the most satisfactory explanation of the form given to the theory of the precession. The system having been first constructed on the assumption of the equality of the tropical and sidereal years, when it began later to appear, too plainly to be disregarded, that the equinox had changed its place, the question was how to introduce the new element. Now to assign to the equinox a complete revolution would derange the whole system, acknowledging a different number of solar from sidereal years in the chronological periods; if, however, a libratory motion were assumed, the equilibrium would be maintained, since what the solar year lost in one part of the revolution of libration it would gain in another, and so the tropical and sidereal years would coincide, in number and in limits, in each great period. The circumstance which determined the limit to be assigned to the libration we conceive to have been, as suggested by Bentley (Hind. Ast., p. 132), that the earliest recorded Hindu year had been made to begin when the sun entered the asterism Kṛttikâ, or was 26° 40′ west of the point fixed upon as the commencement of the sidereal sphere for all time (see above, under i. 27), on which account it was desirable to make the arc of libration include the beginning of Kṛttikâ.

Besides these considerations, drawn from the general history of the Hindu astronomy, and the position of the element of the precession in the system of the Sûrya-Siddhânta, we have still to urge the blind and incoherent, as well as unusual, form of statement of the phenomenon, as fully exposed above. There is nothing to compare with it in this respect in any other part of the treatise, and we are unwilling to believe that in the original composition of the Siddhânta a clearer explanation, and one more consistent in its method and language with those of the treatise generally, would not have been found for the subject. We even discover evidences of more than one revision of the passage. The first half of verse 9 so distinctly teaches, if read independently of what follows it, a complete revolution of the equinoxes, that, especially when taken in connection with Bhâskara's statement, as cited above, it almost amounts to proof that the theory put forth in the Sûrya-Siddhânta was at one time that of a complete revolution. The same conclusion is not a little strengthened, farther, by the impossibility of deducing from verse 9, through the processes prescribed in the following verses, a true expression for the direction of the movement at present: we can see no reason why, if the whole passage came from the same hand, at the same time, this difficulty should not have been avoided; while it is readily explainable upon the supposition that the libratory theory of verse 10 was added as an amendment to the theory of verse 9, while at the same time the language of the latter was left as nearly unaltered as possible.

There seems, accordingly, sufficient ground for suspecting that in the Sûrya-Siddhânta, as originally constituted, no account was taken of the precession; that its recognition is a later interpolation, and was made at first in the form of a theory of complete revolution, being afterward altered to its present shape. Whether the statement of Bhâskara truly represents the earlier theory, as displayed in the Sûrya-Siddhânta of his time, we must leave an undetermined question. The very slow rate assigned by it to the movement of the equinox—only 9″ a year—throws a doubt upon the matter: but it must be borne in mind that, so far as we can see, the actual amount of the precession since about A. D. 570 (see above, under i. 27) might by that first theory have been distributed over the whole duration of the present Age, since B. C. 3102.

In his own astronomy, Bhâskara teaches the complete revolution of the equinoxes, giving the number of revolutions in an Æon (of 4,320,000,000 years) as 199,669; this makes the time of a single revolution to be 21,635.8073 years, and the yearly rate of precession 59″.9007. It is not to be supposed that he considered himself to have determined the rate with such exactness as would give precisely the odd number of 199,669 revolutions to the Æon; the number doubtless stands in some relation which we do not at present comprehend to the other elements of his astronomical system. Bhâskara's own commentators, however, reject his theory, and hold to that of a libration, which has been and is altogether the prevailing doctrine throughout India, and seems to have made its way thence into the Arabian, and even into the early European astronomy (see Colebrooke, as above).

Bentley, it may be remarked here, altogether denies (Hind. Ast., p. 130, etc.) that the libration of the equinoxes is taught in the Sûrya-Siddhânta, maintaining, with arrogant and unbecoming depreciation of those who venture to hold a different opinion, that its theory is that of a continuous revolution in an epicycle, of which the circumference is equal to 108° of the zodiac. In truth, however, Bentley's own theory derives no color of support from the text of the Siddhânta, and is besides in itself utterly untenable. It is not a little strange that he should not

have perceived that, if the precession were to be explained by a revolution in an epicycle, its rate of increase would not be equable, but as the increment of the sine of the arc in the epicycle traversed by the movable point, farther varied by the varying distance at which it would be seen from the centre in different parts of the revolution; and also that, the dimensions of the epicycle being 108°, the amount of precession would never come to equal 27°, but would, when greatest, fall short of 18°, being determined by the radius of the epicycle. Bentley's whole treatment of the passage shows a thorough misapprehension of its meaning and relations: he even commits the blunder of understanding the first half of verse 9 to refer to the motion of the equinox, instead of to that of the initial point of the sidereal sphere.

Among the Greek astronomers, Hipparchus is regarded as the first who discovered the precession of the equinoxes; their rate of motion, however, seems not to have been confidently determined by him, although he pronounces it to be at any rate not less than 36″ yearly. For a thorough discussion of the subject of the precession in Greek astronomy see Delambre's History of Ancient Astronomy, ii. 247, etc. From the observations reported as the data whence Hipparchus made his discovery, Delambre deduces very nearly the true rate of the precession. Ptolemy, however, was so unfortunate as to adopt for the true rate Hipparchus's minimum, of 36″ a year: the subject is treated of by him in the seventh book of the Syntaxis. The actual motion of the equinox at the present time is 50″.25; its rate is slowly on the increase, having been, at the epoch of the Greek astronomy, somewhat less than 50″. How the Hindus succeeded in arriving at a determination of it so much more accurate than was made by the great Greek astronomer, or whether it was anything more than a lucky hit on their part, we will not attempt here to discuss.

The term by which the precession is designated in this passage is *ayanânça*, "degrees of the *ayana*." The latter word is employed in different senses: by derivation, it means simply "going, progress," and it seems to have been first introduced into the astronomical language to designate the half-revolutions of the sun, from solstice to solstice; these being called respectively (see xiv. 9) the *uttarâyaṇa* and *dakshiṇâyana*, "northern progress" and "southern progress." From this use the word was transferred to denote also the solstices themselves, as we have translated it in the first half of verse 11. In the latter sense we conceive it to be employed in the compound *ayanânça;* although why the name of the precession should be derived from the solstice we are unable clearly to see. The term *krântipâtagati*, "movement of the node of declination," which is often met with in modern works on Hindu astronomy, does not occur in the Sûrya-Siddhânta.

12. . . . In like manner, the equatorial shadow which is cast at mid-day at one's place of observation

13. Upon the north and south line of the dial—that is the equinoctial shadow (*vishuvatprabhâ*) of that place. . . .

The equinoctial shadow has been already sufficiently explained, in connection with a preceding passage (above, v. 7). In this treatise it is known only by names formed by combining one of the words for shadow (*châyâ*, *bhâ*, *prabhâ*), with *vishuvat*, "equinox" (see above, under v. 6). In modern Hindu astronomy it is also called *akshabhâ*, "shadow of latitude"—i. e., which determines the latitude—and *palabhâ*, of which, as used in this sense, the meaning is obscure.

13. . . . Radius, multiplied respectively by gnomon and shadow, and divided by the equinoctial hypothenuse,

14. Gives the sines of co-latitude (*lamba*) and of latitude (*aksha*): the corresponding arcs are co-latitude and latitude, always south. . . .

The proportions upon which these rules are founded are illustrated by the following figure (Fig. 11), in which, as in a previous figure (Fig. 8, p. 88), Z S represents a quadrant of the meridian, Z being the zenith and S the south point, C being the centre, and E C the projection of the plane of the equator. In order to illustrate the corresponding relations of the dial, we have conceived the gnomon, C *b*, to be placed at the centre. Then, when the sun is on the meridian and in the equator, at E, the shadow cast, which is the equinoctial shadow, is *b e*, while C *e* is the corresponding hypothenuse. But, by similarity of triangles,

Fig. 11.

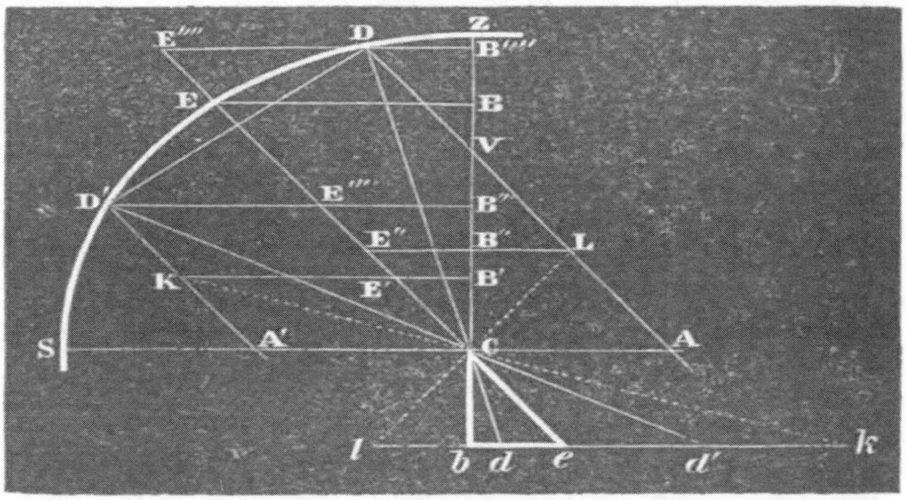

$$Ce : be :: CE : BE$$

and $$Ce : Cb :: CE : CB$$

and as B E is the sine of E Z, which equals the latitude, and C B the sine of E S, its complement, the reduction of these proportions to the form of equations gives the rules of the text.

14. . . . The mid-day shadow is the base (*bhuja*); if radius be multiplied by that,

15. And the product divided by the corresponding hypothenuse, the result, converted to arc, is the sun's zenith-distance (*nata*), in minutes: this, when the base is south, is north, and when the base is north, is south. Of the sun's zenith-distance and his declination, in minutes,

16. Take the sum, when their direction is different—the difference, when it is the same; the result is the latitude, in minutes. From this find the sine of latitude; subtract its square from the square of radius, and the square-root of the remainder

17. Is the sine of co-latitude. . . .

This passage applies to cases in which the sun is not upon the equator, but has a certain declination, of which the amount and direction are known. Then, from the shadow cast at noon, may be derived his zenith-distance when upon the meridian, and the latitude. Thus, supposing the sun, having north declination ED (Fig. 11), to be upon the meridian, at D: the shadow of the gnomon will be *b d*, and the proportion

$$Cd : db :: CD : DB''''$$

gives DB'''', the sine of the sun's zenith-distance, ZD, which is found from it by the conversion of sine into arc by a rule previously given (ii. 33). ZD in this case being south, and ED being north, their sum, EZ, is the latitude: if, the declination being south, the sun were at D', the difference of ED' and ZD' would be EZ, the latitude. The figure does not give an illustration of north zenith-distance, being drawn for the latitude of Washington, where that is impossible. The latitude being thus ascertained, it is easy to find its sine and cosine: the only thing which deserves to be noted in the process is that, to find the cosine from the sine, resort is had to the laborious method of squares, instead of taking from the table the sine of the complementary arc, or the *koṭijyâ*.

The sun's distance from the zenith when he is upon the meridian is called *natâs*, "deflected," an adjective belonging to the noun *liptâs*, "minutes," or *bhâgâs*, *ançâs*, "degrees." The same term is also employed, as will be seen farther on (vv. 34–36), to designate the hour-angle. For zenith-distance off the meridian another term is used (see below, v. 33).

17. . . . The sine of latitude, multiplied by twelve, and divided by the sine of co-latitude, gives the equinoctial shadow. . . .

That is (Fig. 11),

$$BC : BE :: Cb : be$$

the value of the gnomon in digits being substituted in the rule for the gnomon itself.

17. . . . The difference of the latitude of the place of observation and the sun's meridian zenith-distance in degrees (*natabhâgâs*), if their direction be the same, or their sum,

18. If their direction be different, is the sun's declination: if the sine of this latter be multiplied by radius and divided by the sine of greatest declination, the result, converted to arc, will be the sun's longitude, if he is in the quadrant commencing with Aries;

19. If in that commencing with Cancer, subtract from a half-circle; if in that commencing with Libra, add a half-circle; if in that commencing with Capricorn, subtract from a circle: the result, in each case, is the true (*sphuṭa*) longitude of the sun at mid-day.

20. To this if the equation of the apsis (*mânda phala*) be repeatedly applied, with a contrary sign, the sun's mean longitude will be found. . . .

This passage teaches how, when the latitude of the observer is known, the sun's declination, and his true and mean longitudes, may be found by observing his zenith-distance at noon. The several parts of the process are all of them the converse of processes previously given, and require no explanation. To find the sun's declination from his meridian zenith-distance and the latitude (reckoned as south, by v. 14), the rule given above, in verses 15 and 16, is inverted; the true longitude is found from the declination by the inversion of the method taught in ii. 28, account being taken of the quadrant in which the sun may be according to the principle of ii. 30: and finally, the mean may be derived from the true longitude by a method of successive approximation, applying in reverse the equation of the centre, as calculated by ii. 39.

It is hardly necessary to remark that this is a very rough process for ascertaining the sun's longitude, and could give, especially in the hands of Hindu observers, results only distantly approaching to accuracy.

20. . . . The sum of the latitude of the place of observation and the sun's declination, if their direction is the same, or, in the contrary case, their difference,

21. Is the sun's meridian zenith-distance (*natânçâs*); of that find the base-sine (*bahujyâ*) and the perpendicular-sine (*koṭijyâ*). If, then, the base-sine and radius be multiplied respectively by the measure of the gnomon in digits,

22. And divided by the perpendicular-sine, the results are the shadow and hypothenuse at mid-day. . . .

The problem here is to determine the length of the shadow which will be cast at mid-day when the sun has a given declination, the latitude of the observer being also known. First, the sun's meridian zenith-distance is found, by a process the converse of that taught in verses 15 and 16; then, the corresponding sine and cosine having been calculated, a simple proportion gives the desired result. Thus, let us suppose the sun to be at D′ (Fig. 11, p. 106); the sum of his south declination, E D′, and the north latitude, E Z, gives the zenith-distance, Z D′: its sine (*bhujajyâ*) is D′ B‴, and its cosine (*koṭijyâ*) is C B‴. Then

$$CB''' : B'''D' :: Cb : bd'$$

and

$$CB''' : CD' :: Cb : Cd'$$

which proportions, reduced to equations, give the value of bd', the shadow, and Cd', its hypothenuse.

22. . . . The sine of declination, multiplied by the equinoctial hypothenuse, and divided by the gnomon-sine (*çankujîvâ*),

23. Gives, when farther multiplied by the hypothenuse of any given shadow, and divided by radius (*madhyakarṇa*), the sun's measure of amplitude (*arkâgrâ*) corresponding to that shadow. . .

In this passage we are taught, the declination being known, how to find the measure of amplitude (*agrâ*) of any given shadow, as preparatory to determining, by the next following rule, the base (*bhuja*) of the shadow, by calculation instead of measurement. The first step is to find the sine of the sun's amplitude: in order to this, we compare the triangles A B C and C E H (Fig. 13, p. 110), which are similar, since the angles A C B and C E H are each equal to the latitude of the observer. Hence

E H : E C : : B C : A C

But the triangles C E H (Fig. 13) and C *b e* (Fig. 11) are also similar; and

E H : E C : : C *b* : C *e*

Hence, by equality of ratios, C *b* : C *e* : : B C : A C

and A C, the sine of the sun's amplitude, equals B C—which is the sine of declination, being equal to D F—multiplied by C *e*, the equinoctial hypothenuse, and divided by C *b*, the gnomon. The remaining part of the process depends upon the principle which we have demonstrated above, under verse 7, that the measure of amplitude is to the hypothenuse of the shadow as the sine of amplitude to radius.

Why the gnomon is in this passage called the "gnomon-sine," it is not easy to discover. Verse 23 presents also a name for radius, *madhyakarṇa*, "half-diameter," which is not found again; nor is *karṇa* often employed in the sense of "diameter" in this treatise.

23. . . . The sum of the equinoctial shadow and the sun's measure of amplitude (*arkâgrâ*), when the sun is in the southern hemisphere, is the base, north;

24. When the sun is in the northern hemisphere, the base is found, if north, by subtracting the measure of amplitude from the equinoctial shadow; if south, by a contrary process—according to the direction of the interval between the end of the shadow and the east and west axis.

25. The mid-day base is invariably the midday shadow. . . .

We have already had occasion to notice, in connection with the first verses of this chapter, that the base (*bhuja*) of the shadow is its projection upon a north and south line, or the distance of its extremity from the east and west axis of the dial. It is that line which, as shown above (under v. 7), corresponds to and represents the perpendicular let fall from the sun upon the plane of the prime vertical. Thus, if (Fig. 11, p. 106) K, L, D′, D be different positions of the sun—K and L being conceived to be upon the surface of the sphere—the perpendiculars KB′, L B″, D′ B‴, D B⁗ are represented upon the dial by *k b*, *l b*, *d′ b*, *d b*, or, in Fig. 9 (p. 241), by *k b′*, *l b″*, *d′ b*, *d b*. Of these, the two latter coincide with their respective shadows, the shadow cast at noon being always itself upon a north and south line. The base of any shadow may be found by combining its measure of amplitude (*agrâ*) with the equinoctial shadow. When the sun is in the southern hemisphere, as at D′ or K (Fig. 11), the measure of amplitude, *e d′* or *ek*, is to be added always to the equinoctial shadow, *b e*, in order to give the base, *b d′* or *b k*. If, on the contrary, the sun's declination be north, a different method of procedure will be necessary, according as he is north or south from the prime vertical. If he be south, as at D, the shadow, *b d*, will be thrown northward, and the base will be found by subtracting the measure of amplitude, *d e*, from the equinoctial shadow, *b e*: if he be north, as at L, the extremity of the shadow, *l*, will be south from the east and west axis, and the base, *b l*, will be obtained by subtracting the equinoctial shadow, *b e*, from the measure of amplitude, *l e*.

25. . . . Multiply the sines of co-latitude and of latitude respectively by the equinoctial shadow and by twelve,

26. And divide by the sine of declination; the results are the hypothenuse when the sun is on the prime vertical (*samamaṇḍala*). When north declination is less than the latitude, then the mid-day hypothenuse (*çrava*),

27. Multiplied by the equinoctial shadow, and divided by the mid-day measure of amplitude (*agrâ*), is the hypothenuse. . . .

Here we have two separate and independent methods of finding the hypothenuse of the east and west shadow cast by the sun at the moment when he is upon the prime vertical. In connection with the second of the two are stated the circumstances under which alone a transit of the sun across the prime vertical will take place: if his declination is south, or if, being north, it is greater than the latitude, his diurnal revolution will be wholly to the south, or wholly to the north, of that circle.

The first method is illustrated by the following figures. Let V C″ (Fig. 12) be an arc of the prime vertical, V being the point at which the sun crosses it in his daily revolution; and let C′ be the centre; then V C′ is radius, and V C the sine of the sun's altitude; and, C′ *b* being the gnomon, *b v* will be the shadow, and C′ *v* its hypothenuse. But, by similarity of triangles,

Fig. 12.

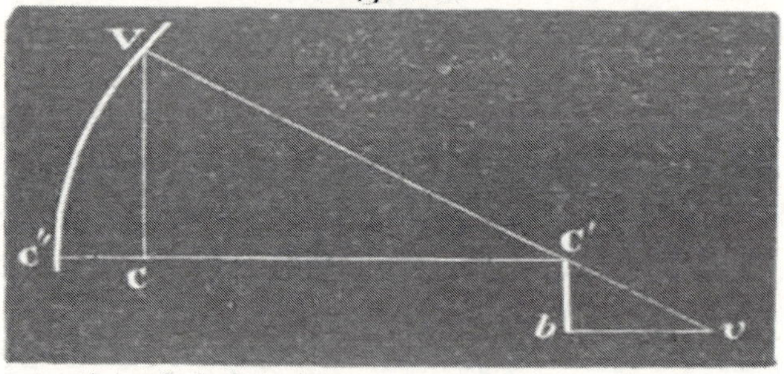

$$VC : VC' :: C'b : C'v$$

Again, in the other figure (Fig. 13)—of which the general relations are those of Fig. 8 (p. 88)—A D being the projection of the circle of the sun's diurnal revolution, and the point at which it crosses the prime vertical being seen projected in V, V C is the sine of the sun's altitude at that point. But V C B and E C H are similar triangles, the angles B V C and C E H being each equal to the latitude; hence

Fig. 13.

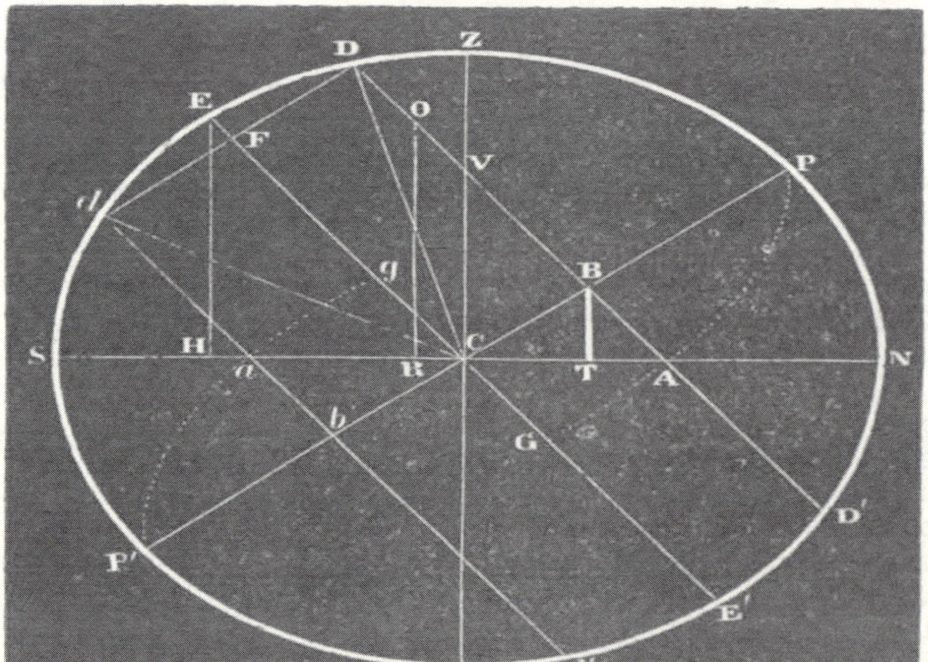

$$VC : EC :: BC : CH$$

Now the first of these ratios is—since E C equals V C′, both being radius—the same with the first

in the former proportion; and therefore

BC : CH : : C′*b* : C′*v*

or sin decl. : sin lat. : : gnom. : hyp. pr. vert. shad.

but sin lat. : cos. lat. : : eq. shad. : gnom.

therefore, by combining terms,

sin. decl. : cos. lat. : : eq. shad. : hyp. pr. vert. shad.

and the reduction of the first and third of these proportions to the form of equations gives the rules of the text.

The other method of finding the same quantity is an application of the principle demonstrated above, under verse 7, that, with a given declination, the measure of amplitude of any shadow is to that of any other shadow as the hypothenuse of the former to that of the latter. Now when the sun is upon the prime vertical, the shadow falls directly eastward or directly westward, and hence its extremity lies in the east and west axis of the dial, and its measure of amplitude is equal to the equinoctial shadow. The noon measure of amplitude is, accordingly, to the hypothenuse of the noon shadow as the equinoctial shadow to the hypothenuse of the shadow cast when the sun is upon the prime vertical.

27. . . . If the sine of declination of a given time be multiplied by radius and divided by the sine of co-latitude, the result is the sine of amplitude (*agramâurvikâ*).

28. And this, being farther multiplied by the hypothenuse of a given shadow at that time, and divided by radius, gives the measure of amplitude (*agrâ*), in digits (*angula*), etc. . . .

The sine of the sun's amplitude is found—his declination and the latitude being known—by a comparison of the similar triangles ABC and CEH (Fig. 13), in which

HE : EC : : BC : CA

or cos. lat. : R : : sin. decl. : sin. ampl.

And the proportion upon which is founded the rule in verse 28—namely, that radius is to the sine of amplitude as the hypothenuse of a given shadow to the corresponding measure of amplitude—has been demonstrated under verse 7, above.

28. . . . If from half the square of radius the square of the sine of amplitude (*agrajyâ*) be subtracted, and the remainder multiplied by twelve,

29. And again multiplied by twelve, and then farther divided by the square of the equinoctial shadow increased by half the square of the gnomon—the result obtained by the wise

30. Is called the "surd" (*karaṇî*): this let the wise man set down in two places. Then multiply the equinoctial shadow by twelve, and again by the sine of amplitude,

31. And divide as before: the result is styled the "fruit" (*phala*). Add its square to the "surd," and take the square root of their sum; this, diminished and increased by the "fruit," for the southern and northern hemispheres,

32. Is the sine of altitude (*çanku*) of the southern intermediate directions (*vidiç*); and equally, whether the sun's revolution take place to the south or to the north of the gnomon (*çanku*)—only, in the latter case, the sine of altitude is that of the northern intermediate directions.

33. The square root of the difference of the squares of that and of radius is styled the sine of zenith-distance (*drç.*) If, then, the sine of zenith-distance and radius be multiplied respectively by twelve, and divided by the sine of altitude,

34. The results are the shadow and hypothenuse at the angles (*koṇa*), under the given circumstances of time and place. . . .

The method taught in this passage of finding, with a given declination and latitude, the sine of the sun's altitude at the moment when he crosses the south-east and south-west vertical circles, or when the shadow of the gnomon is thrown toward the angles (*koṇa*) of the circumscribing square of the dial, is, when stated algebraically, as follows:

$$\frac{(\frac{1}{2}\text{R}^2 - \text{sin}^2\text{ ampl.}) \times \text{gn.}^2}{\frac{1}{2}\text{gn.}^2 + \text{eq. sh.}^2} = \text{surd.}$$

$$\frac{\text{eq. sh.} \times \text{gn.} \times \text{sin ampl.}}{\frac{1}{2}\text{gn.}^2 + \text{eq. sh.}^2} = \text{fruit.}$$

$$\sqrt{\text{surd} + \text{fruit}^2} + \text{fruit} = \text{sin alt., declination being north.}$$

$$\sqrt{\text{surd} + \text{fruit}^2} - \text{fruit} = \text{sin alt., declination being south.}$$

It is at once apparent that a problem is here presented more complicated and difficult of solution than any with which we have heretofore had to do in the treatise. The commentary gives a demonstration of it, in which, for the first time, the notation and processes of the Hindu algebra are introduced, and with these we are not sufficiently familiar to be able to follow the course of the demonstration. The problem, however, admits of solution without the aid of mathematical knowledge of a higher character than has been displayed in the processes already explained; by means, namely, of the consideration of right-angled triangles, situated in the same plane, and capable of being represented by a single figure. We give below such a solution, which, we are persuaded, agrees in all its main features with the process by which the formulas of the text were originally deduced.

Fig. 14.

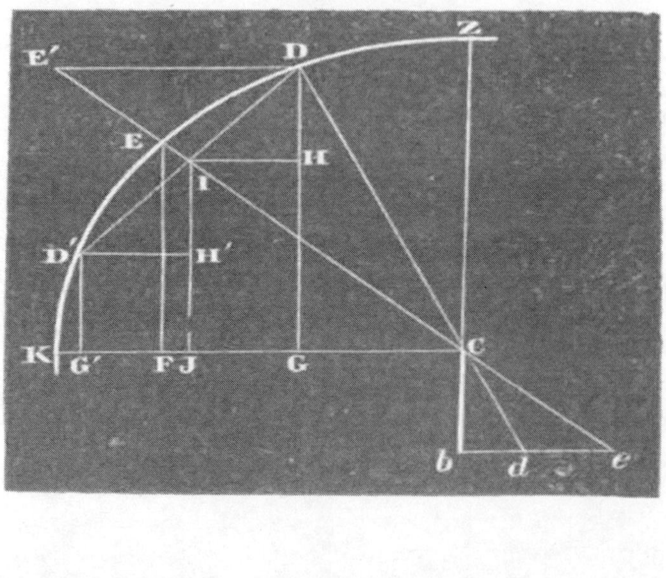

Let Z E K be the south-eastern circle of altitude, from the zenith, Z, to the horizon, K; let E be its intersection with the equator, and D the position of the sun; and let C *b* represent the gnomon.

Since *e* is in the line of the equinoctial shadow (see above, v. 7), and since *be* makes an angle of 45° with either axis of the dial, we have $be^2 = 2$ eq. sh.^{2}, and $\text{C}e^2 = \text{C}b^2 + be^2 =$ gn.$^2 + 2$ eq. sh.^{2}

In like manner, $de^2 = 2$ meas. ampl.^{2} But the similar triangles Cde and CDE' give $Cd^2 : de^2 :: CD^2 : DE'^2$; which, by halving the two consequents, and observing the constant relation of Cd to the measure of amplitude (see above, under v. 7), gives R^2 : sin ampl.^{2} :: $R^2 : \frac{1}{2}DE'^2$: whence $\frac{1}{2}DE'^2 =$ sin ampl.^{2}, or $DE'^2 = 2$ sin ampl.^{2}

Now the required sine of altitude is DG, and $DG = DH + HG = DH + IJ$. And, obviously, the triangles DHI, DIE', EFC, IJC, and Cbe are all similar. Then, from DHI and Cbe, we derive

$$DH : DI :: be : Ce$$

from DIE' and Cbe, $\quad DI : DE' :: Cb : Ce$

and, by combining terms, $DH : DE' :: be \times Cb : Ce^2$

whence $\quad DH = \dfrac{\sqrt{2}\,.\,\text{eq. sh.} \times \text{gn.} \times \sqrt{2}\,.\,\text{sin ampl.}}{\text{gn.}^2 + 2\,\text{eq. sh.}^2} = \dfrac{\text{eq. sh.} \times \text{gn.} \times \text{sin ampl.}}{\frac{1}{2}\text{gn.}^2 + \text{eq. sh.}^2} = \text{fruit.}$

Again, from DHI and EFC, we derive

$$IH^2 : DI^2 :: EF^2 : EC^2$$

from IJC and EFC, $\quad IJ^2 : IC^2 :: EF^2 : EC^2$

whence, by adding the terms of the equal ratios, and observing that $IH^2 + IJ^2 = JH^2$, and $DI^2 + IC^2 = DC^2 = EC^2$, we have

$$JH^2 : EC^2 :: EF^2 : EC^2$$

or $JH^2 = EF^2$. Hence $IJ^2 = JH^2 - IH^2 = EF^2 - IH^2 = EF^2 - DI^2 + DH^2$

But from EFC and Cbe are derived

$$Ce^2 : Cb^2 :: EC^2 : EF^2$$

from DIE' and Cbe, $\quad Ce^2 : Cb^2 :: DE'^2 : DI^2$

whence $EF^2 = \dfrac{EC^2 . Cb^2}{Ce^2}$, and $DI^2 = \dfrac{DE'^2 . Cb^2}{Ce^2}$, and $EF^2 - DI^2 = \dfrac{(EC^2 - DE'^2)Cb^2}{Ce^2}$

that is to say,

$$EF^2 - DI^2 = \frac{(R^2 - 2\,\text{sin ampl.}^2) \times \text{gn.}^2}{\text{gn.}^2 + 2\,\text{eq. sh.}^2} = \frac{(\frac{1}{2}R^2 - \text{sin ampl.}^2) \times \text{gn.}^2}{\frac{1}{2}\text{gn.}^2 + \text{eq. sh.}^2} = \text{surd.}$$

But, as was shown above, $IJ^2 = EF^2 - DI^2 + DH^2 =$ surd + fruit^2 and $\sqrt{\text{surd} + \text{fruit}^2} + \text{fruit} = IJ + DH = DG =$ sine of altitude.

When declination is south, so that the sun crosses the circle of altitude at D', IH', the equivalent of DH, is to be subtracted from IJ, to give $D'G'$, the sine of altitude.

The correctness of the Hindu formulas may likewise be briefly and succinctly demonstrated by means of our modern methods. Thus, let PZS (Fig. 15) be a spherical triangle, of which the three angular points are P, the pole, Z, the zenith, and S, the place of the sun when upon the south-east or the south-west vertical circles; PZ, then, is the co-latitude, ZS the zenith-distance, or co-altitude, and PS the co-declination; and the angle PZS is 135°; the problem is, to find the sine of the complement of ZS, or of the sun's altitude. By spherical trigonometry, cos SP = cos ZS cos ZP + sin ZS sin ZP cos Z. Dividing by sin ZP, and observing that cos SP ÷ sin ZP = sin decl. ÷ cos lat. = sine of amplitude, we have sin ampl. = sin alt. tan lat. + cos alt. cos 135°. If, now, we represent sin ampl. by a, tan lat. by b, cos 135° by

Fig. 15.

$-\sqrt{\tfrac{1}{2}}$, sin alt. by x, and cos alt. by $\sqrt{1-x^2}$, we have $a^2-2abx+b^2x^2=\tfrac{1}{2}(1-x^2)$; and, by reduction, $x^2-\dfrac{2ab}{\tfrac{1}{2}+b^2}x=\dfrac{\tfrac{1}{2}-a^2}{\tfrac{1}{2}+b^2}$. Representing, again, $\dfrac{ab}{\tfrac{1}{2}+b^2}$ by f, and $\dfrac{\tfrac{1}{2}-a^2}{\tfrac{1}{2}+b^2}$ by s, and reducing, we have $x=f+\sqrt{f^2+s}$. But f is evidently the same with the "fruit," since b, or tan lat., equals eq. sh.$\div$gnom., and therefore $\dfrac{ab}{\tfrac{1}{2}+b^2}=\dfrac{\text{eq. sh.}\times\text{gn.}\times\text{sin. ampl.}}{\tfrac{1}{2}\text{ gnom.}^2+\text{eq. sh.}^2}$: and s is also the same with the "surd," for $\dfrac{\tfrac{1}{2}-a^2}{\tfrac{1}{2}+b^2}=\dfrac{(\tfrac{1}{2}\text{R}^2-\sin^2\text{ampl.})\times\text{gn.}^2}{\tfrac{1}{2}\text{gnom.}^2+\text{eq. sh.}^2}$.

If, the latitude being north, we consider the north direction as positive, b will be positive. The value of f, given above, will then evidently be positive or negative as the sign of a is plus or minus. But a, the sine of amplitude, is positive when declination is north, and negative when declination is south. Hence f is to be added to or subtracted from the radical, according as the sun is north or south of the equator, as prescribed by the Hindu rule. A minus sign before the radical would correspond to a second passage of the sun across the south-east and north-west vertical circle; which, except in a high latitude, would take place always below the horizon.

The construction of the last part of verse 32 is by no means clear, yet we cannot question that the meaning intended to be conveyed by it is truly represented by our translation. When declination is greater than north latitude, the sun's revolution is made wholly to the north of the prime vertical, and the vertical circles which he crosses are the north-east and the north-west. The process prescribed in the text, however, gives the correct value for the sine of altitude in this case also. For, in the triangle S Z P (Fig. 15), all the parts remain the same, excepting that the angle P Z S becomes 45°, instead of 135°: but the cosine of the former is the same as that of the latter arc, with a difference only of sign, which disappears in the process, the cosine being squared.

The sine of altitude being found, that of its complement, or of zenith-distance, is readily derived from it by the method of squares (as above, in vv. 16, 17). To ascertain, farther, the length of the corresponding shadow and of its hypothenuse, we make the proportions

sin alt. : sin zen. dist. : : gnom. : shad.

and sin alt. : R : : gnom. : hyp. shad.

In this passage, as in those that follow, the sine of altitude is called by the same name, *çanku*, "staff," which is elsewhere given to the gnomon: the gnomon, in fact, representing in all cases, if the hypothenuse be made radius, the sine of the sun's altitude. The word is frequently used in this sense in the modern astronomical language: thus V C (Fig. 13, p. 110), the sine of the sun's altitude when upon the prime vertical, is called the *samamaṇḍalaçanku*, "prime vertical staff," and B T, the sine of altitude when the sun crosses the *unmaṇḍala*, or east and west hour-circle, is styled the *unmaṇḍalaçanku*: of the latter line, however, the Sûrya-Siddhânta makes no account. We are surprised, however, not to find a distinct name for the altitude, as for its complement, the zenith-distance: the sine of the latter might with very

nearly the same propriety be called the "shadow," as that of the former the "gnomon." The particular sine of altitude which is the result of the present process is commonly known as the *koṇaçanku*, from the word *koṇa*, which, signifying originally "angle," is used, in connection with the dial, to indicate the angles of the circumscribing square (see Fig. 9, p. 97), and then the directions in which those angles lie from the gnomon. The word itself is doubtless borrowed from the Greek *γωνία*, the form given to it being that in which it appears in the compounds *τρίγωνον* (Sanskrit *trikoṇa*), etc. Lest it seem strange that the Hindus should have derived from abroad the name for so familiar and elementary a quantity as an angle, we would direct attention to the striking fact that in that stage of their mathematical science, at least, which is represented by the Sûrya-Siddhânta, they appear to have made no use whatever in their calculations of the angle: for, excepting in this passage (v. 34) and in the term for "square" employed in a previous verse (v. 5) of this chapter, no word meaning "angle" is to be met with anywhere in the text of this treatise. The term *dṛç*, used to signify "zenith-distance"—excepting when this is measured upon the meridian; see above, under vv. 14–16—means literally "sight:" in this sense, it occurs here for the first time: we have had it more than once above with the signification of "observed place," as distinguished from a position obtained by calculation. In verse 32, *çanku* might be understood as used in the sense of "zenith," yet it has there, in truth, its own proper signification of "gnomon;" the meaning being, that the sun, in the cases supposed, makes his revolution to the south or to the north of the gnomon itself, or in such a manner as to cast the shadow of the latter, at noon, northward or southward. One of the factors in the calculation is styled *karaṇî*, "surd" (see Colebrooke's Hind. Alg., p. 145), rather, apparently, as being a quantity of which the root is not required to be taken, than one of which an integral root is always impossible; or, it may be, as being the square of a line which is not, and cannot be, drawn. The term translated "fruit" (*phala*) is one of very frequent occurrence elsewhere, as denoting "quotient, result, corrective equation," etc.

The form of statement and of injunction employed in verses 29 and 30, in the phrases "the result obtained by the wise," and "let the wise man set down," etc., is so little in accordance with the style of our treatise elsewhere, while it is also frequent and familiar in other works of a kindred character, that it furnishes ground for suspicion that this passage, relating to the *koṇaçanku*, is a later interpolation into the body of the text; and the suspicion is strengthened by the fact that the process prescribed here is so much more complicated than those elsewhere presented in this chapter.

34. . . . If radius be increased by the sine of ascensional difference (*cara*) when declination is north, or diminished by the same, when declination is south,

35. The result is the day-measure (*antyâ*); this, diminished by the versed sine (*utkramajyâ*) of the hour-angle (*nata*), then multiplied by the day-radius and divided by radius, is the "divisor" (*cheda*); the latter, again, being multiplied by the sine of co-latitude (*lamba*), and divided

36. By radius, gives the sine of altitude (*çanku*): subtract its sine from that of radius, and the square root of the remainder is the sine of zenith-distance (*drç*): the shadow and its hypothenuse are found as in the preceding process.

The object of this process is, to find the sine of the sun's altitude at any given hour of the day, when his distance from the meridian, his declination, and the latitude, are known. The sun's angular distance from the meridian, or the hour-angle, is found, as explained by the commentary, by subtracting the time elapsed since sunrise, or which is to elapse before sunset, from the half day, as calculated by a rule previously given (ii. 61–63). From the declination and the latitude the sine of ascensional difference (*carajyâ*) is supposed to have been already derived, by the method taught in the same passage; as also, from the declination (by ii. 60), the radius of the diurnal circle. The successive steps of the process of calculation will be made clear by a reference to the annexed figure (Fig. 16), taken in connection with Fig. 13 (p. 110), with which it corresponds in dimensions and lettering. Let G G′ C′ E represent a portion of the plane of the equator, C being its centre, and G E its intersection with the plane of the meridian; and let A A′ B′ D represent a corresponding portion of the plane of the diurnal circle, as seen projected upon the other, its centre and its line of intersection with the meridian coinciding with those of the latter. Let C G equal the sine of ascensional difference, and A B its correspondent in the lesser circle, or the earth-sine (*kujyâ* or *kshitijyâ ;* see above, ii. 61). Now let O′ be the place of the sun at a given time; the angle O′ C D, measured by the arc of the equator Q′ E, is the hour-angle: from Q′ draw Q′ Q perpendicular to C E; then Q′ Q is the sine, and Q E is the versed sine, of Q′ E. Add to radius, E C, the sine of ascensional difference, C G; their sum, E G—which is the equivalent, in terms of a great circle, of D A, that part of the diameter of the circle of diurnal revolution which is above the horizon, and which consequently measures the length of the day—is the day-measure (*antyâ*). From E G deduct E Q, the versed sine of the hour-angle; the remainder, G Q, is the same quantity in terms of a great circle which A O is in terms of the diurnal circle: hence the reduction of G Q to the dimensions of the lesser circle, by the proportion

Fig. 16.

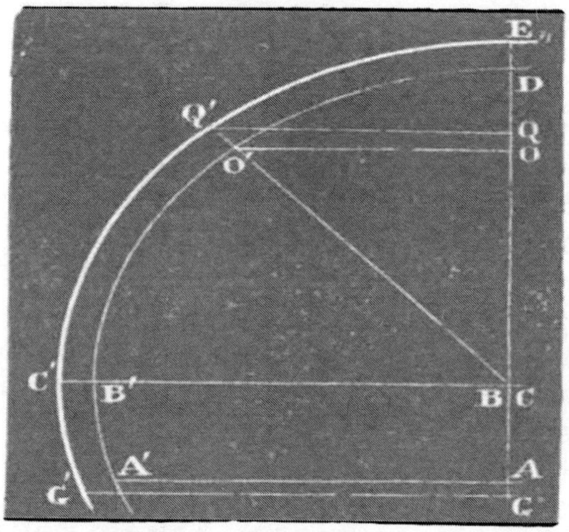

$$CE : BD :: GQ : AO$$

gives us the value of A O; to this the text gives the technical name of "divisor" (*cheda*). But, by Fig. 13,

$$CE : EH :: AO : OR$$

hence O R, which is the sine of the sun's altitude at the given time, equals A O, the "divisor," multiplied by E H, the cosine of latitude, and divided by C E, or radius.

The processes for deriving from the sine of altitude that of zenith-distance, and from both the length of the corresponding shadow and its hypothenuse, are precisely the same as in the last problem.

For the meaning of *antyâ*—which, for lack of a better term, we have translated "day-measure"—see above, under verse 7. The word *nata*, by which the hour-angle is designated, is the same with that employed above with the signification of "meridian zenith-distance;" see the note to verses 14–17.

37. If radius be multiplied by a given shadow, and divided by the corresponding hypothenuse, the result is the sine of zenith-distance (*dṛc*): the square root of the difference between the square of that and the square of radius

38. Is the sine of altitude (*çanku*); which, multiplied by radius and divided by the sine of co-latitude (*lamba*), gives the "divisor" (*cheda*); multiply the latter by radius, and divide by the radius of the diurnal circle,

39. And the quotient is the sine of the sun's distance from the horizon (*unnata*); this, then, being subtracted from the day-measure (*antyâ*), and the remainder turned into arc by means of the table of versed sines, the final result is the hour-angle (*nata*), in respirations (*asu*), east or west.

The process taught in these verses is precisely the converse of the one described in the preceding passage. The only point which calls for farther remark in connection with it is, that the line G Q (Fig. 16) is in verse 39 called the "sine of the *unnata*." By this latter term is designated the opposite of the hour-angle (*nata*)—that is to say, the sun's angular distance from the horizon upon his own circle, O′ A′, reduced to time, or to the measure of a great circle. Thus, when the sun is at O′, his hour-angle (*nata*), or the time till noon, is Q′ E; his distance from the horizon (*unnata*), or the time since sunrise, is Q′ G′. But G Q is with no propriety styled the sine of G′ Q′; it is not itself a sine at all, and the actual sine of the arc in question would have a very different value.

40. Multiply the sine of co-latitude by any given measure of amplitude (*agrâ*), and divide by the corresponding hypothenuse in digits; the result is the sine of declination. This, again, is to be multiplied by radius, and divided by the sine of greatest declination;

41. The quotient, converted into arc, is, in signs, etc., the sun's place in the quadrant; by means of the quadrants is then found the actual longitude of the sun at that point. . . .

By the method taught in this passage, the sun's declination, and, through that, his true and mean longitude, may, the latitude of the observer being known, be found from a single observation upon the shadow at any hour in the day. The declination is obtained from the measure of amplitude and the hypothenuse of the shadow, in the following manner: first, as was shown in connection with verse 7 of this chapter,

	hyp. shad. : meas. ampl. : : E C : C A (Fig. 13, p. 110)
but	E C : C A : : E H : B C
therefore	hyp. shad. : meas. ampl. : : E H : B C

B C, the sine of declination, being thus ascertained, the longitude is deduced from it as in a previous process (see above, vv. 17–20).

41. . . . Upon a given day, the distances of three bases, at noon, in the forenoon, and in the afternoon, being laid off,

42. From the point of intersection of the lines drawn between them by means of two fish-figures, (*matsya*), and with a radius touching the three points, is described the path of the shadow. . .

This method of drawing upon the face of the dial the path which will be described by the extremity of the shadow upon a given day proceeds upon the assumption that that path will be an arc of a circle—an erroneous assumption, since, excepting within the polar circles, the path of the shadow is always a hyperbola, when the sun is not in the equator. In low latitudes, however, the difference between the arc of the hyperbola, at any point not too far from the gnomon, and the arc of a circle, is so small, that it is not very surprising that the Hindus should have overlooked it. The path being regarded as a true circle, of course it can be drawn if any three points in it can be found by calculation: and this is not difficult, since the rules above given furnish means of ascertaining, if the sun's declination and the observer's latitude be known, the length of the shadow and the length of its base, or the distance of its extremity from the east and west axis of the dial, at different times during the day. One part of the process, however, has not been provided for in the rules hitherto given. Thus (Fig. 9, p. 97), supposing *d*, *m*, and *l* to be three points in the same daily path of the shadow, we require, in order to lay down *l* and *m*, to know not only the bases $l\,b''$, $m\,b'''$, but also the distances $b\,b''$, $b\,b'''$. But these are readily found when the shadow and the base corresponding to each are known, or they may be calculated from the sines of the respective hour-angles.

The three points being determined, the mode of describing a circle through them is virtually the same with that which we should employ: lines are drawn from the noon-point to each of the others, which are then, by fish-figures (see above, under vv. 1–5), bisected by other lines at right angles to them, and the intersection of the latter is the centre of the required circle.

42. . . . Multiply by the day-radius of three signs, and divide by their own respective day-radii,

43. In succession, the sines of one, of two, and of three signs; the quotients, converted into arc, being subtracted, each from the one following, give, beginning with Aries, the times of rising (*udayâsavas*) at Lankâ;

44. Namely sixteen hundred and seventy, seventeen hundred and ninety-five, and nineteen hundred and thirty-five respirations. And these, diminished each by its portion of ascensional

difference (*carakhaṇḍa*), as calculated for a given place, are the times of rising at that place.

45. Invert them, and add their own portions of ascensional difference inverted, and the sums are the three signs beginning with Cancer: and these same six, in inverse order, are the other six, commencing with Libra.

The problem here is to determine the "times of rising" (*udayâsavas*) of the different signs of the ecliptic—that is to say, the part of the 5400 respirations (*asavas*) constituting a quarter of the sidereal day, which each of the three signs making up a quadrant of the ecliptic will occupy in rising (*udaya*) above the horizon. And in the first place, the times of rising at the equator, or in the right sphere—which are the equivalents of the signs in right ascension—are found as follows:

Let Z N (Fig. 17) be a quadrant of the solstitial colure, A N the projection upon its plane of the equinoctial colure, A Z of the equator, and A C of the ecliptic; and let A, T, G, and C be the projections upon A C of the initial points of the first four signs; then A T is the sine of one sign, or of 30°, A G of two signs, or of 60°, and A C, which is radius, the sine of three signs, or of 90°. From T, G, and C, draw T t, G g, C c, perpendicular to A N. Then A T t and A C c are similar triangles, and, since A C equals radius,

Fig. 17.

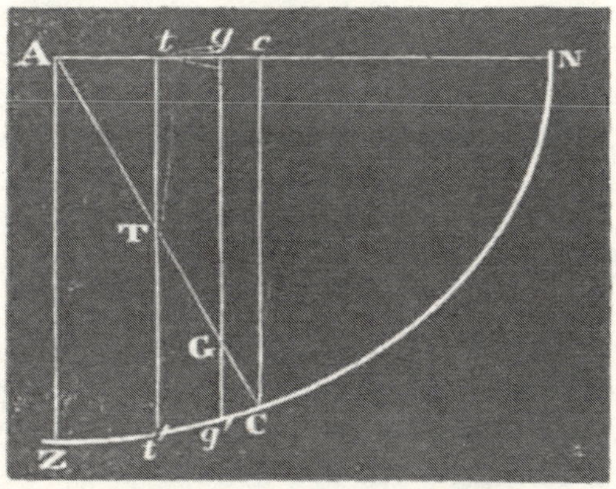

$$R : Cc :: AT : Tt$$

But the arc of which T t is sine, is the same part of the circle of diurnal revolution of which the radius is $t\,t'$, as the required ascensional equivalent of one sign is of the equator: hence the sine of the latter, which we may call x, is found by reducing T t to the measure of a great circle, which is done by the proportion

$$tt' : R :: Tt : \sin x$$

Combining this with the preceding proportion, we have,

$$tt' : Cc :: AT : \sin x$$

Again, to find the ascensional equivalent of two signs, which we will call y, we have first, by comparison of the triangles A G g and A C c,

$$R : Cc :: AG : Gg$$

and

$$gg' : R :: Gg : \sin y$$

therefore, as before,

$$gg' : Cc :: AG : \sin y$$

Hence, the sines of the ascensional equivalents of one and of two signs respectively are equal to the sines of one and of two signs, A T and A G, multiplied by the day-radius of three signs, C c, and divided each by its own day-radius, $t\,t'$ and $g\,g'$; and the conversion of the sines thus obtained into arc gives the ascensional equivalents themselves. The rule of the text includes also the equivalent of three signs, but this is so obviously equal to a quadrant that it is unnecessary to draw out the process, all the terms in the proportions disappearing except radius.

Upon working out the process, by means of the table of sines given in the second chapter (vv. 15–22), and assuming the inclination of the plane of the ecliptic to be 24° (ii. 28), we find, by the rule given above (ii. 60), that the day-radii of one, of two, and of three sines, or $t\,t'$, $g\,g'$, C c, are 3366′, 3216′, and 3140′ respectively, and that the sines of x and y are 1604′ and 2907′, to which the corresponding arcs are 27° 50′ and 57° 45′, or 1670′ and 3465′. The former is the ascensional equivalent of the first sign; subtracting it from the latter gives that of the second sign, which is 1795′, and subtracting 3465′ from a quadrant, 5400′, gives the equivalent of the third sign, which is 1935′—all as stated in the text.

These, then, are the periods of sidereal time which the first three signs of the ecliptic will occupy in rising above the horizon at the equator, or in passing the meridian of any latitude. It is obvious that the same quantities, in inverse order, will be the equivalents in right ascension of the three following signs also, and that the series of six equivalents thus found will belong also to the six signs of the other half of the ecliptic. In order, now, to ascertain the equivalents of the signs in oblique ascension, or the periods of sidereal time which they will occupy in rising above the horizon of a given latitude, it is necessary first to calculate, for that latitude, the ascensional difference (*cara*) of the three points T, G, and C (Fig. 17), which is done by the rule given in the last chapter (vv. 61, 62). We have calculated these quantities, in the Hindu method, for the latitude of Washington, 38° 54′, and find the ascensional difference of T to be 578′, that of G 1061′, and that of C 1263′. The manner in which these are combined with the equivalents in right ascension to produce the equivalents in oblique ascension may be explained by the following figure (Fig. 18), which, although not a true projection, is sufficient for the purpose of illustration. Let A C S be a semicircle of the ecliptic, divided into its successive signs, and A S a semicircle of the equator, upon which A T′, T′ G′, etc., are the equivalents of those signs in right ascension; and let t, g, etc., be the points which rise simultaneously with T, G, etc. Then t T′ and v V′, the ascensional difference of T and V, are 578′, g G′ and l L′ are 1061′, and c C′ is 1263′. Then A t, the equivalent in oblique ascension of A T, equals A T′ − t T′, or 1092′. To find, again, the value of $t\,g$, the second equivalent, the text directs to subtract from T′ G′ the difference between t T′ and g G′, which is called the *carakhaṇḍa*, "portion of ascensional difference"—that is to say, the increment or decrement of ascensional difference at the point G as compared with T. Thus

Fig. 18.

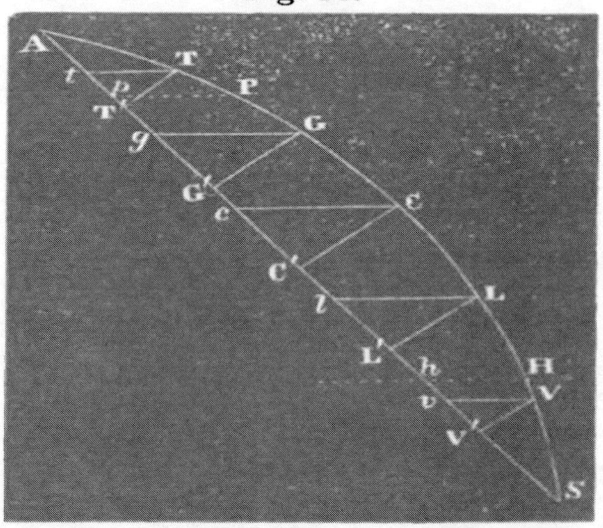

$$t\,g = \mathrm{T}'\,\mathrm{G}' - (g\,\mathrm{G}' - t\,\mathrm{T}') = \mathrm{T}'\,\mathrm{G}' + t\,\mathrm{T}' - g\,\mathrm{G}' = t\,\mathrm{G}' - g\,\mathrm{G}' = 1312'$$

and

$$g\,c = \mathrm{G}'\,\mathrm{C}' - (c\,\mathrm{C}' - g\,\mathrm{G}') = \mathrm{G}'\,\mathrm{C}' + g\,\mathrm{G}' - c\,\mathrm{C}' = g\,\mathrm{C}' - c\,\mathrm{C}' = 1733'$$

Farther, to find the oblique equivalents in the second quadrant, we are directed to invert the right equivalents, and to add to each its own *carakhaṇḍa*, decrement of ascensional difference. Thus

$$c\,l = \mathrm{C}'\,\mathrm{L}' + (c\,\mathrm{C}' - l\,\mathrm{L}') = c\,\mathrm{L}' - l\,\mathrm{L}' = 2137'$$
$$l\,v = \mathrm{L}'\,\mathrm{V}' + (l\,\mathrm{L}' - v\,\mathrm{V}') = l\,\mathrm{V}' - v\,\mathrm{V}' = 2278'$$

and finally, $v\,\mathrm{S} = \mathrm{V}'\,\mathrm{S} + v\,\mathrm{V}' = 2248'.$

It is obvious without particular explanation that the arcs of oblique ascension thus found as the equivalents, in a given latitude, of the first six signs of the ecliptic, are likewise, in inverse order, the equivalents of the other six. We have, then, the following table of times of rising (*udayâsavas*), for the equator and for the latitude of Washington, of all the divisions of the ecliptic:

Equivalents in Right and Oblique Ascension of the Signs of the Ecliptic.

Sign.		Equivalent in	Lat. of Washington.		Sign.	
No.	Name.	Right Ascension.	Ascens. Diff.	Equiv. in Obl. Ascension.	Name.	No.
		′ or p.	′ or p.	′ or p.		
1.	Aries, *mesha,*	1670	578	1092	Pisces, *mîna,*	12.
2.	Taurus, *vṛshan,*	1795	1061	1312	Aquarius, *kumbha,*	11.
3.	Gemini, *mithuna,*	1935	1263	1733	Capricornus, *makara,*	10.
4.	Cancer, *karkaṭa,*	1935	1061	2137	Sagittarius, *dhanus,*	9
5.	Leo, *sinha,*	1795	578	2278	Scorpio, *âli,*	8.
6.	Virgo, *kanyâ,*	1670		2248	Libra, *tulâ,*	7.

For the expression "at Lankâ," employed in verse 43 to designate the equator, see above, under i. 62.

46. From the longitude of the sun at a given time are to be calculated the ascensional equivalents of the parts past and to come of the sign in which he is: they are equal to the number of degrees traversed and to be traversed, multiplied by the ascensional equivalent (*udayâsavas*) of the sign, and divided by thirty;

47. Then, from the given time, reduced to respirations, subtract the equivalent, in respirations, of the part of the sign to come, and also the ascensional equivalents (*lagnâsavas*) of the following signs, in succession—so likewise, subtract the equivalents of the part past, and of the signs past, in inverse order;

48. If there be a remainder, multiply it by thirty and divide by the equivalent of the unsubtracted sign; add the quotient, in degrees, to the whole signs, or subtract it from them: the result is the point of the ecliptic (*lagna*) which is at that time upon the horizon (*kshitija*).

49. So, from the east or west hour-angle (*nata*) of the sun, in nâḍîs, having made a similar calculation, by means of the equivalents in right ascension (*lankodayâsavas*), apply the result as an additive or subtractive equation to the sun's longitude: the result is the point of the ecliptic then upon the meridian (*madhyalagna*).

The word *lagna* means literally "attached to, connected with," and hence, "corresponding, equivalent to." It is, then, most properly, and likewise most usually, employed to designate the point or the arc of the equator which corresponds to a given point or arc of the ecliptic. In such a sense it occurs in this passage, in verse 47, where *lagnâsavas* is precisely equivalent to *udayâsavas,* explained in connection with the next preceding passage; also below, in verse 50, and in several other places. In verses 48 and 49, however, it receives a different signification, being taken to indicate the point of the ecliptic which, at a given time, is upon the meridian or at the horizon; the former being called *lagnam kshitije,* "*lagna* at the horizon"—or, in one or two cases elsewhere, simply *lagna*—the other receiving the name of *madhyalagna,* "meridian-*lagna.*"

The rules by which, the sun's longitude and the hour of the day being known, the points of the ecliptic at the horizon and upon the meridian are found, are very elliptically and obscurely stated in the text; our translation itself has been necessarily made in part also a paraphrase and explication of them. Their farther illustration may be best effected by means of an example, with reference to the last figure (Fig. 18).

At a given place of observation, as Washington, let the moment of local time—reckoned in the usual Hindu manner, from sunrise—be 18^n 12^v 3^p, and let the longitude of the sun, as corrected by the precession, be, by calculation, 42°, or 1^s 12°: it is required to know the longitude of the point of the ecliptic (*lagna*) then upon the eastern horizon.

Let P (Fig. 18) be the place of the sun, and H *h* the line of the horizon, at the given time; and let *p* be the point of the equator which rose with the sun; then the arc *p h* is equivalent to the time since sunrise, 18^n 12^v 3^p, or 6555^p. The value of *t g,* the equivalent in oblique ascension of the second sign T G, in which the sun is, is given in the table presented at the end of the note upon the preceding passage as 1312^p. To find the value of the part of it *p g* we make the proportion

$$\mathrm{TG} : \mathrm{PG} :: tg : pg$$

or

$$30^\circ : 18^\circ :: 1312^p : 787^p$$

From *p h,* or 6555^p, we now subtract *p g,* 787^p, and then, in succession, the ascensional equivalents of the following signs, G C and C L—that is, *g c,* or 1733^p, and *c l,* or 2137^p—until there is left a remainder, *l h,* or 1898^p, which is less than the equivalent of the next sign. To this remainder of oblique ascension the corresponding arc of longitude is then found by a proportion the reverse of that formerly made, namely

$$lv : lh :: \mathrm{LV} : \mathrm{LH}$$

or

$$2278^p : 1898^p :: 30^\circ : 25^\circ$$

The result thus obtained being added to A L, or 4^s, the sum, 4^s 25°, or 145°, is the longitude of H.

The arc *p g* is called in the text *bhogyâsavas,* "the equivalent in respirations of the part of the sign to be traversed," while *t p* is styled *bhuktâsavas,* "the respirations of the part traversed."

If, on the other hand, it were desired to arrive at the same result by reckoning in the opposite direction from the sun to the horizon, either on account of the greater proximity of the two in that direction, or for any other reason, the manner of proceeding would be somewhat different. Thus, if A H (Fig. 18) were the sun's longitude, and *p* P the line of the eastern horizon, we should first find *h p*, by subtracting the part of the day already elapsed from the calculated length of the day (this step is, in the text, omitted to be specified); from it we should then subtract the *bhuktâsavas*, *l h*, and then the equivalents of the signs through which the sun has already passed, in inverse order, until there remained only the part of an equivalent, *p g*, which would be converted into the corresponding arc of longitude, P G, in the same manner as before: and the subtraction of P G from A G would give A P, the longitude of the point P.

But again, if it be required to determine the point of the ecliptic which is at any given time upon the meridian, the general process is the same as already explained, excepting that for the time from sunrise is substituted the time until or since noon, and also for the equivalents in oblique ascension those in right ascension, or, in the language of the text, the "times of rising at Lankâ" (*lankodâyasavas*); since the meridian, like the equatorial horizon, cuts the equator at right angles.

It will be observed that all these calculations assume the increments longitude of to be proportional to those of ascension throughout each sign: in a process of greater pretensions to accuracy, this would lead to errors of some consequence.

The use and value of the methods here taught, and of the quantities found as their results, will appear in the sequel (see ch. v. 1–9; vii. 7; ix. 5–11; x. 2).

The term *kshitija*, by which the horizon is designated, may be understood, according to the meaning attributed to *kshiti* (see above, under ii. 61–63), either as the "circle of situation"—that is, the one which is dependent upon the situation of the observer, varying with every change of place on his part—or as the "earth-circle," the one produced by the intervention of the earth below the observer, or drawn by the earth upon the sky. Probably the latter is its true interpretation.

50. Add together the ascensional equivalents, in respirations, of the part of the sign to be traversed by the point having less longitude, of the part traversed by that having greater longitude, and of the intervening signs—thus is made the ascertainment of time (*kâlasâdhana*).

51. When the longitude of the point of the ecliptic upon the horizon (*lagna*) is less than that of the sun, the time is in the latter part of the night; when greater, it is in the day-time; when greater than the longitude of the sun increased by half a revolution, it is after sunset.

The process taught in these verses is, in a manner, the converse of that which is explained in the preceding passage, its object being to find the instant of local time when a given point of the ecliptic will be upon the horizon, the longitude of the sun being also known. Thus (Fig. 18), supposing the sun's longitude, A P, to be, at a given time, $1^s\ 12°$; it is required to know at what time the point H, of which the longitude is

4^s 25°, will rise. The problem, is, virtually, to ascertain the arc of the equator intercepted between *p*, the point which rose with the sun, and *h*, which will rise with H, since that arc determines the time elapsed between sunrise and the rise of H, or the time in the day at which the latter will take place. In order to this, we ascertain, by a process similar to that illustrated in connection with the last passage, the *bhogyâsavas*, "ascensional equivalent of the part of the sign to be traversed," of the point having less longitude—or *p g*—and the *bhuktâsavas*, "ascensional equivalent of the part traversed," belonging to H, the point having greater longitude—or *l h*—and add the sum of both to that of the ascensional equivalents of the intervening whole signs, *g c* and *c l*, which the text calls *antaralagnâsavas*, "equivalent respirations of the interval;" the total is, in respirations of time, corresponding to minutes of arc, the interval of time required: it will be found to be 6555^p, or 18^n 12^v 3^p: and this, in the case assumed, is the time in the day at which the rise of H takes place: were H, on the other hand, the position of the sun, 18^n 12^v 3^p would be the time before sunrise of the same event, and would require to be subtracted from the calculated length of the day to give the instant of local time.

It is evident that the main use of this process must be to determine the hour at which a given planet, or a star of which the longitude is known, will pass the horizon, or at which its "day" (see above, ii. 59–63) will commence. A like method—substituting only the equivalents in right for those in oblique ascension—might be employed in determining at what instant of local time the complete day, *ahorâtra*, of any of the heavenly bodies, reckoned from transit to transit across the lower meridian, would commence: and this is perhaps to be regarded as included also in the terms of verse 50; even though the following verse plainly has reference to the time of rising, and the word *lagna*, as used in it, means only the point upon the horizon.

The last verse we take to be simply an obvious and convenient rule for determining at a glance in which part of the civil day will take place the rising of any given point of the ecliptic, or of a planet occupying that point. If the longitude of a planet be less than that of the sun, while at the same time they are not more than three signs apart—this and the other corresponding restrictions in point of distance are plainly implied in the different specifications of the verse as compared with one another, and are accordingly explicitly stated by the commentator—the hour when that planet comes to assume the position called in the text *lagna*, or to pass the eastern horizon, will evidently be between midnight and sunrise, or in the after part (*çesha*, literally "remainder") of the night: if, again, it be more than three and less than six signs behind the sun, or, which is the same thing, more than six and less than nine signs in advance of him, its time of rising will be between sunset and midnight: if, once more, it be in advance of the sun by less than six signs, it will rise while the sun is above the horizon.

The next three chapters treat of the eclipses of the sun and moon, the fourth being devoted to lunar eclipses, and the fifth to solar, and the sixth containing directions for projecting an eclipse.

CHAPTER IV.

OF ECLIPSES, AND ESPECIALLY OF LUNAR ECLIPSES.

Contents:—1, dimensions of the sun and moon; 2–3, measurement of their apparent dimensions; 4–5, measurement of the earth's shadow; 6, conditions of the occurrence of an eclipse; 7–8, ascertainment of longitude at the time of conjunction or of opposition; 9, causes of eclipses; 10–11, to determine whether there will be an eclipse, and the amount of obscuration; 12–15, to find half the time of duration of the eclipse, and half that of total obscuration; 16–17, to ascertain the times of contact and of separation, and, in a total eclipse, of immersion and emergence; 18–21, to determine the amount of obscuration at a given time; 22–23, to find the time corresponding to a given amount of obscuration; 24–25, measurement of the deflection of the ecliptic, at the point occupied by the eclipsed body, from an east and west line; 26, correction of the scale of projection for difference of altitude.

1. The diameter of the sun's disk is six thousand five hundred yojanas; of the moon's, four hundred and eighty.

We shall see, in connection with the next passage, that the diameters of the sun and moon, as thus stated, are subject to a curious modification, dependent upon and representing the greater or less distance of those bodies from the earth; so that, in a certain sense, we have here only their mean diameters. These represent, however, in the Hindu theory—which affects to reject the supposition of other orbits than such as are circular, and described at equal distances about the earth—the true absolute dimensions of the sun and moon.

Of the two, only that for the moon is obtained by a legitimate process, or presents any near approximation to the truth. The diameter of the earth being, as stated above (i. 59), 1600 yojanas, that of the moon, 480 yojanas, is .3 of it: while the true value of the moon's diameter in terms of the earth's is .2716, or only about a tenth less. An estimate so nearly correct supposes, of course, an equally correct determination of the moon's horizontal parallax, distance from the earth, and mean apparent diameter. The Hindu valuation of the parallax may be deduced from the value given just below (v. 3), of a minute on the moon's orbit, as 15 yojanas. Since the moon's horizontal parallax is equal to the angle subtended at her centre by the earth's radius, and since, at the moon's mean distance, 1′ of arc equals 15 yojanas, and the earth's radius, 800 yojanas, would accordingly subtend an angle of 53′ 20″—the latter angle, 53′ 20″, is, according to the system of the Sûrya-Siddhânta, the moon's parallax, when in the horizon and at her mean distance. This is considerably less than the actual value of the quantity, as determined by modern science, namely 57″ 1′; and it is practically, in the calculation of solar eclipses, still farther lessened by 3′ 51″, the excess of the value assigned to the sun's horizontal parallax, as we shall see farther on. Of the variation in the parallax, due to the varying distance of the moon, the Hindu system makes no account: the variation is actually nearly 8′, being from 53′ 48″, at the apogee, to 61′ 24″, at the perigee.

How the amount of the parallax was determined by the Hindus—if, indeed, they had the instruments and the skill in observation requisite for making themselves an independent determination of it—we are not informed. It is not to be supposed, however, that an actual estimate of the mean horizontal parallax as precisely 53′ 20″ lies at the foundation of the other elements which seem to rest upon it; for, in the making up of the artificial Hindu system, all these elements have been modified and adapted to one another in such a manner as to produce certain whole numbers as their results, and so to be of more convenient use.

From this parallax the moon's distance may be deduced by the proportion

sin 53′ 20″ : R : : earth's rad. : moon's dist.

or 53′⅓ : 3438′ : : 800y : 51,570y

The radius of the moon's orbit, then, is 51,570 yojanas, or, in terms of the earth's radius, 64.47. The true value of the moon's mean distance is 59.96 radii of the earth.

The farther proportion

3438′ : 5400′ : : 57,570y : 81,000y

would give, as the value of a quadrant of the moon's orbit, 81,000 yojanas, and, as the whole orbit, 324,000 yojanas. This is, in fact, the circumference of the orbit assumed by the system, and stated in another place (xii. 85). Since, however, the moon's distance is nowhere assumed as an element in any of the processes of the system, and is even directed (xii. 84) to be found from the circumference of the orbit by the false ratio of 1 : $\sqrt{10}$, it is probable that it was also made no account of in constructing the system, and that the relations of the moon's parallax and orbit were fixed by some such proportion as

53′ 20″ : 360° : : 800y : 324,000y

The moon's orbit being 324,000 yojanas, the assignment of 480 yojanas as her diameter implies a determination of her apparent diameter at her mean distance as 32′; since

360° : 32′ : : 324,000y : 480y

The moon's mean apparent diameter is actually 31′ 7″.

In order to understand, farther, how the dimensions of the sun's orbit and of the sun himself are determined by the Hindus, we have to notice that, the moon's orbit being 324,000 yojanas, and her time of sidereal revolution $27^{d}.32167416$, the amount of her mean daily motion is $11{,}858^{y}.717$. The Hindu system now assumes that this is the precise amount of the actual mean daily motion, in space, of all the planets, and ascertains the dimensions of their several orbits by multiplying it by the periodic time of revolution of each (see below, xii. 80–90). The length of the sidereal year being $365^{d}.25875648$, the sun's orbit is, as stated elsewhere (xii. 86), 4,331,500 yojanas. From a quadrant of this, by the ratio 5400′ : 3438′, we derive the sun's distance from the earth, 689,430 yojanas, or 861.8 radii of the earth. This is vastly less than his true distance, which is about 24,000 radii. His horizontal parallax is, of course, proportionally over-estimated, being made to be nearly 4′ (more exactly, 3′ 59″.4), instead of 8″.6, its true value, an amount so small that it should properly have been neglected as inappreciable.

It is an important property of the parallaxes of the sun and moon, resulting from the manner in which the relative distances of the latter from the earth are determined, that they are to one another as the mean daily motions of the planets respectively: that is to say,

53′ 20″ : 3′ 59″ : : 790′ 35″ : 59′ 8″

Each is likewise very nearly one fifteenth of the whole mean daily motion, or equivalent to the amount of arc traversed by each planet in 4 nâḍîs; the difference being, for the moon, about 38″, for the sun, about 3″. We shall see that, in the calculations of the next chapter, these differences are neglected, and the parallax taken as equal, in each case, to the mean motion during 4 nâḍîs.

The circumference of the sun's orbit being 4,331,500 yojanas, the assignment of 6500 yojanas as his diameter implies that his mean apparent diameter was considered to be 32′ 24″.8; for

360° : 32′ 24″.8 : : 4,331,500y : 6500y

The true value of the sun's apparent diameter at his mean distance is 32′ 3″.6.

The results arrived at by the Greek astronomers relative to the parallax, distance, and magnitude of the sun and moon are not greatly discordant with those here presented. Hipparchus found the moon's horizontal parallax to be 57′: Aristarchus had previously, by observation upon the angular distance of the sun and moon when the latter is half-illuminated, made their relative distances to be as 19 to 1; this gave Hipparchus 3′ as the sun's parallax. Ptolemy makes the mean distances of the sun and moon from the earth equal to 1210 and 59 radii of the earth, and their parallaxes 2′ 51″ and 58′ 14″ respectively: he also states the diameter of the moon, earth, and sun to be as 1, $3\frac{2}{5}$, $18\frac{4}{5}$, while the Hindus make them as 1, $3\frac{1}{8}$, and $13\frac{13}{24}$, and their true values, as determined by modern science, are as 1, $3\frac{2}{3}$, and $412\frac{3}{4}$, nearly.

2. These diameters, each multiplied by the true motion, and divided by the mean motion, of its own planet, give the corrected (*sphuṭa*) diameters. If that of the sun be multiplied by the number of the sun's revolutions in an Age, and divided by that of the moon's,

3. Or if it be multiplied by the moon's orbit (*kakshâ*), and divided by the sun's orbit, the result will be its diameter upon the moon's orbit: all these, divided by fifteen, give the measures of the diameters in minutes.

The absolute values of the diameters of the sun and moon being stated in yojanas, it is required to find their apparent values, in minutes of arc. In order to this, they are projected upon the moon's orbit, or upon a circle described about the earth at the moon's mean distance, of which circle—since 324,000 ÷ 21,600 = 15—one minute is equivalent to fifteen yojanas.

The method of the process will be made clear by the annexed figure (Fig. 19). Let E be the earth's place, E M or E *m* the mean distance of

Fig. 19.

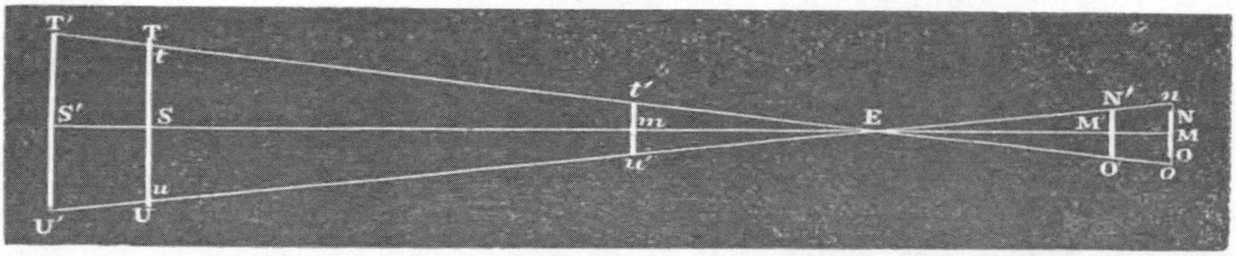

the moon, and E S the mean distance of the sun. Let T U equal the sun's diameter, 6500y. But now let the sun be at the greater distance E S′: the part of his mean orbit which his disk will cover will no longer be T U, but a less quantity, *t u*, and *t u* will be to T U, or T′ U′, as E S to E S′. But the text is not willing to acknowledge here, any more than in the second chapter, an actual inequality in the distance of the sun from the earth at different times, even though that inequality be most unequivocally implied in the processes it prescribes: so, instead of calculating E S′ as well as E S, which the method of epicycles affords full facilities for doing, it substitutes, for the ratio of E S to E S′, the inverse ratio of the daily motion at the mean distance E S to that at the true distance E S′. The ratios, however, are not precisely equal. The arc *a m* (Fig. 4, p. 67) of the eccentric circle is supposed to be traversed by the sun or moon with a uniform velocity. If, then, the motion at any given point, as *m*, were perpendicular to E *m*, the apparent motion would be inversely as the distance. But the motion at *m* is perpendicular to *e m* instead of E *m*. The resulting error, it is true, and especially in the case of the sun, is not very great. It may be added that the eccentric circle which best represents the apparent motions of the sun and moon in their elliptic orbits, gives much more imperfectly the distances and apparent diameters of those bodies. The value of *t u*, however, being thus at least approximately determined, *t′ u′*, the arc of the moon's mean orbit subtended by it, is then found by the proportion ES : E *m* (or E M) : : *t u* : *t′ u′*—excepting that here, again, for the ratio of the distances, E S and E M, is substituted either that of the whole circumferences of which they are respectively the radii, or the inverse ratio of the number of revolutions in a given time of the two planets, which, as shown in the note to the preceding passage, is the same thing. Having thus ascertained the value of *t′ u′* in yojanas, division by 15 gives us the number of minutes in the arc *t′ u′*, or in the angle *t′* E *u′*.

In like manner, if the moon be at less than her mean distance from the earth, as E M′, she will subtend an arc of her mean orbit *n o*, greater than N O, her true diameter; the value of *n o*, in yojanas and in minutes, is found by a method precisely similar to that already described.

There is hardly in the whole treatise a more curious instance than this of the mingling together of true theory and false assumption in the same process, and of the concealment of the real character of a process by substituting other and equivalent data for its true elements.

We meet for the first time, in this passage, the term employed in the treatise to designate a planetary orbit, namely *kakshâ*, literally "border, girdle, periphery." The value finally obtained for the apparent diameter of the sun or moon, as later of the shadow, is styled its *mâna*, "measure."

In order to furnish a practical illustration of the processes taught in this chapter, we have calculated in full, by the methods and elements of the Sûrya-Siddhânta, the lunar eclipse of Feb. 6th, 1860. Rather, however, than present the calculation piecemeal, and with its different processes severed from their natural connection, and arranged under the passages to which they severally belong, we have preferred to give it entire in the Appendix, whither the reader is referred for it.

4. Multiply the earth's diameter by the true daily motion of the moon, and divide by her mean motion: the result is the earth's corrected diameter (*sûcî*). The difference between the earth's diameter and the corrected diameter of the sun

5. Is to be multiplied by the moon's mean diameter, and divided by the sun's mean diameter: subtract the result from the earth's corrected diameter (*sûcî*), and the remainder is the diameter of the shadow; which is reduced to minutes as before.

The method employed in this process for finding the diameter of the earth's shadow upon the moon's mean orbit may be explained by the aid of the following figure (Fig. 20).

As in the last figure, let E represent the earth's place, S and M points in the mean orbits of the sun and moon, and M′ the moon's actual place. Let *t u* be the sun's corrected diameter, or the part of his mean orbit which his disk at its actual distance covers, ascertained as directed in the preceding passage, and let F G be the earth's diameter. Through

Fig. 20.

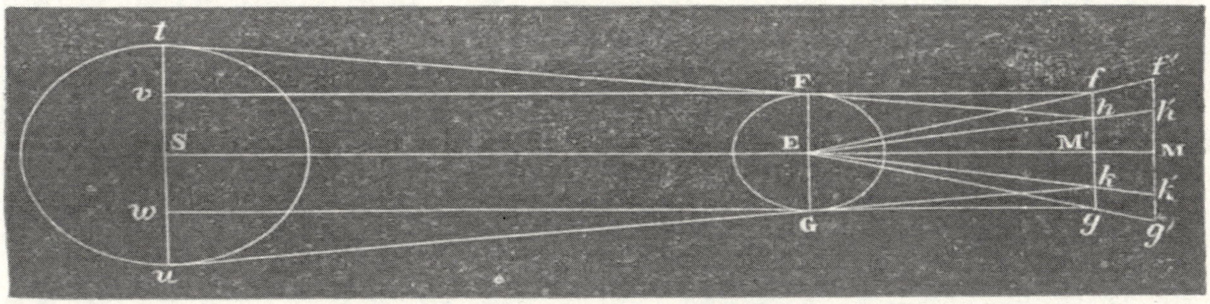

F and G draw *v* F *f* and *w* G *g* parallel to S M, and also *t* F *h* and *u* G *k*: then *h k* will be the diameter of the shadow where the moon actually enters it. The value of *h k* evidently equals *fg* (or F G) − (*f h* + *g k*); and the value of *f h* + *g k* may be found by the proportion

$$Fv \text{ (or ES)} : tv + wu \text{ (or } tu - FG) :: Ff \text{ (or EM}') : fh + gk$$

But the Hindu system provides no method of measuring the angular value of quantities at the distance E M′, nor does it ascertain the value of E M′ itself: and as, in the last process, the diameter of the moon was reduced, for measurement, to its value at the distance E M′, so, to be made commensurate with it, all the data of this process must be similarly modified. That is to say, the proportion

$$EM' : EM :: fg : f'g'$$

—substituting, as before, the ratio of the moon's mean to her true motion for that of E M′ to E M—gives $f'g'$, which the text calls the *sûcî:* the word means literally "needle, pyramid; we do not see precisely how it comes to be employed to designate the quantity $f'g'$, and have translated it, for lack of a better term, and in analogy with the language of the text respecting the diameters of the sun and moon, "corrected diameter of the earth." It is also evident that

$$EM' : fh+gk :: EM : f'h'+g'k'$$

hence, substituting the latter of these ratios for the former in our first proportion, and inverting the middle terms, we have

$$ES : EM :: tu-FG : f'h'+g'k'$$

Once more, now, we have a substitution of ratios, E S : E M being replaced by the ratio of the sun's mean diameter to that of the moon. In this there is a slight inaccuracy. The substitution proceeds upon the assumption that the mean apparent values of the diameters of the sun and moon are precisely equal, in which case, of course, their absolute diameters would be as their distances; but we have seen, in the note to the first verse of this chapter, that the moon's mean angular diameter is made a little less than the sun's, the former being 32′, the latter 32′ 24″.8. The error is evidently neglected as being too small to impair sensibly the correctness of the result obtained: it is not easy to see, however, why we do not have the ratio of the mean distances represented here, as in verses 2 and 3, by that of the orbits, or by that of the revolutions in an Age taken inversely. The substitution being made, we have the final proportion on which the rule in the text is based, viz., the sun's mean diameter is to the moon's mean diameter as the excess of the sun's corrected diameter over the actual diameter of the earth is to a quantity which, being subtracted from the *sûcî*, or corrected diameter of the earth, leaves as a remainder the diameter of the shadow as projected upon the moon's mean orbit: it is expressed in yojanas, but is reduced to minutes, as before, by dividing by fifteen. The earth's penumbra is not taken into account in the Hindu process of calculation of an eclipse.

The lines fg, $f'g'$, etc., are treated here as if they were straight lines, instead of arcs of the moon's orbit: but the inaccuracy never comes to be of any account practically, since the value of these lines always falls inside of the limits within which the Hindu methods of calculation recognize no difference between an arc and its sine.

6. The earth's shadow is distant half the signs from the sun: when the longitude of the moon's node is the same with that of the shadow, or with that of the sun, or when it is a few degrees greater or less, there will be an eclipse.

To the specifications of this verse we need to add, of course, "at the time of conjunction or of opposition."

It will be noticed that no attempt is made here to define the lunar and solar ecliptic limits, or the distances from the moon's node within which eclipses are possible. Those limits are, for the moon, nearly 12°; for the sun, more than 17°.

The word used to designate "eclipse," *grahaṇa*, means literally "seizure": it, with other kindred terms, to be noticed later, exhibits the influence of the primitive theory of eclipses, as seizures of the heavenly bodies by the monster Râhu. In verses 17 and 19, below, instead of *grahaṇa* we have *graha*, another derivative from the same root *grah* or *grabh*, "grasp, seize." Elsewhere *graha* never occurs except as signifying "planet," and it is the only word which the Sûrya-Siddhânta employs with that signification: as so used, it is an active instead of a passive derivative, meaning "seizer," and its application to the planets is due to the astrological conception of them, as powers which "lay hold upon" the fates of men with their supernatural influences.

7. The longitudes of the sun and moon, at the moment of the end of the day of new moon (*amâvâsyâ*), are equal, in signs, etc.; at the end of the day of full moon (*paurṇamâsî*) they are equal in degrees, etc., at a distance of half the signs.

8. When diminished or increased by the proper equation of motion for the time, past or to come, of opposition or conjunction, they are made to agree, to minutes: the place of the node at the same time is treated in the contrary manner.

The very general directions and explanations contained in verses 6, 7, and 9 seem out of place here in the middle of the chapter, and would have more properly constituted its introduction. The process prescribed in verse 8, also, which has for its object the determination of the longitudes of the sun, moon, and moon's node, at the moment of opposition or conjunction, ought no less, it would appear, to precede the ascertainment of the true motions, and of the measures of the disks and shadow, already explained. Verse 8, indeed, by the lack of connection in which it stands, and by the obscurity of its language, furnishes a striking instance of the want of precision and intelligibility so often characteristic of the treatise. The subject of the verse, which requires to be supplied, is, "the longitudes of the sun and moon at the instant of midnight next preceding or following the given opposition or conjunction"; that being the time for which the true longitudes and motions are first calculated, in order to test the question of the probability of an eclipse. If there appears to be such a probability, the next step is to ascertain the interval between midnight and the moment of opposition or conjunction, past or to come: this is done by the method taught in ii. 66, or by some other analogous process: the instant of the occurrence of opposition or conjunction, in local time, counted from sunrise of the place of observation, must also be determined, by ascertaining the interval between mean and apparent midnight (ii. 46), the length of the complete day (ii. 59), and of its parts (ii. 60–63), etc.; the whole process is sufficiently illustrated by the two examples of the calculation of eclipses given in the Appendix. When we have thus found the interval between midnight and the moment of opposition or conjunction, verse 8 teaches us how to ascertain the true longitudes for that moment: it is by calculating—in the manner taught in i. 67, but with the true daily motions—the amount of motion of the sun, moon, and node during the interval, and applying it as a corrective equation to the longitude of each at midnight, subtracting in the case of the sun and moon, and adding in the case of the node, if the moment was then already past; and the contrary, if it was still to come. Then, if the process has been correctly performed, the longitudes of the sun and moon will be found to correspond, in the manner required by verse 7.

For the days of new and full moon, and their appellations, see the note to ii. 66, above. The technical expression employed here, as in one or two other passages, to designate the "moment of opposition or conjunction" is *parvanâdyas*, "nâḍîs of the *parvan*," or "time of the *parvan* in nâḍîs, etc.:" *parvan* means literally "knob, joint," and is frequently applied, as in this term, to denote a conjuncture, the moment that distinguishes and separates two intervals, and especially one that is of prominence and importance.

9. The moon is the eclipser of the sun, coming to stand underneath it, like a cloud: the moon, moving eastward, enters the earth's shadow, and the latter becomes its eclipser.

The names given to the eclipsed and eclipsing bodies are either *châdya* and, as here, *châdaka*, "the body to be obscured" and "the obscurer," or *grâhya* and *grâhaka*, "the body to be seized" and "the seizer." The latter terms are akin with *grahaṇa* and *graha*, spoken of above (note to v. 6), and represent the ancient theory of the phenomena, while the others are derived from their modern and scientific explanation, as given in this verse.

10. Subtract the moon's latitude at the time of opposition or conjunction from half the sum of the measures of the eclipsed and eclipsing bodies: whatever the remainder is, that is said to be the amount obscured.

11. When that remainder is greater than the eclipsed body, the eclipse is total; when the contrary, it is partial; when the latitude is greater than the half sum, there takes place no obscuration (*grâsa*).

It is sufficiently evident that when, at the moment of opposition, the moon's latitude—which is the distance of her centre from the ecliptic, where is the centre of the shadow—is equal to the sum of the radii of her disk and of the shadow, the disk and the shadow will just touch one another; and that, on the other hand, the moon will, at the moment of opposition, be so far immersed in the shadow as her latitude is less than the sum of the radii: and so in like manner for the sun, with due allowance for parallax. The Hindu mode of reckoning the amount eclipsed is not by digits, or twelfths of the diameter of the eclipsed body, which method we have inherited from the Greeks, but by minutes.

The word *grâsa*, used in verse 11 for obscuration or eclipse, means literally "eating, devouring," and so speaks more distinctly than any other term we have had of the old theory of the physical cause of eclipses.

12. Divide by two the sum and difference respectively of the eclipsed and eclipsing bodies: from the square of each of the resulting quantities subtract the square of the latitude, and take the square roots of the two remainders.

13. These, multiplied by sixty and divided by the difference of the daily motions of the sun and moon, give, in nâḍîs, etc., half the duration (*sthiti*) of the eclipse, and half the time of total obscuration.

These rules for finding the intervals of time between the moment of opposition or conjunction in longitude, which is regarded as the middle of the eclipse, and the moments of first and last contact, and, in a total eclipse, of the beginning and end of total obscuration, may be illustrated by help of the annexed figure (Fig. 21).

Let E C L represent the ecliptic, the point C being the centre of the shadow, and let C D be the moon's latitude at the moment of opposi-

Fig. 21.

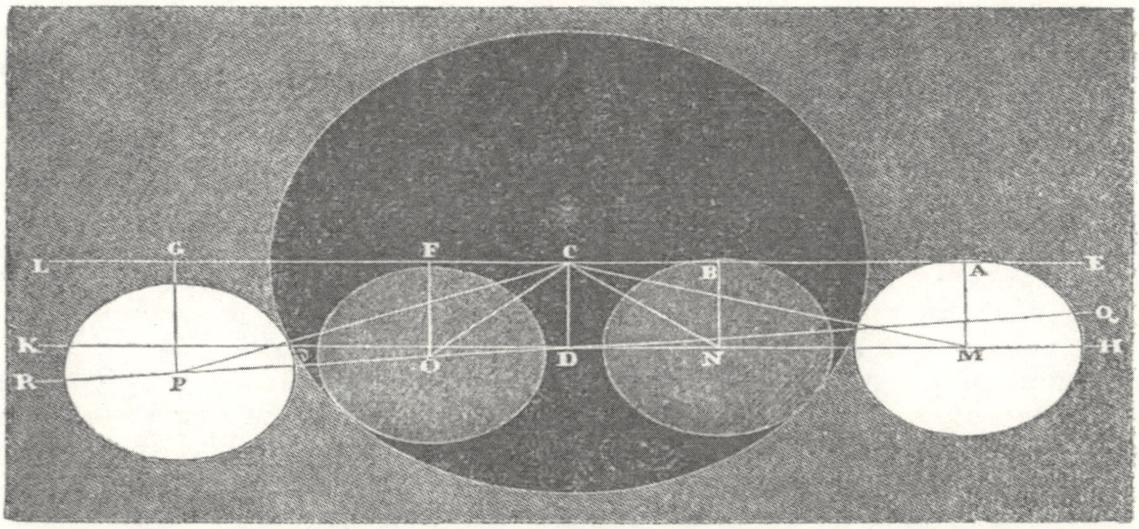

tion; which, for the present, we will suppose to remain unchanged through the whole continuance of the eclipse. It is evident that the first contact of the moon with the shadow will take place when, in the triangle C A M, A C equals the moon's distance in longitude from the centre of the shadow, A M her latitude, and C M the sum of her radius and that of the shadow. In like manner, the moon will disappear entirely within the shadow when B C equals her distance in longitude from the centre of the shadow, B N her latitude, and C N the difference of the two radii. Upon subtracting, then, the square of A M or B N from those of C M and C N respectively, and taking the square roots of the remainders, we shall have the values of A C and B C in minutes. These may be reduced to time by the following proportion: as the excess at the given time of the moon's true motion in a day over that of the sun is to a day, or sixty nâḍîs, so are A C and B C, the amounts which the moon has to gain in longitude upon the sun between the moments of contact and immersion respectively and the moment of opposition, to the corresponding intervals of time.

But the process, as thus conducted, involves a serious error: the moon's latitude, instead of remaining constant during the eclipse, is constantly and sensibly changing. Thus, in the figure above, of which the conditions are those found by the Hindu processes for the eclipse of Feb. 6th, 1860, the moon's path, instead of being upon the line H K, parallel to the ecliptic, is really upon Q R. The object of the process next taught is to get rid of this error.

14. Multiply the daily motions by the half-duration, in nâḍîs, and divide by sixty: the result, in minutes, subtract for the time of contact (*pragraha*), and add for that of separation (*moksha*), respectively;

15. By the latitudes thence derived, the half-duration, and likewise the half-time of total obscuration, are to be calculated anew, and the process repeated. In the case of the node, the proper correction, in minutes, etc., is to be applied in the contrary manner.

This method of eliminating the error involved in the supposition of a constant latitude, and of obtaining another and more accurate determination of the intervals between the moment of opposition and those of first and last contact, and of immersion and emergence, is by a series of successive approximations. For instance: A C, as already determined, being assumed as the interval between opposition and first contact, a new calculation of the moon's longitude is made for the moment A, and, with this and the sum of the radii, a new value is found for A C. But now, as the position of A is changed, the former determination of its latitude is vitiated and must be made anew, and made to furnish anew a corrected value of A C; and so on, until the position of A is fixed with the degree of accuracy required. The process must be conducted separately, of course, for each of the four quantities affected; since, where latitude is increasing, as in the case illustrated, the true values of A C and B C will be greater than their mean values, while G C and F C, the true intervals in the after part of the eclipse, will be less than A C and B C: and the contrary when latitude is decreasing.

We have illustrated these processes by reference only to a lunar eclipse: their application to the conditions of a solar eclipse requires the introduction of another element, that of the parallax, and will be explained in the notes upon the next chapter.

The first contact of the eclipsed and eclipsing bodies is styled in this passage *pragraha*, "seizing upon, laying hold of;" elsewhere it is also called *grâsa*, "devouring," and *sparça*, "touching:" the last contact, or separation, is named *moksha*, "release, letting go." The whole duration of the eclipse, from contact to separation, is the *sthiti*, "stay, continuance;" total obscuration is *vimarda*, "crushing out, entire destruction."

16. The middle of the eclipse is to be regarded as occurring at the true close of the lunar day: if from that time the time of half-duration be subtracted, the moment of contact (*grâsa*) is found; if the same be added, the moment of separation.

17. In like manner also, if from and to it there be subtracted and added, in the case of a total eclipse, the half-time of total obscuration, the results will be the moments called those of immersion and emergence.

The instant of true opposition, or of apparent conjunction (see below, under ch. v. 9), in longitude, of the sun and moon, is to be taken as the middle of the eclipse, even though, owing to the motion of the moon in latitude, and also, in a solar eclipse, to parallax, that instant is not midway between those of contact and separation, or of immersion and emergence. To ascertain the moment of local time of each of these phases of the eclipse, we subtract and add, from and to the local time of opposition or conjunction, the true intervals found by the processes described in verses 12 to 15.

The total disappearance of the eclipsed body within, or behind, the eclipsing body, is called *nimîlana,* literally the "closure of the eyelids, as in winking:" its first commencement of reappearance is styled *unmîlana,* "parting of the eyelids, peeping." We translate the terms by "immersion" and "emergence" respectively.

18. If from half the duration of the eclipse any given interval be subtracted, and the remainder multiplied by the difference of the daily motions of the sun and moon, and divided by sixty, the result will be the perpendicular (*koṭi*) in minutes.

19. In the case of an eclipse (*graha*) of the sun, the perpendicular in minutes is to be multiplied by the mean half-duration, and divided by the true (*sphuṭa*) half-duration, to give the true perpendicular in minutes.

20. The latitude is the base (*bhuja*): the square root of the sum of their squares is the hypothenuse (*çrava*): subtract this from half the sum of the measures, and the remainder is the amount of obscuration (*grâsa*) at the given time.

21. If that time be after the middle of the eclipse, subtract the interval from the half-duration on the side of separation, and treat the remainder as before: the result is the amount remaining obscured on the side of separation.

The object of the process taught in this passage is to determine the amount of obscuration of the eclipsed body at any given moment during the continuance of the eclipse. It, as well as that prescribed in the following passage, is a variation of that which forms the subject of verses 12 and 13 above, being founded, like the latter, upon a consideration of the right-angled triangle formed by the line joining the centres of the eclipsed and eclipsing bodies as hypothenuse, the difference of their longitudes as perpendicular, and the moon's latitude as base. And whereas, in the former problem, we had the base and hypothenuse given to find the perpendicular, here we have the base and perpendicular given to find the hypothenuse. The perpendicular is furnished us in time, and the rule supposes it to be stated in the form of the interval between the given moment and that of contact or of separation: a form to which, of course, it may readily be reduced from any other mode of statement. The interval of time is reduced to its equivalent as difference of longitude by a proportion the reverse of that given in verse 13, by which difference of longitude was converted into time; the moon's latitude is then calculated; from the two the hypothenuse is deduced; and the comparison of this with the sum of the radii gives the measure of the amount of obscuration.

Verse 21 seems altogether superfluous: it merely states the method of proceeding in case the time given falls anywhere between the middle and the end of the eclipse, as if the specifications of the preceding verses applied only to a time occurring before the middle: whereas they are general in their character, and include the former case no less than the latter.

When the eclipse is one of the sun, allowance needs to be made for the variation of parallax during its continuance; this is done by the process described in verse 19, of which the explanation will be given in the notes to the next chapter (vv. 14–17).

In verse 20, for the first and only time, we have latitude called *kshepa*, instead of *vikshepa*, as elsewhere. In the same verse, the term employed for "hypothenuse" is *çrava*, "hearing, organ of hearing;" this, as well as the kindred *çravaṇa*, which is also once or twice employed, is a synonym of the ordinary term *karṇa*, which means literally "ear." It is difficult to see upon what conception their employment in this signification is founded.

22. From half the sum of the eclipsed and eclipsing bodies subtract any given amount of obscuration, in minutes: from the square of the remainder subtract the square of the latitude at the time, and take the square root of their difference.

23. The result is the perpendicular (*koṭi*) in minutes—which, in an eclipse of the sun, is to be multiplied by the true, and divided by the mean, half-duration—and this, converted into time by the same manner as when finding the duration of the eclipse, gives the time of the given amount of obscuration (*grâsa*).

The conditions of this problem are precisely the same with those of the problem stated above, in verses 12–15, excepting that here, instead of requiring the instant of time when obscuration commences, or becomes total, we desire to know when it will be of a certain given amount. The solution must be, as before, by a succession of approximative steps, since, the time not being fixed, the corresponding latitude of the moon cannot be otherwise determined.

24. Multiply the sine of the hour-angle (*nata*) by the sine of the latitude (*aksha*), and divide by radius: the arc corresponding to the result is the degrees of deflection (*valanânçâs*), which are north and south in the eastern and western hemispheres (*kapâla*) respectively.

25. From the position of the eclipsed body increased by three signs calculate the degrees of declination: add them to the degrees of deflection, if of like direction; take their difference, if of different direction: the corresponding sine is the deflection (*valana*)—in digits, when divided by seventy.

This process requires to be performed only when it is desired to project an eclipse. In making a projection according to the Hindu method, as will be seen in connection with the sixth chapter, the eclipsed body is represented as fixed in the centre of the figure, with a north and south line, and an east and west line, drawn through it. The absolute position of these lines upon the disk of the eclipsed body is, of course, all the time changing: but the change is, in the case of the sun, not observable, and in the case of the moon it is disregarded: the Sûrya-Siddhânta takes no notice of the figure visible in the moon's face as determining any fixed and natural directions upon her disk. It is desired to represent to the eye, by the figure drawn, where, with reference to the north, south, east, and west points of the moment, the contact, immersion, emergence, separation, or other phases of the eclipse, will take place. In order to this, it is necessary to know what is, at each given moment, the direction of the ecliptic, in which the motions of both eclipsed and eclipsing bodies are made. The east and west direction is represented by a small circle drawn through the eclipsed body, parallel to the prime vertical; the north and south direction, by a great circle passing through the body and through the north and south points of the horizon: and the direction of the ecliptic is determined by ascertaining the angular amount of its deflection from the small east and west circle at the point occupied by the eclipsed body. Thus, in the annexed figure (Fig. 22), if M be the place of the eclipsed body upon the ecliptic, C L, and if E W be the small east and west circle drawn through M parallel with

Fig. 22.

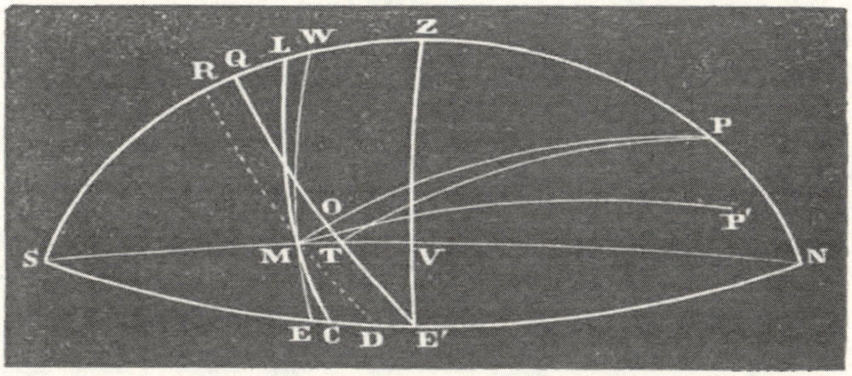

E′ Z, the prime vertical, then the deflection will be the angle made at M by C M and E M, which is equal to P′ M N, the angle made by perpendiculars to the two circles drawn from their respective poles. In order to find the value of this angle, a double process is adopted: first, the angle made at M by the two small circles E M and D M, which is equivalent to P M N, is approximately determined: as this depends for its amount upon the observer's latitude, being nothing in a right sphere, it is called by the commentary *âksha valana*, "the deflection due to latitude:" the text calls it simply *valanânçâs*, "degrees of deflection," since it does not, like the net result of the whole operation, require to be expressed in terms of its sine. Next, the angle made at M by the ecliptic,

C L, and the circle of daily revolution, D R, which angle is equal to P M P′, is also measured: this the commentary calls *âyana valana*, "the deflection due to the deviation of the ecliptic from the equator;" the text has no special name for it. The sum of these two results, or their difference, as the case may be, is the *valana*, or the deflection of the ecliptic from the small east and west circle at M, or the angle P′ M N.

In explaining the method and value of these processes, we will commence with the second one, or with that by which P M P′, the *âyana valana*, is found. In the following figure (Fig. 23), let O Q be the equator, and M L the ecliptic, P and P′ being their respective poles. Let M be the point at which the amount of deflection of M L from the circle of diurnal revolution, D R, is sought. Let M L equal a quadrant; draw P′ L, cutting the equator at Q; as also P L, cutting it at B; then draw P M and Q M. Now P′ M L is a tri-quadrantal triangle, and hence M Q is a quadrant; and therefore Q is a pole of the circle P O M, and Q O is also a quadrant, and Q M O is a right angle. But D R also makes right angles at M with P M; hence Q M and D R are tangents to one another at M, and the spherical angle Q M L is equal to that which the ecliptic makes at M with the circle of declination, or to P M P′: and Q M L is measured by Q L. The rule given in the text produces a result which is a near approach to this, although not entirely accordant with it excepting at the solstice and equinox, the points where the deflection is greatest and where it is nothing. We are directed to reckon forward a quadrant from the position of the eclipsed body—that is, from M to L, in the figure—and then to calculate the declination at that point, which will be the amount of deflection. But the declination at L is B L, and since L B Q is a right-angled triangle, having a right angle at B, and since L Q and L B are always less than quadrants, L B must be less than L Q. The difference between them, however, can never be of more than trifling amount; for, as the angle Q L B increases, Q L diminishes; and the contrary.

Fig. 23.

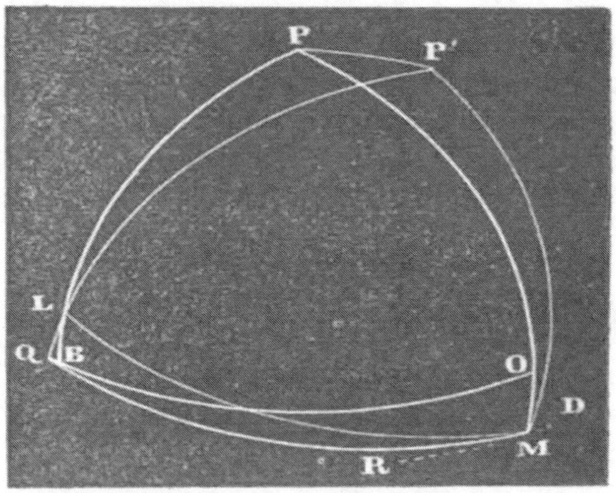

In order to show how the Hindus have arrived at a determination of this part of the deflection so nearly correct, and yet not quite correct, we will cite the commentator's explanation of the process. He says: "The 'east' (*prâcî*) of the equator [i. e., apparently, the point of the equator eastward toward which the small circle must be considered as pointing at M] is a point 90° distant from that where a circle drawn from the pole (*dhruva*) through the planet cuts the equator:" that is to say, it is the point Q (Fig. 23), a quadrant from O: "and the interval by which this is separated from the 'east' of the ecliptic at 90° from the planet, that is the *âyana valana*." This is entirely correct, and would give us Q L, the true measure of the deflection. But the commentator goes on farther to say that since this interval, when the planet is at the solstice, is nothing, and when at the equinox is equal to the greatest declination, it is therefore always equal to the declination at a quadrant's distance from the planet. This is, as we have seen, a false conclusion, and leads to an erroneous result: whether they who made the rule were aware of this, but deemed the process a convenient one, and its result a sufficiently near approximation to the truth, we will not venture to say.

The other part of the operation, to determine the amount of deflection of the circle of declination from the east and west small circle, is considerably more difficult, and the Hindu process correspondingly defective. We will first present the explanation of it which the commentator gives. He states the problem thus: "by whatever interval the directions of the equator are deflected from directions corresponding to those of the prime vertical, northward or southward, that is the deflection due to latitude (*âksha valana*). Now then: if a movable circle be drawn through the pole of the prime vertical (*sama*) and the point occupied by the planet [i. e., the circle N M S, Fig. 22], then the interval of the 'easts,' at the distance of a quadrant upon each of the two circles, the equator and the prime vertical, from the points where they are respectively cut by that circle [i. e., from T and V] will be the deflection. . . . Now when the planet is at the horizon [as at D, referred to E′], then that interval is equal to the latitude [Z Q]; when the planet is upon the meridian (*yâmyottaravṛtta*, "south and north circle") [i. e., when it is at R, referred to Q and Z], there is no interval [as at E′]. Hence, by the following proportion—with a sine of the hour-angle which is equal to radius the sine of deflection for latitude is equal to the sine of latitude; then with any given sine of the hour-angle what is it?—a sine of latitude is found, of which the arc is the required deflection for latitude." This is, in the Hindu form of statement, the proportion represented by the rule in verse 24, viz. R : sin lat. : : sin hour-angle : sin deflection.

It seems to us very questionable, at least, whether the Hindus had any more rigorous demonstration than this of the process they adopted, or knew wherein lay the inaccuracies of the latter. These we will now proceed to point out. In the first place, instead of measuring the angle made at the point in question, M, by the two small circles, the east and west circle and that of daily revolution—which would be the angle P M N—they refer the body to the equator by a circle passing through the north and south points of the horizon, and measure the deflection of the equator from a small east and west circle at its intersection with that circle—which is the angle P T N. Or, if we suppose that, in the process formerly explained, no regard was had to the circle of daily revolution, D R, the intention being to measure the difference in direction of the ecliptic at M and the equator at O, then the two parts of the process are inconsistent in this, that the one takes as its equatorial point of measurement O, and the other T, at which two points the direction of the equator is different. But neither is the value of P T N correctly found. For, in the spherical triangle P N T, to find the angle at T, we should make the proportion

$$\sin PT \text{ (or R)} : \sin PN :: \sin PNT : \sin PTN$$

But, as the third term in this proportion, the Hindus introduce the sine of the hour-angle, Z P M or M P N, although with a certain modification which the commentary prescribes, and which makes of it something very near the angle T P N. The text says simply *natajyâ*, "the sine of the hour-angle" (for *nata*, see notes to iii. 34–36, and 14–16), but the commentary specifies that, to find the desired angle in degrees, we must multiply the hour-angle in time by 90, and divide by the half-day of the planet. This is equivalent to making a quadrant of that part of the circle of diurnal revolution which is between the horizon and the meridian, or to measuring distances upon D R as if they were proportional parts of E′ Q. To make the Hindu process correct, the product of this modification should be the angle P N T, with which, however, it only coincides at the horizon, where both T P N and T N P become right-angles, and at the meridian, where both are reduced to nullity. The error is closely analogous to that involved in the former process, and is of slight account when latitude is small, as is also the error in substituting T for O or M when neither the latitude nor the declination is great.

The direction of the ecliptic deflection (*âyana valana*) is the same, evidently, with that of the declination a quadrant eastward from the point in question; thus, in the case illustrated by the figure, it is south. The direction of the equatorial deflection (*âksha valana*) depends upon the position of the point considered with reference to the meridian, being—in northern latitudes, which alone the Hindu system contemplates—north when that point is east of the meridian, and south when west of it, as specified in verse 24: since, for instance, E′ being the east point of the horizon, the equator at any point between E′ and Q points, eastward, toward a point north of the prime vertical. In the case for which the figure is drawn, then, the difference of the two would be the finally resulting deflection. Since, in making the projection of the eclipse, it is laid off as a straight line (see the illustration given in connection with chapter vi), it must be reduced to its value as a sine; and moreover, since it is laid down in a circle of which the radius is 49 digits (see below, vi. 2), or in which one digit equals 70′—for $3438' \div 49 = 70'$, nearly—that sine is reduced to its value in digits by dividing it by 70.

The general subject of this passage, the determination of directions during an eclipse, for the purpose of establishing the positions, upon the disk of the eclipsed body, of the points of contact, immersion, emergence, and separation, also engaged the attention of the Greeks; Ptolemy devotes to it the eleventh and twelfth chapters of the sixth book of his Syntaxis: his representation of directions, however, and consequently his method of calculation also, are different from those here exposed.

26. To the altitude in time (*unnata*) add a day and a half, and divide by a half-day; by the quotient divide the latitudes and the disks; the results are the measures of those quantities in digits (*angula*).

By this process due account is taken, in the projection of an eclipse, of the apparent increase in magnitude of the heavenly bodies when near the horizon. The theory lying at the foundation of the rule is this:

that three minutes of arc at the horizon, and four at the zenith, are equal to a digit, the difference between the two, or the excess above three minutes of the equivalent of a digit at the zenith, being one minute. To ascertain, then, what will be, at any given altitude, the excess above three minutes of the equivalent of a digit, we ought properly, according to the commentary, to make the proportion

R : 1′ : : sin altitude : corresp. excess

Since, however, it would be a long and tedious process to find the altitude and its sine, another and approximative proportion is substituted for this "by the blessed Sun," as the commentary phrases it, "through compassion for mankind, and out of regard to the very slight difference between the two." It is assumed that the scale of four minutes to the digit will be always the true one at the noon of the planet in question, or whenever it crosses the meridian, although not at the zenith; and so likewise, that the relation of the altitude to 90° may be measured by that of the time since rising or until setting (*unnata*—see above, iii. 37–39) to a half-day. Hence the proportion becomes

half-day : 1′ : : altitude in time : corresp. excess

and the excess of the digital equivalent above 3′ equals $\frac{\text{alt. in time}}{\text{half-day}}$.

Adding, now, the three minutes, and bringing them into the fractional expression, we have

$$\text{equiv. of digit in minutes at given time} = \frac{\text{alt. in time} + 3\ \text{half-days}}{\text{half-day}}$$

The title of the fourth chapter is *candragrahaṇâdhikâra*, "chapter of lunar eclipses," as that of the fifth is *sûryagrahaṇâdhikâra*, "chapter of solar eclipses." In truth, however, the processes and explanations of this chapter apply not less to solar than to lunar eclipses, while the next treats only of parallax, as entering into the calculation of a solar eclipse. We have taken the liberty, therefore, of modifying accordingly the headings which we have prefixed to the chapters.

CHAPTER V.

OF PARALLAX IN A SOLAR ECLIPSE.

CONTENTS:—1, when there is no parallax in longitude, or no parallax in latitude; 2, causes of parallax; 3, to find the orient-sine; 4–5, the meridian-sine; 5–7, and the sines of ecliptic zenith-distance and altitude; 7–8, to find the amount, in time, of the parallax in longitude; 9, its application in determining the moment of apparent conjunction; 10–11, to find the amount, in arc, of the parallax in latitude; 12–13, its application in calculating an eclipse; 14–17, application of the parallax in longitude in determining the moments of contact, of separation, etc.

1. When the sun's place is coincident with the meridian ecliptic-point (*madhyalagna*), there takes place no parallax in

longitude (*harija*): farther, when terrestrial latitude (*aksha*) and north declination of the meridian ecliptic-point (*madhyabha*) are the same, there takes place no parallax in latitude (*avanati*).

The latter of these specifications is entirely accurate: when the north declination of that point of the ecliptic which is at the moment upon the meridian (*madhyalagna;* see iii. 49) is equal to the observer's latitude—regarded by the Hindus as always north—the ecliptic itself passes through the zenith, and becomes a vertical circle; of course, then, the effect of parallax would be only to depress the body in that circle, not to throw it out of it. The other is less exact: when the sun is upon the meridian, there is, indeed, no parallax in right ascension, but there is parallax in longitude, unless the ecliptic is also bisected by the meridian. Here, as below, in verses 8 and 9, the text commits the inaccuracy of substituting the meridian ecliptic-point (L in Fig. 26) for the central or highest point of the ecliptic (B in the same figure). The latter point, although we are taught below (vv. 5–7) to calculate the sine and cosine of its zenith-distance, is not once distinctly mentioned in the text; the commentary calls it *tribhonalagna,* "the orient ecliptic-point (*lagna*—see above, iii. 46–48: it is the point C in Fig. 26) less three signs." The commentary points out this inaccuracy on the part of the text.

In order to illustrate the Hindu method of looking at the subject of parallax, we make the following citation from the general exposition of it given by the commentator under this verse: "At the end of the day of new moon (*amâvâsyâ*) the sun and moon have the same longitude; if, now, the moon has no latitude, then a line drawn from the earth's centre [C in the accompanying figure] to the sun's place [S] just touches the moon [M]: hence, at the centre, the moon becomes an eclipsing, and the sun an eclipsed, body. Since, however, men are not at the earth's centre, (*garbha,* "womb") but upon the earth's surface (*pṛshṭha,* "back"), a line drawn from the earth's surface [B] up to the sun does not just touch the moon; but it cuts the moon's sphere above the point occupied by the moon [at *m*], and when the moon arrives at this point, then is she at the earth's surface the eclipser of the sun. But when the sun is at the zenith (*khamadhya,* "mid-heaven"), then the lines drawn up to the sun from the earth's centre and surface, being one and the same, touch the moon, and so the moon becomes an eclipsing body at the end of the day of new moon. Hence, too, the interval [M *m*] of the lines from the earth's centre and surface is the parallax (*lambana*)."

Fig. 24.

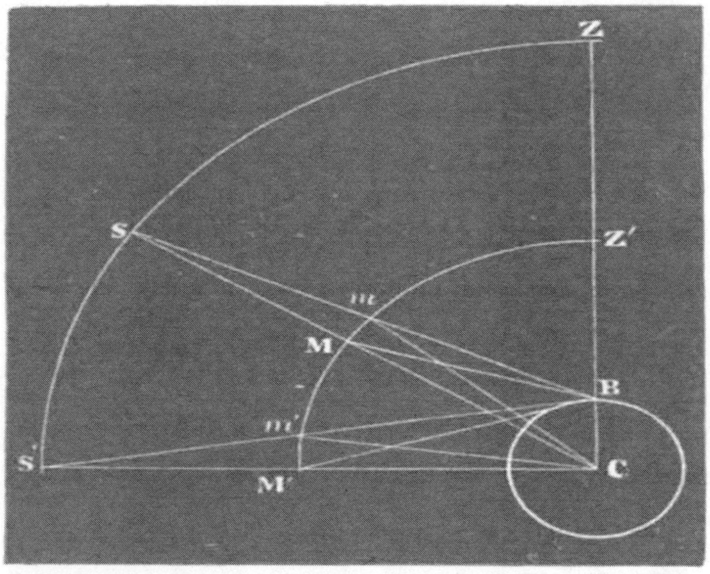

It is evident from this explication how far the Hindu view of parallax is coincident with our own. The principle is the same, but its application is somewhat different. Instead of taking the parallax absolutely, determining that for the sun, which is B S C, and that for the moon, which is B M C, the Hindus look at the subject practically, as it must be taken account of in the calculation of an eclipse, and calculate only the difference of the two parallaxes, which is *m* B M, or, what is virtually the same thing, M C *m*. The Sûrya-Siddhânta, however, as we shall see hereafter more plainly, takes no account of any case in which the line C S would not pass through M, that is to say, the moon's latitude is neglected, and her parallax calculated as if she were in the ecliptic.

We cite farther from the commentary, in illustration of the resolution of the parallax into parallax in longitude and parallax in latitude.

"Now by how many degrees, measured on the moon's sphere (*gola*), the line drawn from the earth's surface up to the sun cuts the moon's vertical circle (*dṛgvṛtta*) above the point occupied by the moon—this is, when the vertical circle and the ecliptic coincide, the moon's parallax in longitude (*lambana*). But when the ecliptic deviates from a vertical circle, then, to the point where the line from the earth's surface cuts the moon's sphere on the moon's vertical circle above the moon [i. e., to *m*, Fig. 25], draw through the pole of the ecliptic (*kadamba*) a circle [P′ *m n*′] north and south to the ecliptic on the moon's sphere [M *n*′]: and then the east and west interval [M *n*′] on the ecliptic between the point occupied by the moon [M] and the point where the circle as drawn cuts the ecliptic on the moon's sphere [*n*′] is the moon's true (*sphuṭa*) parallax in longitude, in minutes, and is the perpendicular (*koṭi*). And since the moon moves along with the ecliptic, the north and south interval, upon the circle we have drawn, between the ecliptic and the vertical circle [*m n*′] is, in minutes, the parallax in latitude (*nati*); which is the base (*bhuja*). The interval, in minutes, on the vertical circle [Z A], between the lines from the earth's centre and surface [*m* M], is the vertical parallax (*dṛglambana*), and the hypothenuse."

Fig. 25.

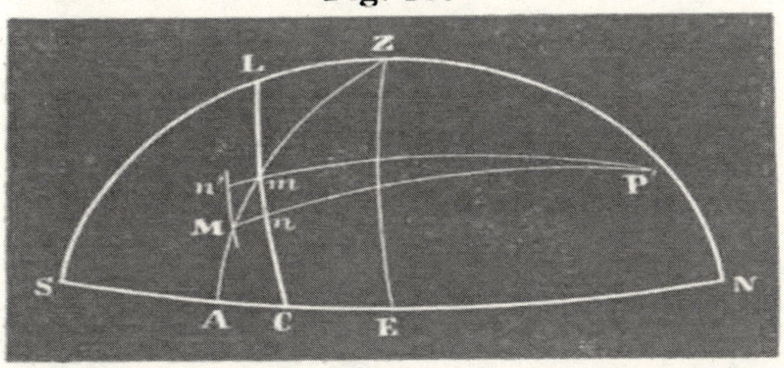

The conception here presented, it will be noticed, is that the moon's path, or the "ecliptic on the moon's sphere," is depressed away from C L, which might be called the "ecliptic on the sun's sphere," to an amount measured as latitude by *m n*′, and as longitude by *n*′ M. To our apprehension, *m n* M, rather than *m n*′ M, would be the triangle of resolution: the two are virtually equal.

The commentary then goes on farther to explain that when the vertical circle and the secondary to the ecliptic coincide, the parallax in longitude disappears, the whole vertical parallax becoming parallax in latitude: and again, when the vertical circle and the ecliptic coincide, the parallax in latitude disappears, the whole vertical parallax becoming parallax in longitude.

The term uniformly employed by the commentary, and more usually by the text, to express parallax in longitude, namely *lambana*, is from the same root which we have already more than once had occasion to notice (see above, under i. 25, 60), and means literally "hanging downward." In this verse, as once or twice later (vv. 14, 16), the text uses *harija*, which the commentary explains as equivalent to *kshitija*, "produced by the earth:" this does not seem very plausible, but we have nothing better to suggest. For parallax in latitude the text presents only the term *avanati*, "bending downward, depression:" the commentary always substitutes for it *nati*, which has nearly the same sense, and is the customary modern term.

2. How parallax in latitude arises by reason of the difference of place (*deça*) and time (*kâla*), and also parallax in longitude (*lambana*) from direction (*diç*) eastward or the contrary—that is now to be explained.

This distribution of the three elements of direction, place, and time, as causes respectively of parallax in longitude and in latitude, is somewhat arbitrary. The verse is to be taken, however, rather as a general introduction to the subject of the chapter, than as a systematic statement of the causes of parallax.

3. Calculate, by the equivalents in oblique ascension (*udayâsavas*) of the observer's place, the orient ecliptic-point (*lagna*) for the moment of conjunction (*parvavinâḍyas*): multiply the sine of its longitude by the sine of greatest declination, and divide by the sine of co-latitude (*lamba*): the result is the quantity known as the orient-sine (*udaya*).

The object of this first step in the rather tedious operation of calculating the parallax is to find for a given moment—here the moment of true conjunction—the sine of amplitude of that point of the ecliptic which is then upon the eastern horizon. In the first place the longitude of that point (*lagna*) is determined, by the data and methods taught above, in iii. 46–48, and which are sufficiently explained in the note to that passage: then its sine of amplitude is found, by a process which is a combination of that for finding the declination from the longitude, and that for finding the amplitude from the declination. Thus, by ii. 28,

R : sin gr. decl. : : sin long. : sin decl.

and, by iii. 22–23,

sin co-lat. : R : : sin decl. : sin ampl.

Hence, by combining terms, we have

sin co-lat. : sin gr. decl. : : sin long. : sin ampl.

This sine of amplitude receives the technical name of *udaya*, or *udayajyâ*: the literal meaning of *udaya* is simply "rising."

4. Then, by means of the equivalents in right ascension (*lankodayâsavas*), find the ecliptic-point (*lagna*) called that of the meridian (*madhya*): of the declination of that point and the lati-

tude of the observer take the sum, when their direction is the same; otherwise, take their difference.

5. The result is the meridian zenith-distance, in degrees (*natânçâs*): its sine is denominated the meridian-sine (*madhyajyâ*). . . .

The accompanying figure (Fig. 26) will assist the comprehension of this and the following processes. Let N E S W be a horizontal plane, N S the projection upon it of the meridian, and E W that of the prime vertical, Z being the zenith. Let C L T be the ecliptic. Then C is the orient ecliptic-point (*lagna*), and C D the sine of its amplitude (*udayajyâ*), found by the last process. The meridian ecliptic point (*madhyalagna*) is L: it is ascertained by the method prescribed in iii. 49, above. Its distance from the zenith is found from its declination and the latitude of the place of observation, as taught in iii. 20–22; and the sine of that distance, by which, in the figure, it is seen projected, is Z L: it is called by the technical name *madhyajyâ*, which we have translated "meridian-sine."

Fig. 26.

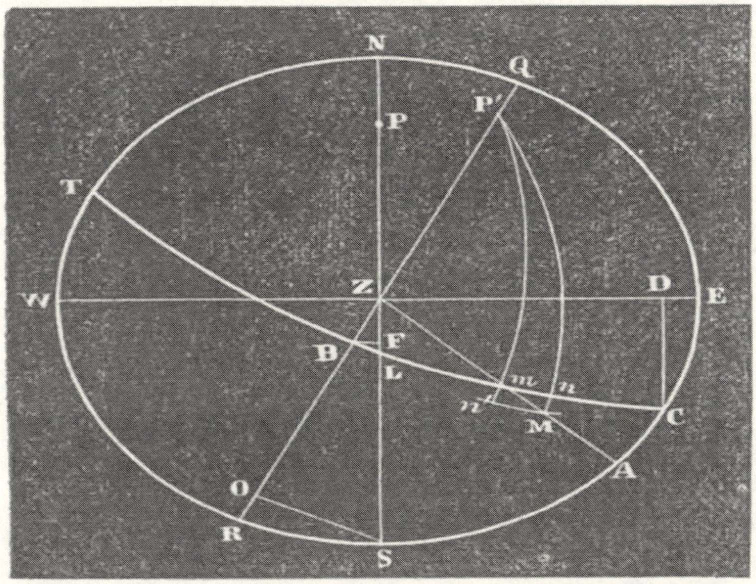

5. . . . Multiply the meridian-sine by the orient-sine, and divide by radius: square the result,

6. And subtract it from the square of the meridian-sine: the square root of the remainder is the sine of ecliptic zenith-distance (*dṛkkshepa*); the square root of the difference of the squares of that and radius is the sine of ecliptic-altitude (*dṛggati*).

Here we are taught how to find the sines of the zenith-distance and altitude respectively of that point of the ecliptic which has greatest altitude, or is nearest to the zenith, and which is also the central point of the portion of the ecliptic above the horizon: it is called by the commentary, as already noticed (see note to v. 1), *tribhonalagna*. Thus, in the last figure, if Q R be the vertical circle passing through the pole of the ecliptic, P′, and cutting the ecliptic, C T, in B, B is the central ecliptic-point (*tribhonalagna*), and the arcs seen projected in Z B and B R are its zenith-distance and altitude respectively. In order, now, to find the sine of Z B, we first find that of B L, and by the following process. C D is the orient-sine, already found. But since C Z and C P′ are quadrants, C is a pole of the vertical circle Q R, and C R is a quadrant. E S is also a quadrant: take away their common part C S, and C E remains equal to S R, and the sine of the latter, S O, is equal to that of the former, C D, the "orient-sine." Now, then, Z B L is treated as if it were a plane horizontal triangle, and similar to Z O S, and the proportion is made

ZS : SO : ZL : BL

or R : or.-sine : : mer.-sine : B L

This is so far a correct process, that it gives the true sine of the arc B L: for, by spherical trigonometry, in the spherical triangle Z B L, right-angled at B,

sin Z B L : sin B Z L : : sin arc Z L : sin arc B L

or R : S O : : Z L : sin B L

But the third side of a plane right-angled triangle of which the sines of the arcs Z B and Z L are hypothenuse and perpendicular, is not the sine of B L. If we conceive the two former sines to be drawn from Z, meeting in *b* and *l* respectively the lines drawn from B and L to the centre, then the line joining *b l* will be the third side, being plainly less than sin B L. Hence, on subtracting $\sin^2$ B L from $\sin^2$ Z L, and taking the square root of the remainder, we obtain, not sin Z B, but a less quantity, which may readily be shown, by spherical trigonometry, to be sin Z B cos B L. The value, then, of the sine of ecliptic zenith-distance (*dṛkkshepa*) as determined by this process, is always less than the truth, and as the corresponding cosine (*dṛggati*) is found by subtracting the square of the sine from that of radius, and taking the square root of the remainder, its value is always proportionally greater than the truth. This inaccuracy is noticed by the commentator, who points out correctly its reason and nature: probably it was also known to those who framed the rule, but disregarded, as not sufficient to vitiate the general character of the process: and it may, indeed, well enough pass unnoticed among all the other inaccuracies involved in the Hindu calculations of the parallax.

As regards the terms employed to express the sines of ecliptic zenith-distance and altitude, we have already met with the first member of each compound, *dṛç*, literally "sight," in other connected uses: as in *dṛgjyâ*, "sine of zenith-distance" (see above, iii. 33), *dṛgvṛtta*, "vertical-circle" (commentary to the first verse of this chapter): here it is combined with words which seem to be rather arbitrarily chosen, to form technical appellations for quantities used only in this process: the literal meaning of *kshepa* is "throwing, hurling;" of *gati*, "gait, motion."

7. The sine and cosine of meridian zenith-distance (*natânçâs*) are the approximate (*asphuṭa*) sines of ecliptic zenith-distance and altitude (*dṛkkshepa, dṛggati*). . . .

This is intended as an allowable simplification of the above process for finding the sines of ecliptic zenith-distance and altitude, by substituting for them other quantities to which they are nearly equivalent, and which are easier of calculation. These are the sines of zenith-distance and altitude of the meridian ecliptic-point (*madhyalagna*—L in Fig. 26) the former of which has already been made an element in the other process, under the name of "meridian-sine" (*madhyajyâ*). It might, indeed, from the terms of the text, be doubtful of what point the altitude and zenith-distance were to be taken; a passage cited by the commentator from Bhâskara's Siddhânta-Çiromaṇi (found on page 221 of the published edition of the Gaṇitâdhyâya) directs the sines of zenith-distance and altitude of B (*tribhonalagna*) when upon the meridian—that is to say, the sine and cosine of the arc Z F—to be substituted for those of Z B in a hasty process: but the value of the sine would in this case be too small, as in the other it was too great: and as the text nowhere directly recognizes the point B, and as directions have been given in verse 5 for finding the meridian zenith-distance of L, it seems hardly to admit of a doubt that the latter is the point to which the text here intends to refer.

Probably the permission to make this substitution is only meant to apply to cases where Z L is of small amount, or where C has but little amplitude.

7. . . . Divide the square of the sine of one sign by the sine called that of ecliptic-altitude (*dṛggatijîvâ*); the quotient is the "divisor" (*cheda*).

8. By this "divisor" divide the sine of the interval between the meridian ecliptic-point (*madhyalagna*) and the sun's place: the quotient is to be regarded as the parallax in longitude (*lambana*) of the sun and moon, eastward or westward, in nâdîs, etc.

The true nature of the process by which this final rule for finding the parallax in longitude is obtained is altogether hidden from sight under the form in which the rule is stated. Its method is as follows:

We have seen, in connection with the first verse of the preceding chapter, that the greatest parallaxes of the sun and moon are quite nearly equivalent to the mean motion of each during 4 nâḍis. Hence, were both bodies in the horizon, and the ecliptic a vertical circle, the moon would be depressed in her orbit below the sun to an amount equal to her excess in motion during 4 nâḍis. This, then, is the moon's greatest horizontal parallax in longitude. To find what it would be at any other point in the ecliptic, still considered as a vertical circle, we make the proportion

R : 4 (hor. par.) : : sin zen.-dist. : vert. parallax

This proportion is entirely correct, and in accordance with our modern rule that, with a given distance, the parallax of a body varies as the sine of its zenith-distance: whether the Hindus had made a rigorous demonstration of its truth, or whether, as in so many other cases, seeing that the parallax was greatest when the sine of zenith-distance was greatest, and nothing when this was nothing, they assumed it to vary in the interval as the sine of zenith-distance, saying "if, with a sine of zenith-distance which is equal to radius, the parallax is four nâḍis, with a given sine of zenith-distance what is it?"—this we will not venture to determine.

But now is to be considered the farther case in which the ecliptic is not a vertical circle, but is depressed below the zenith a certain distance, measured by the sine of ecliptic zenith-distance (*dṛkkshepa*), already found. Here again, noting that the parallax is all to be reckoned as parallax in longitude when the ecliptic is a vertical circle, or when the sine of ecliptic-altitude is greatest, and that it would be only parallax in latitude when the ecliptic should be a horizontal circle, or when the sine of ecliptic-altitude should be reduced to nothing, the Hindus assume it to vary in the interval as that sine, and accordingly make the proportion: "if, with a sine of ecliptic-altitude that is equal to radius, the parallax in longitude is equal to the vertical parallax, with any given sine of ecliptic-altitude what is it?"—or, inverting the middle terms,

R : sin ecl.-alt. : : vert. parallax : parallax in long.

But we had before

R : 4 : : sin zen.-dist. : vert. parallax

hence, by combining terms,

R^2 : 4 sin ecl.-alt. : : sin zen.-dist. : parallax in long.

For the third term of this proportion, now, is substituted the sine of the distance of the given point from the central ecliptic-point: that is to say, B *m* (Fig. 26) is substituted for Z *m*; the two are in fact of equal value only when they coincide, or else at the horizon, when each becomes a quadrant; but the error involved in the substitution is greatly lessened by the circumstance that, as it increases in proportional amount, the parallax in longitude itself decreases, until at B the latter is reduced to nullity, as is the vertical parallax at Z. The text, indeed, as in verses 1 and 9, puts *madhyalagna*, L, for *tribhonalagna*, B, in reckoning this distance: but the commentary, without ceremony or apology, reads the latter for the former. These substitutions being made, and the proportion being reduced to the form of an equation, we have

$$\text{par. in long.} = \frac{\text{sin dist.} \times 4 \text{ sin ecl.-alt.}}{R^2}$$

which reduces to

$$\frac{\text{sin dist.}}{R^2 \div 4 \text{ sin ecl.-alt.}} \quad \text{or} \quad \frac{\text{sin dist.}}{\frac{1}{4}R^2 \div \text{sin ecl.-alt.}}$$

and since $\frac{1}{4}R^2 = (\frac{1}{2}R)^2$, and $\frac{1}{2}R = \sin 30°$, we have finally

$$\text{par. in long.} = \frac{\text{sin dist.}}{\sin^2 30° \div \text{sin ecl.-alt.}}$$

which is the rule given in the text. To the denominator of the fraction, in its final form, is given the technical name of *cheda*, "divisor," which word we have had before similarly used, to designate one of the factors in a complicated operation (see above, iii. 35, 38).

We will now examine the correctness of the second principal proportion from which the rule is deduced. It is, in terms of the last figure (Fig. 26),

R : sin Z P′ (= B R) : : *m* M : *m n*

Assuming the equality of the little triangles M *m n* and M *m n′*, and accordingly that of the angles *m* M *n* and M *m n′*, which latter equals Z *m* P′, we have, by spherical trigonometry, as a true proportion,

sin *m n′* M : sin M *m n′* : : *m* M : *m n′*

or

R : sin Z *m* P′ : : *m* M : *m n*

Hence the former proportion is correct only when sin Z P′ and sin Z *m* P′ are equal; that is to say, when Z P′ measures the angle Z *m* P′;

and this can be the case only when **Z *m***, as well as **P′ *m***, is a quadrant, or when *m* is on the horizon. Here again, however, precisely as in the case last noticed, the importance of the error is kept within very narrow limits by the fact that, as its relative consequence increases, the amount of the parallax in longitude affected by it diminishes.

9. When the sun's longitude is greater than that of the meridian ecliptic-point (*madhyalagna*), subtract the parallax in longitude from the end of the lunar day; when less, add the same: repeat the process until all is fixed.

The text so pertinaciously reads "meridian ecliptic-point" (*madhyalagna*) where we should expect, and ought to have, "central ecliptic-point" (*tribhonalagna*), that we are almost ready to suspect it of meaning to designate the latter point by the former name. It is sufficiently clear that, whenever the sun and moon are to the eastward of the central ecliptic-point, the effect of the parallax in longitude will be to throw the moon forward on her orbit beyond the sun, and so to cause the time of apparent to precede that of real conjunction; and the contrary. Hence, in the eastern hemisphere, the parallax, in time, is subtractive, while in the western it is additive. But a single calculation and application of the correction for parallax is not enough; the moment of apparent conjunction must be found by a series of successive approximations: since if, for instance, the moment of true conjunction is $25^n\ 2^v$, and the calculated parallax in longitude for that moment is $2^n\ 21^v$, the apparent end of the lunar day will not be at $27^n\ 23^v$, because at the latter time the parallax will be greater than $2^n\ 21^v$, deferring accordingly still farther the time of conjunction; and so on. The commentary explains the method of procedure more fully, as follows: for the moment of true conjunction in longitude calculate the parallax in longitude, and apply it to that moment: for the time thus found calculate the parallax anew, and apply it to the moment of true conjunction: again, for the time found as the result of this process, calculate the parallax, and apply it as before; and so proceed, until a moment is arrived at, at which the difference in actual longitude, according to the motions of the two planets, will just equal and counterbalance the parallax in longitude.

The accuracy of this approximative process cannot but be somewhat impaired by the circumstance that, while the parallax is reckoned in difference of mean motions, the corrections of longitude must be made in true motions. Indeed, the reckoning of the horizontal parallax in time as 4 nâḍîs, whatever be the rate of motion of the sun and moon, is one of the most palpable among the many errors which the Hindu process involves.

To ascertain the moment of apparent conjunction in longitude, only the parallax in longitude requires to be known; but to determine the time of occurrence of the other phases of the eclipse, it is necessary to take into account the parallax in latitude, the ascertainment of which is accordingly made the subject of the next rule.

10. If the sine of ecliptic zenith-distance (*dṛkkshepa*) be multiplied by the difference of the mean motions of the sun and

moon, and divided by fifteen times radius, the result will be the parallax in latitude (*avanati*).

As the sun's greatest parallax is equal to the fifteenth part of his mean daily motion, and that of the moon to the fifteenth part of hers (see note to iv. 1, above), the excess of the moon's parallax over that of the sun is equal, when greatest, to one fifteenth of the difference of their respective mean daily motions. This will be the value of the parallax in latitude when the ecliptic coincides with the horizon, or when the sine of ecliptic zenith-distance becomes equal to radius. On the other hand, the parallax in latitude disappears when this same sine is reduced to nullity. Hence it is to be regarded as varying with the sine of ecliptic zenith-distance, and, in order to find its value at any given point, we say "if, with a sine of ecliptic zenith-distance which is equal to radius, the parallax in latitude is one fifteenth of the difference of mean daily motions, with a given sine of ecliptic zenith-distance what is it?" or

R : diff. of mean m. $\div$ 15 : : sin ecl. zen.-dist. : parallax in lat.

This proportion, it is evident, would give with entire correctness the parallax at the central ecliptic-point (B in Fig. 26), where the whole vertical parallax is to be reckoned as parallax in latitude. But the rule given in the text also assumes that, with a given position of the ecliptic, the parallax in latitude is the same at any point in the ecliptic. Of this the commentary offers no demonstration, but it is essentially true. For, regarding the little triangle M m n as a plane triangle, right-angled at n, and with its angle n m M equal to the angle Z m B, we have

$$\mathrm{R} : \sin \mathrm{Z}m\mathrm{B} :: \mathrm{M}m : \mathrm{M}n$$

But, in the spherical triangle Z m B, right-angled at B,

$$\mathrm{R} : \sin \mathrm{Z}m\mathrm{B} :: \sin \mathrm{Z}m : \sin \mathrm{ZB}$$

Hence, by equality of ratios,

$$\sin \mathrm{Z}m : \sin \mathrm{ZB} :: \mathrm{M}m : \mathrm{M}n$$

But, as before shown,

$$\mathrm{R} : \sin \mathrm{Z}m :: \text{gr. parallax} : \mathrm{M}m$$

Hence, by combining terms,

$$\mathrm{R} : \sin \mathrm{ZB} :: \text{gr. parallax} : \mathrm{M}n$$

That is to say, whatever be the position of m, the point for which the parallax in latitude is sought, this will be equal to the product of the greatest parallax into the sine of ecliptic zenith-distance, divided by radius: or, as the greatest parallax equals the difference of mean motions divided by fifteen,

$$\text{par. in lat.} = \frac{\text{sin ecl. zen.-dist.} \times \text{diff. of m. m.} \div 15}{\mathrm{R}} \quad \text{or} \quad \frac{\text{sin ecl. zen.-dist.} \times \text{diff. of m. m.}}{\mathrm{R} \times 15}$$

The next verse teaches more summary methods of arriving at the same quantity.

11. Or, the parallax in latitude is the quotient arising from dividing the sine of ecliptic zenith-distance (*dṛkkshepa*) by sev-

enty, or, from multiplying it by forty-nine, and dividing it by radius.

In the expression given above for the value of the parallax in latitude, all the terms are constant excepting the sine of ecliptic zenith-distance. The difference of the mean daily motions is 731′ 27″, and fifteen times radius is 51,570′. Now 731′ 27″÷51,570′ equals $\frac{1}{70.50}$ or 48.77÷R; to which the expressions given in the text are sufficiently near approximations.

12. The parallax in latitude is to be regarded as south or north according to the direction of the meridian-sine (*madhyajyâ*). When it and the moon's latitude are of like direction, take their sum; otherwise, their difference:

13. With this calculate the half-duration (*sthiti*), half total obscuration (*vimarda*), amount of obscuration (*grâsa*), etc., in the manner already taught; likewise the scale of projection (*pramâṇa*), the deflection (*valana*), the required amount of obscuration, etc., as in the case of a lunar eclipse.

In ascertaining the true time of occurrence of the various phases of a solar eclipse, as determined by the parallax of the given point of observation, we are taught first to make the whole correction for parallax in latitude, and then afterward to apply that for parallax in longitude. The former part of the process is succinctly taught in verses 12 and 13: the rules for the other follow in the next passage. The language of the text, as usual, is by no means so clear and explicit as could be wished. Thus, in the case before us, we are not taught whether, as the first step in this process of correction, we are to calculate the moon's parallax in latitude for the time of true conjunction (*tithyanta*, "end of the lunar day"), or for that of apparent conjunction (*madhyagrahaṇa*, "middle of the eclipse"). It might be supposed that, as we have thus far only had in the text directions for finding the sine and cosine of ecliptic zenith-distance at the moment of true conjunction, the former of them was to be used in the calculations of verses 10 and 11, and the result from it, which would be the parallax at the moment of true conjunction, applied here as the correction needed. Nor, so far as we have been able to discover, does the commentator expound what is the true meaning of the text upon this point. It is sufficiently evident, however, that the moment of apparent conjunction is the time required. We have found, by a process of successive approximation, at what time (see Fig. 25), the moon (her latitude being neglected) being at *m* and the sun at *n*, the parallax in longitude and the difference of true longitude will both be the same quantity, *m n*, and so, when apparent conjunction will take place. Now, to know the distance of the two centres at that moment, we require to ascertain the parallax in latitude, *n* M, for the moon at *m*, and to apply it to the moon's latitude when in the same position, taking their sum when their direction is the same, and their difference when their direction is different, as prescribed by the text; the net result will be the distance required. The commentary, it may be remarked, expressly states that the moon's latitude is to be calculated in this operation for the time of apparent conjunction (*madhyagrahaṇa*). The distance thus found will determine the amount of greatest obscuration, and the character of the eclipse, as taught in verse 10 of the preceding chapter. It is then farther to be taken as the foundation of precisely such a process as that described in verses 12–15 of the same chapter, in order to ascertain the half-time of duration, or of total obscuration: that is to say, the distance in latitude of the two centres being first assumed as invariable through the whole duration of the eclipse, the half-time of duration, and the resulting moments of contact and separation are to be ascertained: for these moments the latitude and parallax in latitude are to be calculated anew, and by them a new determination of the times of contact and separation is to be made, and so on, until these are fixed with the degree of accuracy required. If the eclipse be total, a similar operation must be gone through with to ascertain the moments of immersion and emergence. No account is made, it will be noticed, of the possible occurrence of an annular eclipse.

The intervals thus found, after correction for parallax in latitude only, between the middle of the eclipse and the moments of contact and separation respectively, are those which are called in the last chapter (vv. 19, 23), the "mean half-duration" (*madhyasthityardha*).

In this process for finding the net result, as apparent latitude, of the actual latitude and the parallax in latitude, is brought out with distinctness the inaccuracy already alluded to; that, whatever be the moon's actual latitude, her parallax is always calculated as if she were in the ecliptic. In an eclipse, however, to which case alone the Hindu processes are intended to be applied, the moon's latitude can never be of any considerable amount.

The propriety of determining the direction of the parallax in latitude by means of that of the meridian-sine (Z L in Fig. 26), of which the direction is established as south or north by the process of its calculation, is too evident to call for remark.

In verse 13 is given a somewhat confused specification of matters which are, indeed, affected by the parallax in latitude, but in different modes and degrees. The amount of greatest obscuration, and the (mean) half-times of duration and total obscuration, are the quantities directly dependent upon the calculation of that parallax, as here presented: to find the amount of obscuration at a given moment—as also the time corresponding to a given amount of obscuration—we require to know also the true half-duration, as found by the rules stated in the following passage: while the scale of projection and the deflection are affected by parallax only so far as this alters the time of occurrence of the phases of the eclipse.

14. For the end of the lunar day, diminished and increased by the half-duration, as formerly, calculate again the parallax in longitude for the times of contact (*grâsa*) and of separation (*moksha*), and find the difference between these and the parallax in longitude (*harija*) for the middle of the eclipse.

15. If, in the eastern hemisphere, the parallax in longitude for the contact is greater than that for the middle, and that for

the separation less; and if, in the western hemisphere, the contrary is the case—

16. Then the difference of parallax in longitude is to be added to the half-duration on the side of separation, and likewise on that of contact (*pragrahaṇa*); when the contrary is true, it is to be subtracted.

17. These rules are given for cases where the two parallaxes are in the same hemisphere: where they are in different hemispheres, the sum of the parallaxes in longitude is to be added to the corresponding half-duration. The principles here stated apply also to the half-time of total obscuration.

We are supposed to have ascertained, by the preceding process, the true amount of apparent latitude at the moments of first and last contact of the eclipsed and eclipsing bodies, and consequently to have determined the dimensions of the triangle—corresponding, in a solar eclipse, to C G P, Fig. 21, in a lunar—made up of the latitude, the distance in longitude, and the sum of the two radii. The question now is how the duration of the eclipse will be affected by the parallax in longitude. If this parallax remained constant during the continuance of the eclipse, its effect would be nothing; and, having once determined by it the time of apparent conjunction, we should not need to take it farther into account. But it varies from moment to moment, and the effect of its variation is to prolong the duration of every part of a visible eclipse. For, to the east of the central ecliptic-point, it throws the moon's disk forward upon that of the sun, thus hastening the occurrence of all the phases of the eclipse, but by an amount which is all the time decreasing, so that it hastens the beginning of the eclipse more than the middle, and the middle more than the close: to the west of that same point, on the other hand, it depresses the moon's disk away from the sun's, but by an amount constantly increasing, so that it retards the end of the eclipse more than its middle, and its middle more than its beginning. The effect of the parallax in longitude, then, upon each half-duration of the eclipse, will be measured by the difference between its retarding and accelerating effects upon contact and conjunction, and upon conjunction and separation, respectively: and the amount of this difference will always be additive to the time of half-duration as otherwise determined. If, however, contact and conjunction, or conjunction and separation, take place upon opposite sides of the point of no parallax in longitude, then the sum of the two parallactic effects, instead of their difference, will be to be added to the corresponding half-duration: since the one, on the east, will hasten the occurrence of the former phase, while the other, on the west, will defer the occurrence of the latter phase. The amount of the parallax in longitude for the middle of the eclipse has already been found; if, now, we farther determine its amount—reckoned, it will be remembered, always in time—for the moments of contact and separation, and add the difference or the sum of each of these and the parallax for the moment of conjunction to the corresponding half-duration as previously determined, we shall have the true times of half-duration. In order to find the parallax for contact and separation, we repeat the same process (see above, v. 9) by which that for conjunction was found: as we then started from the moment of true conjunction, and, by a series of successive approximations, ascertained the time when the difference of longitude would equal the parallax in longitude, so now we start from two moments removed from that of true conjunction by the equivalents in time of the two distances in longitude obtained by the last process, and, by a similar series of successive approximations, ascertain the times when the differences of longitude, together with the parallax, will equal those distances in longitude.

In the process, as thus conducted, there is an evident inaccuracy. It is not enough to apply the whole correction for parallax in latitude, and then that for parallax in longitude, since, by reason of the change effected by the latter in the times of contact and separation, a new calculation of the former becomes necessary, and then again a new calculation of the latter, and so on, until, by a series of doubly compounded approximations, the true value of each is determined. This was doubtless known to the framers of the system, but passed over by them, on account of the excessively laborious character of the complete calculation, and because the accuracy of such results as they could obtain was not sensibly affected by its neglect.

The question naturally arises, why the specifications of verse 15 are made hypothetical instead of positive, and why, in the latter half of verse 16, a case is supposed which never arises. The commentator anticipates this objection, and takes much pains to remove it: it is not worth while to follow his different pleas, which amount to no real explanation, saving to notice his last suggestion, that, in case an eclipse begins before sunrise, the parallax for its earlier phase or phases, as calculated according to the distance in time from the lower meridian, may be less than for its later phases—and the contrary, when the eclipse ends after sunset. This may possibly be the true explanation, although we are justly surprised at finding a case of so little practical consequence, and to which no allusion has been made in the previous processes, here taken into account.

The text, it may be remarked, by its use of the terms "eastern and western hemispheres" (*kapâla,* literally "cup, vessel"), repeats once more its substitution of the meridian ecliptic-point (*madhyalagna*) for the central ecliptic-point (*tribhonalagna*), as that of no parallax in longitude; the meridian forming the only proper and recognized division of the heavens into an eastern and a western hemisphere.

We are now prepared to see the reason of the special directions given in verses 19 and 23 of the last chapter, respecting the reduction, in a solar eclipse, of distance in time from the middle of the eclipse to distance in longitude of the two centres. The "mean half-duration" (*madhyasthityardha*) of the eclipse is the time during which the true distance of the centres at the moments of contact or separation, as found by the process prescribed in verses 12 and 13 of this chapter, would be gained by the moon with her actual excess of motion, leaving out of account the variation of parallax in longitude: the "true half-duration" (*sphuṭasthityardha*) is the increased time in which, owing to that variation, the same distance in longitude is actually gained by the moon;

the effect of the parallax being equivalent either to a diminution of the moon's excess of motion, or to a protraction of the distance of the two centers—both of them in the ratio of the true to the mean half-duration. If then, for instance, it be required to know what will be the amount of obscuration of the sun half an hour after the first contact, we shall first subtract this interval from the true half-duration before conjunction; the remainder will be the actual interval to the middle of the eclipse: this interval, then, we shall reduce to its value as distance in longitude by diminishing it, either before or after its reduction to minutes of arc, in the ratio of the true to the mean half-duration. The rest of the process will be performed precisely as in the case of an eclipse of the moon.

Notwithstanding the ingenuity and approximate correctness of many of the rules and methods of calculation taught in this chapter, the whole process for the ascertainment of parallax contains so many elements of error that it hardly deserves to be called otherwise than cumbrous and bungling. The false estimate of the difference between the sun's and moon's horizontal parallax—the neglect, in determining it, of the variation of the moon's distance—the estimation of its value in time made always according to mean motions, whatever be the true motions of the planets at the moment—the neglect, in calculating the amount of parallax, of the moon's latitude—these, with all the other inaccuracies of the processes of calculation which have been pointed out in the notes, render it impossible that the results obtained should ever be more than a rude approximation to the truth.

In farther illustration of the subject of solar eclipses, as exposed in this and the preceding chapters, we present, in the Appendix, a full calculation of the eclipse of May 26th, 1854, mainly as made for the translator, during his residence in India, by a native astronomer.

CHAPTER VI.

OF THE PROJECTION OF ECLIPSES.

CONTENTS:—1, value of a projection; 2–4, general directions; 5–6, how to lay off the deflection and latitude for the beginning and end of the eclipse; 7, to exhibit the points of contact and separation; 8–10, how to lay off the deflection and latitude for the middle of the eclipse; 11, to show the amount of greatest obscuration; 12, reversal of directions in the western hemisphere; 13, least amount of obscuration observable; 14–16, to draw the path of the eclipsing body; 17–19, to show the amount of obscuration at a given time; 20–22, to exhibit the points of immersion and emergence in a total eclipse; 23, color of the part of the moon obscured; 24, caution as to communicating a knowledge of these matters.

1. Since, without a projection (*chedyaka*), the precise (*sphuṭa*) differences of the two eclipses are not understood, I shall proceed to explain the exalted doctrine of the projection.

The term *chedyaka* is from the root *chid*, "split, divide, sunder," and indicates, as here applied, the instrumentality by which distinctive differences are rendered evident. The name of the chapter, *parilekhâdhikâra*, is not taken from this word, but from *parilekha*, "delineation, figure," which occurs once below, in the eighth verse.

2. Having fixed, upon a well prepared surface, a point, describe from it, in the first place, with a radius of forty-nine digits (*angula*), a circle for the deflection (*valana*):

3. Then a second circle, with a radius equal to half the sum of the eclipsed and eclipsing bodies; this is called the aggregate-circle (*samâsa*); then a third, with a radius equal to half the eclipsed body.

4. The determination of the directions, north, south, east, and west, is as formerly. In a lunar eclipse, contact (*grahaṇa*) takes place on the east, and separation (*moksha*) on the west; in a solar eclipse, the contrary.

The larger circle, drawn with a radius of about three feet, is used solely in laying off the deflection (*valana*) of the ecliptic from an east and west circle. We have seen above (iv. 24, 25) that the sine of this deflection was reduced to its value in a circle of forty-nine digits' radius, by dividing by seventy its value in minutes. The second circle is employed (see below, vv. 6, 7) in determining the points of contact and separation. The third represents the eclipsed body itself, always maintaining a fixed position in the centre of the figure, even though, in a lunar eclipse, it is the body which itself moves, relatively to the eclipsing shadow. For the scale by which the measures of the eclipsed and eclipsing bodies, the latitudes, etc., are determined, see above, iv. 26.

The method of laying down the cardinal directions is the same with that used in constructing a dial; it is described in the first passage of the third chapter (iii. 1–4).

The specifications of the latter half of verse 4 apply to the eclipsed body, designating upon which side of it obscuration will commence and terminate.

5. In a lunar eclipse, the deflection (*valana*) for the contact is to be laid off in its own proper direction, but that for the separation in reverse; in an eclipse of the sun, the contrary is the case.

The accompanying figure (Fig. 27) will illustrate the Hindu method of exhibiting, by a projection, the various phases of an eclipse. Its conditions are those of the lunar eclipse of Feb. 6th, 1860, as determined by the data and methods of this treatise: for the calculation see the Appendix. Let M be the centre of the figure and the place of the moon, and let N S and E W be the circles of direction drawn through the moon's centre; the former representing (see above, under iv. 24, 25) a great circle drawn through the north and south points of the horizon, the latter a small circle parallel to the prime vertical. In explanation of the manner in which these directions are presented by the figure, we would remark that we have adapted it to a supposed position of the

observer on the north side of his projection, as at N, and looking southward—a position which, in our latitude, he would naturally assume, for

Fig. 27.

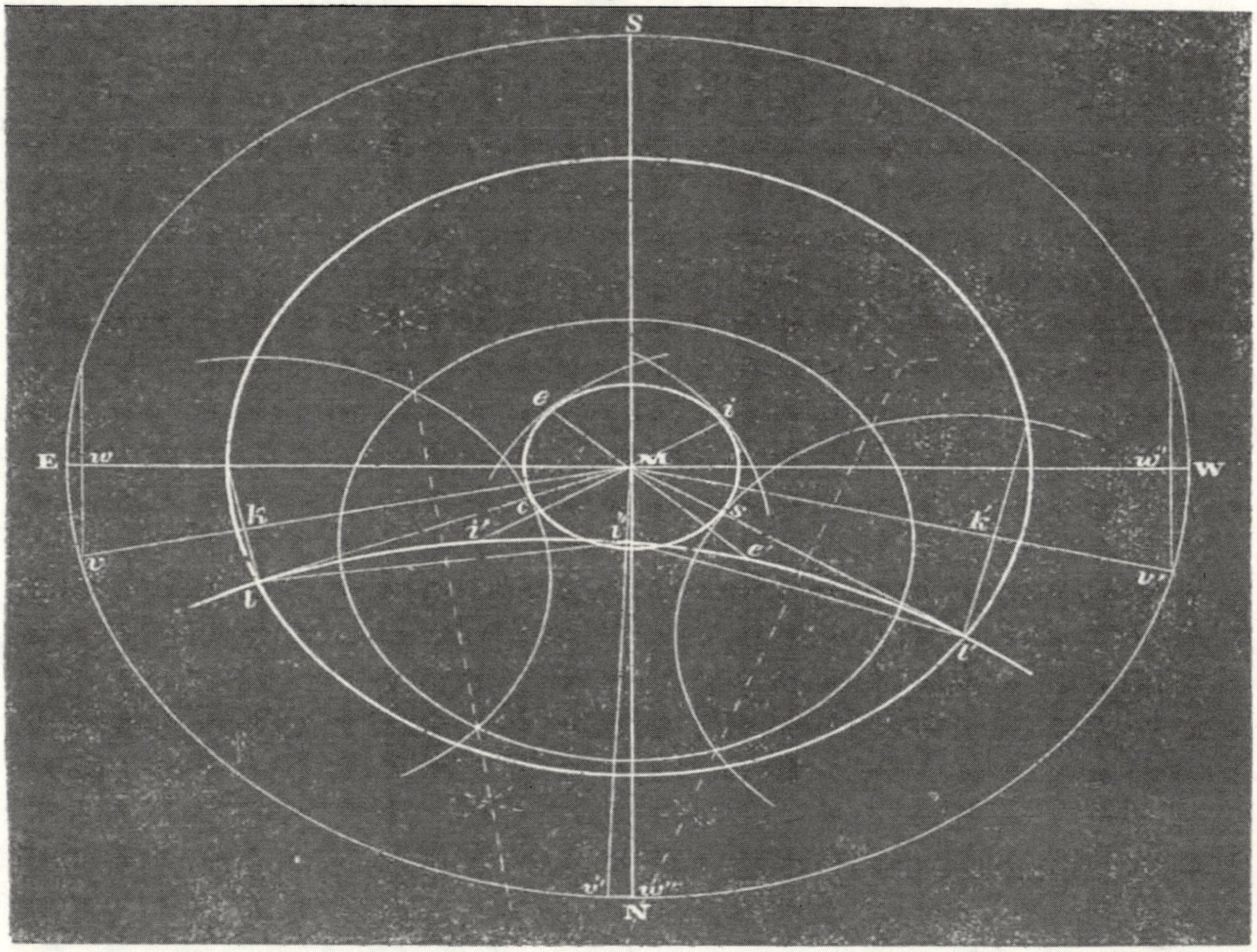

the purpose of comparing the actual phases of the eclipse, as they occurred, with his delineation of them. The heavier circle, *l l'*, is that drawn with the sum of the semi-diameters, or the "aggregate-circle;" while the outer one, N E S W, is that for the deflection. This, in order to reduce the size of the whole figure, we have drawn upon a scale very much smaller than that prescribed; its relative dimensions being a matter of no consequence whatever, provided the sine of the deflection be made commensurate with its radius. In our own, or the Greek, method of laying off an arc, by its angular value, the radius of the circle of deflection would also be a matter of indifference: the Hindus, ignoring angular measurements, adopt the more awkward and bungling method of laying off the arc by means of its sine. Let *v w* equal the deflection, calculated for the moment of contact, expressed as a sine, and in terms of a circle in which E M is radius. Now, as the moon's contact with the shadow takes place upon her eastern limb, the deflection for the contact must be laid off from the east point of the circle; and, as the calculated direction of the deflection indicates in what way the ecliptic is pointing eastwardly, it must be laid off from E in its own proper di-

rection. In the case illustrated, the deflection for the contact is north: hence we lay it off northward from E, and then the line drawn from M to *v*, its extremity—which line represents the direction of the ecliptic at the moment—points northward. Again, upon the side of separation—which, for the moon, is the western side—we lay off the deflection for the moment of separation: but we lay it off from W in the reverse of its true direction, in order that the line from its extremity to the centre may truly represent the direction of the ecliptic. Thus, in the eclipse figured, the deflection for separation is south; we lay it off northward from W, and then the line *v′* M points, toward M, southward. In a solar eclipse, in which, since the sun's western limb is the first eclipsed, the deflection for contact must be laid off from W, and that for separation from E, the direction of the former requires to be reversed, and that of the latter to be maintained as calculated.

6. From the extremity of either deflection draw a line to the centre: from the point where that cuts the aggregate-circle (*samâsa*) are to be laid off the latitudes of contact and of separation.

7. From the extremity of the latitude, again, draw a line to the central point: where that, in either case, touches the eclipsed body, there point out the contact and separation.

8. Always, in a solar eclipse, the latitudes are to be drawn in the figure (*parilekha*) in their proper direction; in a lunar eclipse, in the opposite direction. . . .

The lines *v* M and *v′* M, drawn from *v* and *v′*, the extremities of the sines or arcs which measure the deflection, to the centre of the figure, represent, as already noticed, the direction of the ecliptic with reference to an east and west line at the moments of contact and separation. From them, accordingly, and at right angles to them, are to be laid off the values of the moon's latitude at those moments. Owing, however, to the principle adopted in the projection, of regarding the eclipsed body as fixed in the centre of the figure, and the eclipsing body as passing over it, the lines *v* M and *v′* M do not, in the case of a lunar eclipse, represent the ecliptic itself, in which is the centre of the shadow, but the small circle of latitude, in which is the moon's centre: hence, in laying off the moon's latitude to determine the centre of the shadow, we reverse its direction. Thus, in the case illustrated, the moon's latitude is always south: we lay off, then, the lines *k l* and *k′ l′*, representing its value at the moments of contact and separation, northward: they are, like the deflection, drawn as sines, and in such manner that their extremities, *l* and *l′*, are in the aggregate-circle: then, since *l* M and *l′* M are each equal to the sum of the two semi-diameters, and *l k* and *l′ k′* to the latitudes, *k* M and *k′* M will represent the distances of the centres in longitude, and *l* and *l′* the places of the centre of the shadow, at contact and separation: and upon describing circles from *l* and *l′*, with radii equal to the semi-diameter of the shadow, the points *c* and *s*, where these touch the disk of the moon, will be the points of first and last contact: *c* and *s* being also, as stated in the text, the points where *l* M and *l′* M meet the circumference of the disk of the eclipsed body.

8\. . . . In accordance with this, then, for the middle of the eclipse,

9\. The deflection is to be laid off—eastward, when it and the latitude are of the same direction; when they are of different directions, it is to be laid off westward: this is for a lunar eclipse; in a solar, the contrary is the case.

10\. From the end of the deflection, again, draw a line to the central point, and upon this line of the middle lay off the latitude, in the direction of the deflection.

11\. From the extremity of the latitude describe a circle with a radius equal to half the measure of the eclipsing body: whatever of the disk of the eclipsed body is enclosed within that circle, so much is swallowed up by the darkness (*tamas*).

The phraseology of the text in this passage is somewhat intricate and obscure; it is fully explained by the commentary, as, indeed, its meaning is also deducible with sufficient clearness from the conditions of the problem sought to be solved. It is required to represent the deflection of the ecliptic from an east and west line at the moment of greatest obscuration, and to fix the position of the centre of the eclipsing body at that moment. The deflection is this time to be determined by a secondary to the ecliptic, drawn from near the north or south point of the figure. The first question is, from which of these two points shall the deflection be laid off, and the line to the centre drawn. Now since, according to verse 10, the latitude itself is to be measured upon the line of deflection, the latter must be drawn southward or northward according to the direction in which the latitude is to be laid off. And this is the meaning of the last part of verse 8; "in accordance," namely, with the direction in which, according to the previous part of the verse, the latitude is to be drawn. But again, in which direction from the north or south point, as thus determined, shall the deflection be measured? This must, of course, be determined by the direction of the deflection itself: if south, it must obviously be measured east from the north point and west from the south point; if north, the contrary. The rules of the text are in accordance with this, although the determining circumstance is made to be the agreement or non-agreement, in respect to direction, of the deflection with the moon's latitude—the latter being this time reckoned in its own proper direction, and not, in a lunar eclipse, reversed. Thus, in the case for which the figure is drawn, as the moon's latitude is south, and must be laid off northward from M, the deflection, $v'' w''$, is measured from the north point; as deflection and latitude are both south, it is measured east from N. In an eclipse of the sun, on the other hand, the moon's latitude would, if north, be laid off northward, as in the figure, and hence also, the deflection would be measured from the north point: but it would be measured eastward, if its own direction were south, or disagreed with that of the latitude.

The line of deflection, which is M v'' in the figure, being drawn, and having the direction of a perpendicular to the ecliptic at the moment of opposition, the moon's latitude for that moment, M l'', is laid off directly upon it. The point l'' is, accordingly, the position of the centre of the shadow at the middle of the eclipse, and if from that centre, with a radius equal to the semi-diameter of the eclipsing body, a circle be drawn, it will include so much of the disk of the eclipsed body as is covered when the obscuration is greatest. In the figure the eclipse is shown as total, the Hindu calculations making it so, although, in fact, it is only a partial eclipse.

12. By the wise man who draws the projection (*chedyaka*), upon the ground or upon a board, a reversal of directions is to be made in the eastern and western hemispheres.

This verse is inserted here in order to remove the objection that, in the eastern hemisphere, indeed, all takes place as stated, but, if the eclipse occurs west of the meridian, the stated directions require to be all of them reversed. In order to understand this objection, we must take notice of the origin and literal meaning of the Sanskrit words which designate the cardinal directions. The face of the observer is supposed always to be eastward: then "east" is *prâñc*, "forward, toward the front"; "west" is *paçcât*, "backward, toward the rear": "south" is *dakshiṇa*, "on the right"; "north" is *uttara*, "upward" (i. e., probably, toward the mountains, or up the course of the rivers in north-western India). These words apply, then, in etymological strictness, only when one is looking eastward—and so, in the present case, only when the eclipse is taking place in the eastern hemisphere, and the projector is watching it from the west side of his projection, with the latter before him: if, on the other hand, he removes to E, turning his face westward, and comparing the phenomena as they occur in the western hemisphere with his delineation of them, then "forward" (*prâñc*) is no longer east, but west; "right" (*dakshiṇa*) is no longer south, but north, etc.

It is unnecessary to point out that this objection is one of the most frivolous and hair-splitting character, and its removal by the text a waste of trouble: the terms in question have fully acquired in the language an absolute meaning, as indicating directions in space, without regard to the position of the observer.

13. Owing to her clearness, even the twelfth part of the moon, when eclipsed (*grasta*), is observable; but, owing to his piercing brilliancy, even three minutes of the sun, when eclipsed, are not observable.

The commentator regards the negative which is expressed in the latter half of this verse as also implied in the former, the meaning being that an obscuration of the moon's disk extending over only the twelfth part of it does not make itself apparent. We have preferred the interpretation given above, as being better accordant both with the plain and simple construction of the text and with fact.

14. At the extremities of the latitudes make three points, of corresponding names; then, between that of the contact and

that of the middle, and likewise between that of the separation and that of the middle,

15. Describe two fish-figures (*matsya*): from the middle of these having drawn out two lines projecting through the mouth and tail, wherever their intersection takes place,

16. There, with a line touching the three points, describe an arc: that is called the path of the eclipsing body, upon which the latter will move forward.

The deflection and the latitude of three points in the continuance of the eclipse having been determined and laid down upon the projection, it is deemed unnecessary to take the same trouble with regard to any other points, these three being sufficient to determine the path of the eclipsing body: accordingly, an arc of a circle is drawn through them, and is regarded as representing that path. The method of describing the arc is the same with that which has already been more than once employed (see above, iii. 1–4, 41–42): it is explained here with somewhat more fullness than before. Thus, in the figure, l, l'', and l' are the three extremities of the moon's latitude, at the moments of contact, opposition, and separation, respectively: we join $l\,l''$, $l''\,l'$, and upon these lines describe fish-figures (see note to iii. 1–5); their two extremities ("mouth" and "tail") are indicated by the intersecting dotted lines in the figure: then, at the point, not included in the figure, where the lines drawn through them meet one another, is the centre of a circle passing through l, l'', and l'.

17. From half the sum of the eclipsed and eclipsing bodies subtract the amount of obscuration, as calculated for any given time: take a little stick equal to the remainder, in digits, and, from the central point,

18. Lay it off toward the path upon either side—when the time is before that of greatest obscuration, toward the side of contact; when the obscuration is decreasing, in the direction of separation—and where the stick and the path of the eclipsing body

19. Meet one another, from that point describe a circle with a radius equal to half the eclipsing body: whatever of the eclipsed body is included within it, that point out as swallowed up by the darkness (*tamas*).

20. Take a little stick equal to half the difference of the measures (*mâna*), and lay it off in the direction of contact, calling it the stick of immersion (*nimîlana*): where it touches the path,

21. From that point, with a radius equal to half the eclipsing body, draw a circle, as in the former case: where this meets the circle of the eclipsed body, there immersion takes place.

22. So also for the emergence (*unmîlana*), lay it off in the direction of separation, and describe a circle, as before: it will show the point of emergence in the manner explained.

The method of these processes is so clear as to call for no detailed explanation. The centre of the eclipsing body being supposed to be always in the arc *l l″ l′*, drawn as directed in the last passage, we have only to fix a point in this arc which shall be at a distance from M corresponding to the calculated distance of the centres at the given time, and from that point to describe a circle of the dimensions of the eclipsed body, and the result will be a representation of the then phase of the eclipse. If the point thus fixed be distant from M by the difference of the two semi-diameters, as M *i′*, M *e′*, the circles described will touch the disk of the eclipsed body at the points of immersion and emergence, *i* and *e*.

23. The part obscured, when less than half, will be dusky (*sadhûmra*); when more than half, it will be black; when emerging, it is dark copper-color (*kṛshṇatâmra*); when the obscuration is total, it is tawny (*kapila*).

The commentary adds the important circumstance, omitted in the text, that the moon alone is here spoken of; no specification being added with reference to the sun, because, in a solar eclipse, the part obscured is always black.

A more suitable place might have been found for this verse in the fourth chapter, as it has nothing to do with the projection of an eclipse.

24. This mystery of the gods is not to be imparted indiscriminately: it is to be made known to the well-tried pupil, who remains a year under instruction.

The commentary understands by this mystery, which is to be kept with so jealous care, the knowledge of the subject of this chapter, the delineation of an eclipse, and not the general subject of eclipses, as treated in the past three chapters. It seems a little curious to find a matter of so subordinate consequence heralded so pompously in the first verse of the chapter, and guarded so cautiously at its close.

CHAPTER VII.

OF PLANETARY CONJUNCTIONS.

Contents:—1, general classification of planetary conjunctions; 2–6, method of determining at what point on the ecliptic, and at what time, two planets will come to have the same longitude; 7–10, how to find the point on the ecliptic to which a planet, having latitude, will be referred by a circle passing through the north and south points of the horizon; 11, when a planet must be so referred; 12, how to ascertain the interval between two planets when in conjunction upon such a north and south line; 13–14, dimensions of the lesser planets; 15–18, modes of exhibiting the coincidence between the calculated and actual places of the planets; 18–20, definition of different kinds of conjunction; 20–21, when a planet, in con-

junction, is vanquished or victor; 22, farther definition of different kinds of conjunction; 23, usual prevalence of Venus in a conjunction; 23, planetary conjunctions with the moon; 24, conjunctions apparent only; why calculated.

1. Of the star-planets there take place, with one another, encounter (*yuddha*) and conjunction (*samâgama*); with the moon, conjunction (*samâgama*); with the sun, heliacal setting (*astamana*).

The "star-planets" (*târâgraha*) are, of course, the five lesser planets, exclusive of the sun and moon. Their conjunctions with one another and with the moon, with the asterisms (*nakshatra*), and with the sun, are the subjects of this and the two following chapters.

For the general idea of "conjunction" various terms are indifferently employed in this chapter, as *samâgama*, "coming together", *samyoga*, "conjunction," *yoga*, "junction" (in viii. 14, also, *melaka*, "meeting"): the word *yuti*, "union," which is constantly used in the same sense by the commentary, and which enters into the title of the chapter, *graha-yutyadhikâra*, does not occur anywhere in the text. The word which we translate "encounter," *yuddha*, means literally "war, conflict." Verses 18–20, and verse 22, below, give distinctive definitions of some of the different kinds of encounter and conjunction.

2. When the longitude of the swift-moving planet is greater than that of the slow one, the conjunction (*samyoga*) is past; otherwise, it is to come: this is the case when the two are moving eastward; if, however, they are retrograding (*vakrin*), the contrary is true.

3. When the longitude of the one moving eastward is greater, the conjunction (*samâgama*) is past; but when that of the one that is retrograding is greater, it is to come. Multiply the distance in longitude of the planets, in minutes, by the minutes of daily motion of each,

4. And divide the products by the difference of daily motions, if both are moving with direct, or both with retrograde, motion: if one is retrograding, divide by the sum of daily motions.

5. The quotient, in minutes, etc., is to be subtracted when the conjunction is past, and added when it is to come: if the two are retrograding, the contrary: if one is retrograding, the quotients are additive and subtractive respectively.

6. Thus the two planets, situated in the zodiac, are made to be of equal longitude, to minutes. Divide in like manner the distance in longitude, and a quotient is obtained which is the time, in days, etc.

The object of this process is to determine where and when the two planets of which it is desired to calculate the conjunction will have the same longitude. The directions given in the text are in the main so clear as hardly to require explication. The longitude and the rate of motion of the two planets in question is supposed to have been found for some time not far removed from that of their conjunction. Then, in determining whether the conjunction is past or to come, and at what distance, in arc and in time, three separate cases require to be taken into account—when both are advancing, when both are retrograding, and when one is advancing and the other retrograding. In the two former cases, the planets are approaching or receding from one another by the difference of their daily motions; in the latter, by the sum of their daily motions. The point of conjunction will be found by the following proportion: as the daily rate at which the two are approaching or receding from each other is to their distance in longitude, so is the daily motion of each one to the distance which it will have to move before, or which it has moved since, the conjunction in longitude. The time, again, elapsed or to elapse between the given moment and that of the conjunction, will be found by dividing the distance in longitude by the same divisor as was used in the other part of the process, namely the daily rate of approach or separation of the two planets.

The only other matter which seems to call for more special explanation than is to be found in the text is, at what moment the process of calculation, as thus conducted, shall commence. If a time be fixed upon which is too far removed—as, for instance, by an interval of several days—from the moment of actual conjunction, the rate of motion of the two planets will be liable to change in the mean time so much as altogether to vitiate the correctness of the calculation. It is probable that, as in the calculation of an eclipse (see above, note to iv. 7–8), we are supposed, before entering upon the particular process which is the subject of this passage, to have ascertained, by previous tentative calculations, the midnight next preceding or following the conjunction, and to have determined for that time the longitudes and rates of motion of the two planets. If so, the operation will give, without farther repetition, results having the desired degree of accuracy. The commentary, it may be remarked, gives us no light upon this point, as it gave us none in the case of the eclipse.

We have not, however, thus ascertained the time and place of the conjunction. This, to the Hindu apprehension, takes place, not when the two planets are upon the same secondary to the ecliptic, but when they are upon the same secondary to the prime vertical, or upon the same circle passing through the north and south points of the horizon. Upon such a circle two stars rise and set simultaneously; upon such a one they together pass the meridian: such a line, then, determines approximately their relative height above the horizon, each upon its own circle of daily revolution. We have also seen above, when considering the deflection (*valana*—see iv. 24–25), that a secondary to the prime vertical is regarded as determining the north and south directions upon the starry concave. To ascertain what will be the place of each planet upon the ecliptic when referred to it by such a circle is the object of the following processes.

7. Having calculated the measure of the day and night, and likewise the latitude (*vikshepa*), in minutes; having determined the meridian-distance (*nata*) and altitude (*unnata*), in time, according to the corresponding orient ecliptic-point (*lagna*)—

8. Multiply the latitude by the equinoctial shadow, and divide by twelve; the quotient multiply by the meridian-distance in nâḍîs, and divide by the corresponding half-day:

9. The result, when latitude is north, is subtractive in the eastern hemisphere, and additive in the western; when latitude is south, on the other hand, it is additive in the eastern hemisphere, and likewise subtractive in the western.

10. Multiply the minutes of latitude by the degrees of declination of the position of the planet increased by three signs: the result, in seconds (*vikalâ*), is additive or subtractive, according as declination and latitude are of unlike or like direction.

11. In calculating the conjunction (*yoga*) of a planet and an asterism (*nakshatra*), in determining the setting and rising of a planet, and in finding the elevation of the moon's cusps, this operation for apparent longitude (*dṛkkarman*) is first prescribed.

12. Calculate again the longitudes of the two planets for the determined time, and from these their latitudes: when the latter are of the same direction, take their difference; otherwise, their sum: the result is the interval of the planets.

The whole operation for determining the point on the ecliptic to which a planet, having a given latitude, will be referred by a secondary to the prime vertical, is called its *dṛkkarman*. Both parts of this compound we have had before—the latter, signifying "operation, process of calculation," in ii. 37, 42, etc.—for the former, see the notes to iii. 28–34, and v. 5–6: here we are to understand it as signifying the "apparent longitude" of a planet, when referred to the ecliptic in the manner stated, as distinguished from its true or actual longitude, reckoned in the usual way: we accordingly translate the whole term, as in verse 11, "operation for apparent longitude." The operation, like the somewhat analogous one by which the ecliptic-deflection (*valana*) is determined (see above, iv. 24–25), consists of two separate processes, which receive in the commentary distinct names, corresponding with those applied to the two parts of the process for calculating the deflection. The whole subject may be illustrated by reference to the next figure (Fig. 28). This represents the projection of a part of the sphere upon a horizontal plane, N and E being the north and east points of the horizon, and Z the zenith. Let C L be the position of the ecliptic at the moment of conjunction in longitude, C being the orient ecliptic-point (*lagna*); and let M be the point at which the conjunction in longitude of the two planets S and V, each upon its parallel of celestial latitude, *c l* and *c′ l′*, and having latitude equal to S M and V M respectively, will take place. Through V and S draw secondaries to the prime vertical, N V and N S, meeting the ecliptic in *v* and *s*: these latter are the points of apparent longitude of the two planets, which are still removed from a true conjunction by the distance *v s*: in order to the ascertainment of the time of that true conjunction, it is desired to know the positions of *v* and *s*, or their respective distances from M. From P, the pole of the equator, draw also circles through the two planets, meeting the ecliptic in *s′* and *v′*: then, in order to find M *s*, we ascertain the values of *s s′* and M *s′*; and, in like manner, to find M *v*, we ascertain the values of *v v′* and M *v′*. Now at the equator, or in a right sphere, the circles N S and P S would coincide, and the distance *s s′* disappear: hence, the amount of *s s′* being dependent upon the latitude (*aksha*) of the observer, N P, the process by which it is calculated is called the "operation for latitude" (*akshadṛkkarman*, or else *âksha dṛkkarman*). Again, if P and P′ were the same point, or if the ecliptic and equator coincided, P S and P′ S would coincide, and M *s′* would disappear: hence the process of calculation of M *s′* is called the "operation for ecliptic-deviation" (*ayanadṛkkarman*, or *âyana dṛkkarman*). The latter of the two processes, although stated after the other in the text, is the one first explained by the commentary: we will also, as in the case of the deflection (note to iv. 24–25), give to it our first attention.

Fig. 28.

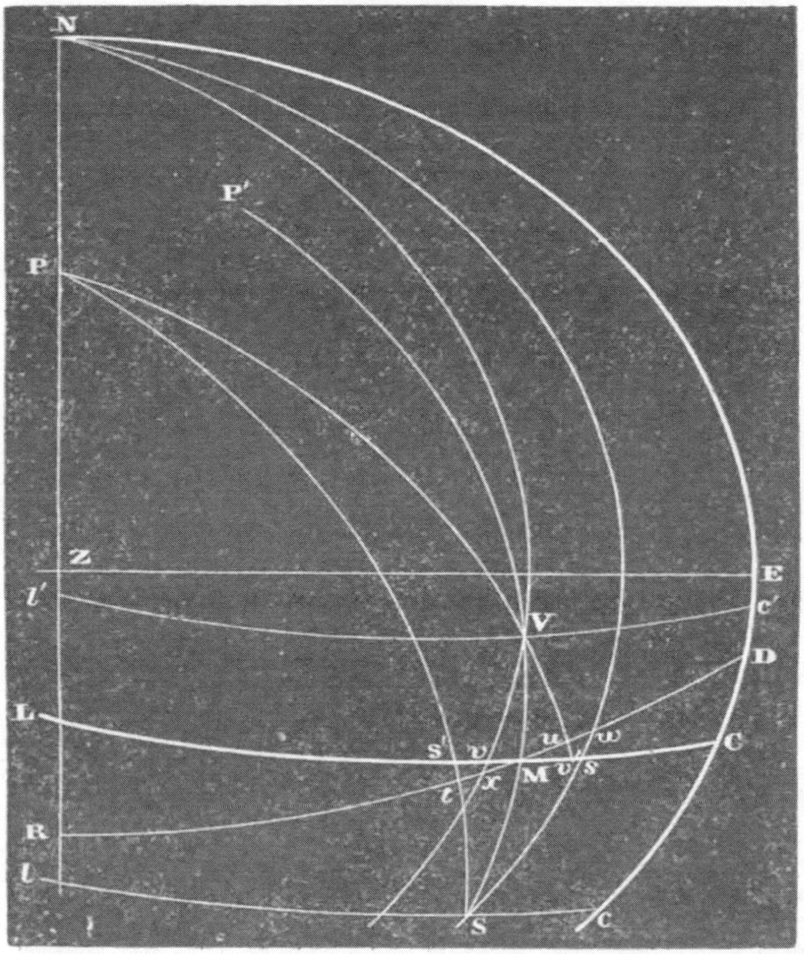

The point *s′*, to which the planet is referred by a circle passing through the pole P, is styled by the commentary *ayanagraha*, "the planet's longitude as corrected for ecliptic-deviation," and the distance M *s′*, which it is desired to ascertain, is called *ayanakalâs*, "the correction, in minutes, for ecliptic-deviation." Instead, however, of finding M *s′*, the process taught in the text finds M *t*, the corresponding distance on the circle of daily revolution, D R, of the point M—which is then assumed equal to M *s′*. The proportion upon which the rule, as stated in verse 10, is ultimately founded, is

$$R : \sin M S t :: M S : M t$$

the triangle M S *t*, which is always very small, being treated as if it were a plane triangle, right-angled at *t*. But now also, as the latitude M S is always a small quantity, the angle P S P′ may be treated as if equal to P M P′ (not drawn in the figure); and this angle is, as was shown in connection with iv. 24–25, the deflection of the ecliptic from the equator (*âyana valana*) at M, which is regarded as equal to the declination of the point 90° in advance of M: this point, for convenience's sake, we will call M′. Our proportion becomes, then

$$R : \sin \text{decl. } M' :: MS : Mt$$

all the quantities which it contains being in terms of minutes. To bring this proportion, now, to the form in which it appears in the text, it is made to undergo a most fantastic and unscientific series of alterations. The greatest declination (ii. 28) being 24°, and its sine 1397′, which is nearly fifty-eight times twenty-four—since $58 \times 24 = 1392$—it is assumed that fifty-eight times the number of degrees in any given arc of declination will be equal to the number of minutes in the sine of that arc. Again, the value of radius, 3438′, admits of being roughly divided into the two factors fifty-eight and sixty—since $58 \times 60 = 3480$. Substituting, then, these values in the proportion as stated, we have

$$58 \times 60 : 58 \times \text{decl. } M' \text{ in degr.} :: \text{latitude in min.} : Mt$$

Cancelling, again, the common factor in the first two terms, and transferring the factor 60 to the fourth term, we obtain finally

$$1 : \text{decl. } M' \text{ in degr.} :: \text{latitude in min.} : Mt \times 60$$

that is to say, if the latitude of the planet, in minutes, be multiplied by the declination, in degrees, of a point 90° in advance of the planet, the result will be a quantity which, after being divided by sixty, or reduced from seconds to minutes, is to be accepted as the required interval on the ecliptic between the real place of the planet and the point to which it is referred by a secondary to the equator.

This explanation of the rule is the one given by the commentator, nor are we able to see that it admits of any other. The reduction of the original proportion to its final form is a process to which we have heretofore found no parallel, and which appears equally absurd and uncalled for. That Mt is taken as equivalent to Ms' has, as will appear from a consideration of the next process, a certain propriety.

The value of the arc Ms' being thus found, the question arises, in which direction it shall be measured from M. This depends upon the position of M with reference to the solstitial colure. At the colure, the lines PS and P′S coincide, so that, whatever be the latitude of a planet, it will, by a secondary to the equator, be referred to the ecliptic at its true point of longitude. From the winter solstice onward to the summer solstice, or when the point M is upon the sun's northward path (*uttarâyaṇa*), a planet having north latitude will be referred backward on the ecliptic by a circle from the pole, and a planet having south latitude will be referred forward. If M, on the other hand, be upon the sun's southward path (*dakshiṇâyana*), a planet having north latitude at that point will be referred forward, and one having south latitude backward: this is the case illustrated by the figure. The statement of the text virtually agrees with this, it being evident that, when M is on the northward path, the declination of the point 90° in advance of it will be north; and the contrary.

We come now to consider the other part of the operation, or the *âksha dṛkkarman*, which forms the subject of verses 7–9. As the first step, we are directed to ascertain the day and the night respectively of the point of the ecliptic at which the two planets are in conjunction in longitude, for the purpose of determining also its distance in time from the horizon and from the meridian. This is accomplished as follows.

Having the longitude of the point in question (M in the last figure), we calculate (by ii. 28) its declination, which gives us (by ii. 60) the radius of its diurnal circle, and (by ii. 61) its ascensional difference; whence, again, is derived (by ii. 62–63) the length of its day and night. Again, having the time of conjunction at M, we easily calculate the sun's longitude at the moment, and this and the time together give us (by iii. 46–48) the longitude of C, the orient ecliptic-point: then (by iii. 50) we ascertain directly the difference between the time when M rose and that when C rises, which is the altitude in time (*unnata*) of M: the difference between this and the half-day is the meridian-distance in time (*nata*) of the same point. If the conjunction takes place when M is below the horizon, or during its night, its distance from the horizon and from the inferior meridian is determined in like manner.

The direct object of this part of the general process being to find the value of *s s′*, we note first that that distance is evidently greatest at the horizon; farther, that it disappears at the meridian, where the lines P S and N S coincide. If, then, it is argued, its value at the horizon can be ascertained, we may assume it to vary as the distance from the meridian. The accompanying figure (Fig. 29) will illustrate the method by which it is attempted to calculate *s s′* at the horizon. Suppose the planet S, being removed in latitude to the distance M S from M, the point of the ecliptic which determines its longitude, to be upon the horizon, and let *s′*, as before, be the point to which it is referred by a circle from the north pole: it is desired to determine the value of *s s′*. Let D R be the circle of diurnal revolution of the point M, meeting S *s′* in *t*, and the horizon in *w*: S *t w* may be regarded as a plane right-angled triangle, having its angles at S and *w* respectively equal to the observer's latitude and co-latitude. In that triangle, to find the value of *t w*, we should make the proportion

Fig. 29.

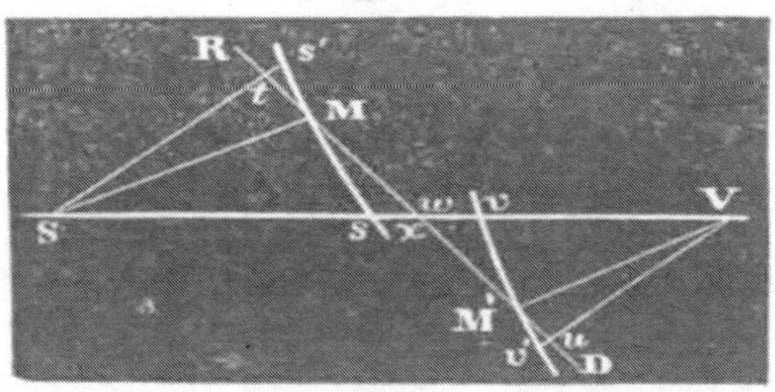

cos *t* S *w* : sin *t* S *w* : : *t* S : *t w*

Now the first of these ratios, that of the cosine to the sine of latitude, is (see above, iii. 17) the same with that of the gnomon to the equinoctial shadow: again, as the difference of M *t* and M *s′* was in the preceding process neglected, so here the difference of S M and S *t*; and finally, *t w*, the true result of the process, is accepted as the equivalent of *s′ s*, the distance sought. The proportion then becomes

gnom. : eq. shad. : : latitude : required dist. at horizon

The value of the required distance at the horizon having been thus ascertained, its value at any given altitude is, as pointed out above, determined by a proportion, as follows: as the planet's distance in time from the meridian when upon the horizon is to the value of this correction at the horizon, so is any given distance from the meridian (*nata*) to the value at that distance; or

half-day : mer.-dist. in time : : result of last proportion : required distance

The direction in which the distance thus found is to be reckoned, starting in each case from the *âyana graha*, or place of the planet on the ecliptic as determined by a secondary to the equator, which was ascertained by the preceding process, is evidently as the text states it in verse 9. In the eastern hemisphere, which is the case illustrated by the figure, *s′ s* is additive to the longitude of *s′*, while *v′ v* is subtractive from the longitude of *v′*: in the western hemisphere, the contrary would be the case. The final result thus arrived at is the longitude of the two points *s* and *v*, to which S and V are referred by the circles N S and N V, drawn through them from the north and south points of the horizon.

The many inaccuracies involved in these calculations are too palpable to require pointing out in detail. The whole operation is a roughly approximative one, of which the errors are kept within limits, and the result rendered sufficiently correct, only by the general minuteness of the quantity entering into it as its main element—namely, the latitude of a planet—and by the absence of any severe practical test of its accuracy. It may be remarked that the commentary is well aware of, and points out, most of the errors of the processes, excusing them by its stereotyped plea of their insignificance, and the merciful disposition of the divine author of the treatise.

Having thus obtained *s* and *v*, the apparent longitudes of the two planets at the time when their true longitude is M, the question arises, how we shall determine the time of apparent conjunction. Upon this point the text gives us no light at all: according to the commentary, we are to repeat the process prescribed in verses 2–6 above, determining, from a consideration of the rate and direction of motion of the planets in connection with their new places, whether the conjunction sought for is past or to come, and then ascertaining, by dividing the distance *v s* by their daily rate of approach or recession, the time of the conjunction. It is evident, however, that one of the elements of the process of correction for latitude (*akshadṛkkarman*), namely the meridian-distance, is changing so rapidly, as compared with the slow motion of the planets in their orbits, that such a process could not yield results at all approaching to accuracy: it also appears that two slow-moving planets might have more than one, and even several apparent conjunctions on successive days, at different times in the day, being found to stand together upon the same secondary to the prime vertical at different altitudes. We do not see how this difficulty is met by anything in the text or in the commentary. The text, assuming the moment of apparent conjunction to have been, by whatever method, already determined, goes on to direct us, in verse 12, to calculate anew, for that moment, the latitudes of the two planets, in order to obtain their distance from one another. Here, again, is a slight inaccuracy: the interval between the two, measured upon a secondary to the prime vertical, is not precisely equal to the sum or difference of their latitudes, which are measured upon secondaries to the ecliptic. The ascertainment of this interval is necessary, in order to determine the name and character of the conjunction, as will appear farther on (vv. 18–20, 22).

The cases mentioned in verse 11, in which, as well as in calculating the conjunctions of two planets with one another, this operation for apparent longitude (*dṛkkarman*) needs to be performed, are the subjects of the three following chapters.

13. The diameters upon the moon's orbit of Mars, Saturn, Mercury, and Jupiter, are declared to be thirty, increased successively by half the half; that of Venus is sixty.

14. These, divided by the sum of radius and the fourth hypothenuse, multiplied by two, and again multiplied by radius, are the respective corrected (*sphuṭa*) diameters: divided by fifteen, they are the measures (*mâna*) in minutes.

We have seen above, in connection with the calculation of eclipses (iv. 2–5), that the diameters of the sun, moon, and shadow had to be reduced, for measurement in minutes, to the moon's mean distance, at which fifteen yojanas make a minute of arc. Here we find the dimensions of the five lesser planets, when at their mean distances from the earth, stated only in the form of the portion of the moon's mean orbit covered by them, their absolute size being left undetermined. We add them below, in a tabular form, both in yojanas and as reduced to minutes, appending also the corresponding estimates of Tycho Brahe (which we take from Delambre), and the true apparent diameters of the planets, as seen from the earth at their greatest and least distances.

Apparent Diameters of the Planets, according to the Sûrya-Siddhânta, to Tycho Brahe, and to Modern Science.

Planet.	Sûrya-Siddhânta:		Tycho Brahe.	Moderns:	
	in yojanas.	in arc.		least.	greatest.
Mars,	30	2′	1′ 40″	4″	27″
Saturn,	37½	2′ 30″	1′ 50″	15″	21″
Mercury,	45	3′	2′ 10″	4″	12″
Jupiter,	52½	3′ 30″	2′ 45″	30″	49″
Venus,	60	4′	3′ 15″	9″	1′ 14″

This table shows how greatly exaggerated are wont to be any determinations of the magnitude of the planetary orbs made by the unassisted eye alone. This effect is due to the well-known phenomenon of the irradiation, which increases the apparent size of a brilliant body when seen at some distance. It will be noticed that the Hindu estimates do not greatly exceed those of Tycho, the most noted and accurate of astronomical observers prior to the invention of the telescope. In respect to order of magnitude they entirely agree, and both accord with the relative apparent size of the planets, except that to Mercury and Venus, whose proportional brilliancy, from their nearness to the sun, is greater, is assigned too high a rank. Tycho also established a scale of apparent diameters for the fixed stars, varying from 2′, for the first magnitude, down to 20″, for the sixth. We do not find that Ptolemy made any similar estimates, either for planets or for fixed stars.

The Hindus, however, push their empiricism one step farther, gravely laying down a rule by which, from these mean values, the true values of the apparent diameters at any given time may be found. The fundamental proportion is, of course,

true dist. : mean dist. : : mean app. diam. : true app. diam.

The second term of this proportion is represented by radius: for the first we have, according to the translation given, one half the sum of radius and the fourth hypothenuse, by which is meant the "variable hypothenuse" (*cala karṇa*) found in the course of the fourth, or last, process for finding the true place of the planet (see above, ii. 43–45). The term, however (*tricatuḥkarṇa*), which is translated "radius and the fourth hypothenuse" is much more naturally rendered "third and fourth hypothenuses"; and the latter interpretation is also mentioned by the commentator as one handed down by tradition (*sâmpradâyika*): but, he adds, owing to the fact that the length of the hypothenuse is not calculated in the third process, that for finding finally the equation of the centre (*mandakarman*), and that that hypothenuse cannot therefore be referred to here as known, modern interpreters understand the first member of the compound (*tri*) as an abbreviation for "radius" (*trijyâ*), and translate it accordingly. We must confess that the other interpretation seems to us to be powerfully supported by both the letter of the text and the reason of the matter. The substitution of *tri* for *trijyâ* in such a connection is quite too violent to be borne, nor do we see why half the sum of radius and the fourth hypothenuse should be taken as representing the planet's true distance, rather than the fourth hypothenuse alone, which was employed (see above, ii. 56–58) in calculating the latitude of the planets. On the other hand, there is reason for adopting, as the relative value of a planet's true distance, the average, or half the sum, of the third hypothenuse, or the planet's distance as affected by the eccentricity of its orbit, and the fourth, or its distance as affected by the motion of the earth in her orbit. There seems to us good reason, therefore, to suspect that verse 14—and with it, probably, also verse 13—is an intrusion into the Sûrya-Siddhânta from some other system, which did not make the grossly erroneous assumption, pointed out under ii. 39, of the equality of the sine of anomaly in the epicycle (*bhujajyâphala*) with the sine of the equation, but in which the hypothenuse and the sine of the equation were duly calculated in the process for finding the equation of the apsis (*mandakarman*), as well as in that for finding the equation of the conjunction (*çîghrakarman*).

15. Exhibit, upon the shadow-ground, the planet at the extremity of its shadow reversed: it is viewed at the apex of the gnomon in its mirror.

As a practical test of the accuracy of his calculations, or as a convincing proof to the pupil or other person of his knowledge and skill, the teacher is here directed to set up a gnomon upon ground properly prepared for exhibiting the shadow, and to calculate and lay off from the base of the gnomon, but in the opposite to the true direction, the shadow which a planet would cast at a given time; upon placing, then, a horizontal mirror at the extremity of the shadow, the reflected image of the planet's disk will be seen in it at the given time by an eye placed at the apex of the gnomon. The principle of the experiment is clearly correct, and the rules and processes taught in the second and third chapters afford the means of carrying it out, since from them the shadow which any star would cast, had it light enough, may be as readily determined as that which the sun actually casts. As no case of precisely this character has hitherto been presented, we will briefly indicate the course of the calculation. The day and night of the planet, and its distance from the meridian, or its hour-angle, are found in the same manner as in the process previously explained (p. 168, above), excepting that here the planet's latitude, and its declination as affected by latitude, must be calculated, by ii. 56–58; and then the hour-angle and the ascensional difference, by iii. 34–36, give the length of the shadow at the given time, together with that of its hypothenuse. The question would next be in what direction to lay off the shadow from the base of the gnomon. This is accomplished by means of the base (*bhuja*) of the shadow, or its value when projected on a north and south line. From the declination is found, by iii. 20–22, the length of the noon-shadow and its hypothenuse, and from the latter, with the declination, comes, by iii. 22–23, the measure of amplitude (*agrâ*) of the given shadow; whence, by iii. 23–25, is derived its base. Having thus both its length and the distance of its extremity from an east and west line running through the base of the gnomon, we lay it off without difficulty.

16. Take two gnomons, five cubits (*hasta*) in height, stationed according to the variation of direction, separated by the interval of the two planets, and buried at the base one cubit.

17. Then fix the two hypothenuses of the shadow, passing from the extremity of the shadow through the apex of each gnomon: and, to a person situated at the point of union of the extremities of the shadow and hypothenuse, exhibit

18. The two planets in the sky, situated at the apex each of its own gnomon, and arrived at a coincidence of observed place (*dṛç*). . . .

This is a proceeding of much the same character with that which forms the subject of the preceding passage. In order to make apprehensible, by observation, the conjunction of two planets, as calculated by the methods of this chapter, two gnomons, of about the height of a man, are set up. At what distance and direction from one another they are to be fixed is not clearly shown. The commentator interprets the expression "interval of the two planets" (v. 16), to mean their distance in minutes on the secondary to the prime vertical, as ascertained according to verse 12, above, reduced to digits by the method taught in iv. 26; while, by "according to the variation of direction," he would understand merely, in the direction from the observer of the hemisphere in which the planets at the moment of conjunction are situated. The latter phrase, however, as thus explained, seems utterly nugatory; nor do we see of what use it would be to make the north and south interval of the bases of the gnomons, in digits, correspond with that of the planets in minutes. We do not think it would be difficult to understand the directions given in the text as meaning, in effect, that the two gnomons should be so stationed as to cast their shadows to the same point: it would be easy to do this, since, at the time in question, the extremities of two shadows cast from one gnomon by the two stars would be in the same north and south line, and it would only be necessary to set the second gnomon as far south of the first as the end of the shadow cast by the southern star was north of that cast by the other. Then, if a hole were sunk in the ground at the point of intersection of the two shadows, and a person enabled to place his eye there, he would, at the proper moment, see both the planets with the same glance, and each at the apex of its own gnomon.

In the eighteenth verse also we have ventured to disregard the authority of the commentator: he translates the words *dṛktulyatâm itâs* "come within the sphere of sight," while we understand by *dṛktulyatâ*, as in other cases (ii. 14, iii. 11), the coincidence between observed and computed position.

Such passages as this and the preceding are not without interest and value, as exhibiting the rudeness of the Hindu methods of observation, and also as showing the unimportant and merely illustrative part which observation was meant to play in their developed system of astronomy.

18. . . . When there is contact of the stars, it is styled "depiction" (*ullekha*); when there is separation, "division" (*bheda*);

19. An encounter (*yuddha*) is called "ray-obliteration" (*ançuvimarda*) when there is mutual mingling of rays: when the interval is less than a degree, the encounter is named "dexter" (*apasavya*)—if, in this case, one be faint (*aṇu*).

20. If the interval be more than a degree, it is "conjunction" (*samâgama*), if both are endued with power (*bala*). One that is vanquished (*jita*) in a dexter encounter (*apasavya yuddha*), one that is covered, faint (*aṇu*), destitute of brilliancy,

21. One that is rough, colorless, struck down (*vidhvasta*), situated to the south, is utterly vanquished (*vijita*). One situated to the north, having brilliancy, large, is victor (*jayin*)—and even in the south, if powerful (*balin*).

22. Even when closely approached, if both are brilliant, it is "conjunction" (*samâgama*): if the two are very small, and struck down, it is "front" (*kûṭa*) and "conflict" (*vigraha*), respectively.

23. Venus is generally victor, whether situated to the north or to the south. . . .

In this passage, as later in a whole chapter (chap. xi), we quit the proper domain of astronomy, and trench upon that of astrology. However intimately connected the two sciences may be in practice, they are, in general, kept distinct in treatment—the Siddhântas, or astronomical text-books, furnishing, as in the present instance, only the scientific basis, the data and methods of calculation of the positions of the heavenly bodies, their eclipses, conjunctions, risings and settings, and the like, while the Sanhitâs, Jâtakas, Tâjikas, etc., the astrological treatises, make the superstitious applications of the science to the explanation of the planetary influences, and their determination of human fates. Thus the celebrated astronomer, Varâha-mihira, besides his astronomies, composed separate astrological works, which are still extant, while the former have become lost. It is by no means impossible that these verses may be an interpolation into the original text of the Sûrya-Siddhânta. They form only a disconnected fragment: it is not to be supposed that they contain a complete statement and definition of all the different kinds of conjunction recognized and distinguished by technical appellations; nor do they fully set forth the circumstances which determine the result of a hostile "encounter" between two planets: while a detailed explanation of some of the distinctions indicated—as, for instance, when a planet is "powerful" or the contrary—could not be given without entering quite deeply into the subject of the Hindu astrology. This we do not regard ourselves as called upon to do here: indeed, it would not be possible to accomplish it satisfactorily without aid from original sources which are not accessible to us. We shall content ourselves with following the example of the commentator, who explains simply the sense and connection of the verses, as given in our translation, citing one or two parallel passages from works of kindred subject. We would only point out farther that it has been shown in the most satisfactory manner (as by Whish, in Trans. Lit. Soc. Madras, 1827; Weber, in his Indische Studien, ii. 236 etc.) that the older Hindu science of astrology, as represented by Varâha-mihira and others, reposes entirely upon the Greek, as its later forms depend also, in part, upon the Arab; the latter connection being indicated even in the common title of the more modern treatises, *tâjika*, which comes from the Persian *tâzî*, "Arab." Weber gives (Ind. Stud. ii. 277 etc.) a translation of a passage from Varâha-mihira's lesser treatise, which states in part the circumstances determining the "power" of a planet in different situations, absolute or relative: partial explanations upon the same subject furnished to the translator in India by his native assistant, agree with these, and both accord closely with the teachings of the Tetrabiblos, the astrological work attributed to Ptolemy.

23. . . . Perform in like manner the calculation of the conjunction (*samyoga*) of the planets with the moon.

This is all that the treatise says respecting the conjunction of the moon with the lesser planets: of the phenomenon, sometimes so striking, of the occultation of the latter by the former, it takes no especial notice. The commentator cites an additional half-verse as sometimes included in the chapter, to the effect that, in calculating a conjunction, the moon's latitude is to be reckoned as corrected by her parallax in latitude (*avanati*), but rejects it, as making the chapter over-full, and as being superfluous, since the nature of the case determines the application here of the general rules for parallax presented in the fifth chapter. Of any parallax of the planets themselves nothing is said: of course, to calculate the moon's parallax by the methods as already given is, in effect, to attribute to them all a horizontal parallax of the same value with that assigned to the sun, or about 4′.

The final verse of the chapter is a caveat against the supposition that, when a "conjunction" of two planets is spoken of, anything more is meant than that they appear to approach one another; while nevertheless, this apparent approach requires to be treated of, on account of its influence upon human fates.

24. Unto the good and evil fortune of men is this system set forth: the planets move on upon their own paths, approaching one another at a distance.

CHAPTER VIII.

OF THE ASTERISMS.

CONTENTS:—1–9, positions of the asterisms; 10–12, of certain fixed stars; 12, direction to test by observation the accuracy of these positions; 13, splitting of Rohiṇî's wain; 14–15, how to determine the conjunction of a planet with an asterism; 16–19, which is the junction-star in each asterism; 20–21, positions of other fixed stars.

1. Now are set forth the positions of the asterisms (*bha*), in minutes. If the share of each one, then, be multiplied by ten, and increased by the minutes in the portions (*bhoga*) of the past asterisms (*dhishṇya*), the result will be the polar longitudes (*dhruva*).

The proper title of this chapter is *nakshatragrahayutyadhikâra*, "chapter of the conjunction of asterisms and planets," but the subject of conjunction occupies but a small space in it, being limited to a direction (vv. 14–15) to apply, with the necessary modifications, the methods taught in the preceding chapter. The chapter is mainly occupied with such a definition of the positions of the asterisms—to which are added also those of a few of the more prominent among the fixed stars—as is necessary in order to render their conjunctions capable of being calculated.

Before proceeding to give the passage which states the positions of the asterisms, we will explain the manner in which these are defined. In the accompanying figure (Fig. 30), let E L represent the equator, and C L the ecliptic, P and P′ being their respective poles. Let S be the position of any given star, and through it draw the circle of declination P S *a*. Then *a* is the point on the ecliptic of which the distance from the first of Aries and from the star respectively are here given as its longitude and latitude. So far as the latitude is concerned, this is not unaccordant with the usage of the treatise hitherto. Latitude (*vikshepa*, "disjection") is the amount by which any body is removed from the declination which it ought to have—that is, from the point of the ecliptic which it ought to occupy—declination (*krânti, apakrama*) being always, according to the Hindu understanding of the term, in the ecliptic itself. In the case of a planet, whose proper path is in the ecliptic, the point of that circle which it ought to occupy is determined by its calculated longitude: in the case of a fixed star, whose only motion is about the pole of the heavens, its point of declination is that to which it is referred by a

Fig. 30.

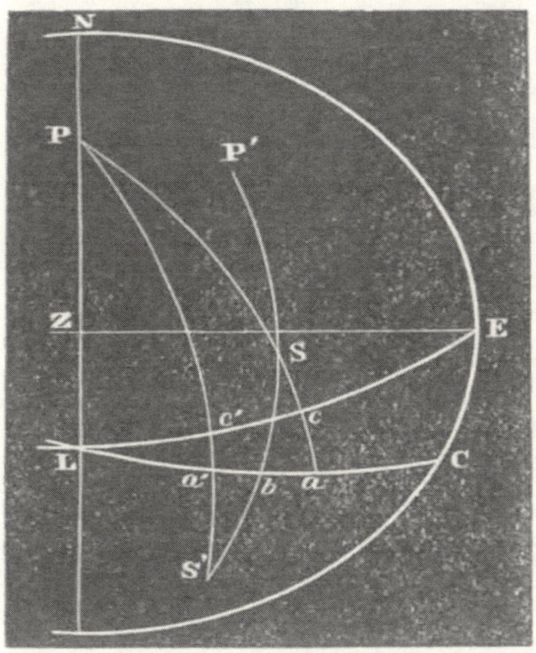

circle through that pole. Thus, in the figure, the declination (*krânti*) of S would be *c a*, or the distance of *a* from the equator at *c* : its latitude (*vikshepa*) is *a* S, or its distance from *a*. We have, accordingly, the same term used here as before. To designate the position in longitude of *a*, on the other hand, we have a new term, *dhruva*, or, as below, (vv. 12, 15), *dhruvaka*. This comes from the adjective *dhruva*, "fixed, immovable," by which the poles of the heaven (see below, xii. 43) are designated; and, if we do not mistake its application, it indicates, as here employed, the longitude of a star as referred to the ecliptic by a circle from the pole. We venture, then, to translate it by "polar longitude," as we also render *vikshepa*, in this connection, by "polar latitude," it being desirable to have for these quantities distinctive names, akin with one another. Colebrooke employs "apparent longitude and latitude," which are objectionable, as being more properly applied to the results of the process taught in the last chapter (vv. 7–10).

The mode of statement of the polar longitudes is highly artificial and arbitrary : a number is mentioned which, when multiplied by ten, will give the position of each asterism, in minutes, in its own "portion" (*bhoga*), or arc of 13° 20′ in the ecliptic (see ii. 64).

This passage presents a name for the asterisms, *dhishnya*, which has not occurred before ; it is found once more below, in xi. 21.

2. Forty-eight, forty, sixty-five, fifty-seven, fifty-eight, four, seventy-eight, seventy-six, fourteen,

3. Fifty-four, sixty-four, fifty, sixty, forty, seventy-four, seventy-eight, sixty-four.

4. Fourteen, six, four: Uttara-Ashâḍhâ, (*vâiçva*) is at the middle of the portion (*bhoga*) of Pûrva-Ashâḍhâ (*âpya*); Abhijit, likewise, is at the end of Pûrva-Ashâḍhâ; the position of Çravaṇa is at the end of Uttara-Ashâḍhâ ;

5. Çravishthâ, on the other hand, is at the point of connection of the third and fourth quarters (*pada*) of Çravaṇa: then, in their own portions, eighty, thirty-six, twenty-two,

6. Seventy-nine. Now their respective latitudes, reckoned from the point of declination (*apakrama*) of each: ten, twelve, five, north; south, five, ten, nine ;

7. North, six ; nothing ; south, seven ; north, nothing, twelve, thirteen ; south, eleven, two ; then thirty-seven, north ;

8. South, one and a half, three, four, nine, five and a half, five ; north, also, sixty, thirty, and also thirty-six ;

9. South, half a degree ; twenty-four, north, twenty-six degrees ; nothing—for Açvinî (*dasra*), etc., in succession.

The text here assumes that the names of the asterisms, and the order of their succession, are so familiarly known as to render it unnecessary to rehearse them. It has been already noticed (see above, i. 48–51, 55, 56–58, etc.) that a similar assumption was made as regards the names and succession of the months, signs of the zodiac, years of Jupiter's cycle, and the like. Many of the asterisms have more than one appellation ; we present in the annexed table those by which they are more generally and familiarly known; the others will be stated farther on. Nearly all these titles are to be found in our text, occurring here and there; a few of the asterisms, however, (the 5th, 6th, 9th, and 17th), are mentioned only by appellations derived from the names of the deities to whom they are regarded as belonging, and one (the 25th) chances not to be once distinctively spoken of. We append to the names, in a tabular form, the data presented in this passage; namely, the position of each asterism (*nakshatra*) in the arc of the ecliptic to which it gives name, and which is styled its "portion" (*bhoga*), the resulting polar longitudes, and the polar latitudes. And since it is probable (see note to the latter half of v. 12, below) that the latter were actually derived by calculation from true declinations and right ascensions, ascertained by observation, we have endeavored to restore those more original data by calculating them back again, according to the data and methods of this Siddhânta—the declinations by ii. 28, the right ascensions by iii. 44–48—and we insert our results in the table, rejecting odd minutes less than ten.

Positions of the Junction-Stars of the Asterisms.

No.	Name.	Position in its Portion.	Polar Longitude.	Polar Latitude.	Right Ascension.	True Declination.	Interval in Longitude.	Interval in R. A.
		° ′	° ′	° ′	° ′	° ′	° ′	° ′
1	Açvinî,	8 0	8 0	10 0 N.	7 30	13 20 N.	12 0	11 0
2	Bharaṇî,	6 40	20 0	12 0 "	18 30	20 0 "	17 30	16 50
3	Kṛttikâ,	10 50	37 30	5 0 "	35 20	19 20 "	12 0	12 0
4	Rohiṇî,	9 30	49 30	5 0 S.	47 20	13 0 "	13 30	13 40
5	Mṛgaçîrsha,	9 40	63 0	10 0 "	61 0	11 20 "	4 20	4 40
6	Ârdrâ,	0 40	67 20	9 0 "	65 40	13 0 "	25 40	27 30
7	Punarvasu,	13 0	93 0	6 0 N.	93 10	30 0 "	13 0	14 0
8	Pushya,	12 40	106 0	0 0	107 10	23 0 "	3 0	3 20
9	Âçleshâ,	2 20	109 0	7 0 S.	110 30	15 40 "	20 0	20 40
10	Maghâ,	9 0	129 0	0 0	131 10	18 20 "	15 0	15 0
11	P.-Phalgunî,	10 40	144 0	12 0 N.	146 10	25 50 "	11 0	10 40
12	U.-Phalgunî,	8 20	155 0	13 0 "	156 50	22 50 "	15 0	13 50
13	Hasta,	10 0	170 0	11 0 S.	170 40	7 0 S.	10 0	9 20
14	Citrâ,	6 40	180 0	2 0 "	180 0	2 0 "	19 0	17 40
15	Svâtî,	12 20	199 0	37 0 N.	197 40	29 20 N.	14 0	13 10
16	Viçâkhâ,	13 0	213 0	1 30 S.	210 50	14 20 S.	11 0	11 0
17	Anurâdhâ,	10 40	224 0	3 0 "	221 50	19 20 "	5 0	5 0
18	Jyeshṭhâ,	2 20	229 0	4 0 "	226 50	21 50 "	12 0	12 0
19	Mûla,	1 0	241 0	9 0 "	238 50	29 50 "	13 0	14 0
20	P.-Ashâḍhâ,	0 40	254 0	5 30 "	252 50	28 30 "	6 0	6 30
21	U.-Ashâḍhâ,	....	260 0	5 0 "	259 20	28 40 "	6 40	7 0
22	Abhijit,	....	266 40	60 0 N.	266 20	36 0 N.	13 20	14 30
23	Çravaṇa,	....	280 0	30 0 "	280 50	6 20 "	10 0	10 40
24	Çravishṭhâ,	....	290 0	36 0 "	291 30	13 30 "	30 0	30 40
25	Çatabhishaj,	13 20	320 0	0 30 S.	322 10	15 40 S.	6 0	6 0
26	P.-Bhâdrapadâ,	6 0	326 0	24 0 N.	328 10	10 50 N.	11 0	10 30
27	U.-Bhâdrapadâ,	3 40	337 0	26 0 "	338 40	16 50 "	22 50	21 10
28	Revatî,	13 10	359 50	0 0	359 50	0 0	8 10	7 40

Fig. 31.

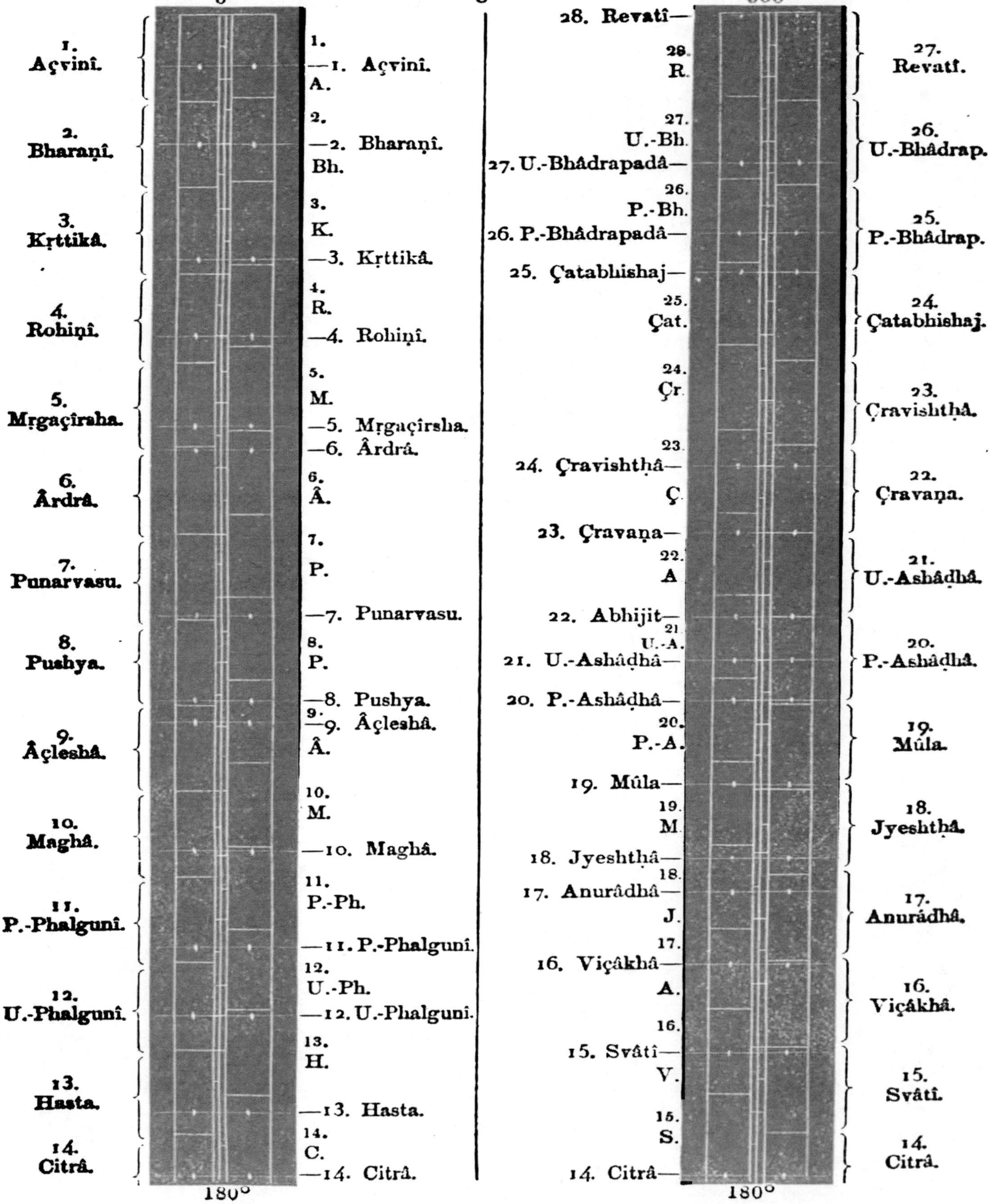
0°
180°
360°
180°
1. Açvinî.
2. Bharaṇî.
3. Kṛttikâ.
4. Rohiṇî.
5. Mṛgaçîrsha.
6. Ârdrâ.
7. Punarvasu.
8. Pushya.
9. Âçleshâ.
10. Maghâ.
11. P.-Phalgunî.
12. U.-Phalgunî.
13. Hasta.
14. Citrâ.
1. —1. Açvinî. A.
2. —2. Bharaṇî. Bh.
3. K. —3. Kṛttikâ.
4. R. —4. Rohiṇî.
5. M. —5. Mṛgaçîrsha. —6. Ârdrâ.
6. Â.
7. P. —7. Punarvasu.
8. P. —8. Pushya.
9. —9. Âçleshâ. Â.
10. M. —10. Maghâ.
11. P.-Ph. —11. P.-Phalgunî.
12. U.-Ph. —12. U.-Phalgunî.
13. H. —13. Hasta.
14. C. —14. Citrâ.
28. Revatî—
28. R.
27. U.-Bh.
27. U.-Bhâdrapadâ—
26. P.-Bh.
26. P.-Bhâdrapadâ—
25. Çatabhishaj—
25. Çat.
24. Çr.
23.
24. Çravishṭhâ—
Ç.
23. Çravaṇa—
22. A.
22. Abhijit—
21. U.-A.
21. U.-Ashâḍhâ—
20. P.-Ashâḍhâ—
20. P.-A.
19. Mûla—
19. M.
18. Jyeshṭhâ—
18.
17. Anurâdhâ—
J.
17.
16. Viçâkhâ—
A.
16.
15. Svâtî—
V.
15. S.
14. Citrâ—
27. Revatî.
26. U.-Bhâdrap.
25. P.-Bhâdrap.
24. Çatabhishaj.
23. Çravishṭhâ.
22. Çravaṇa.
21. U.-Ashâḍhâ.
20. P.-Ashâḍhâ.
19. Mûla.
18. Jyeshṭhâ.
17. Anurâdhâ.
16. Viçâkhâ.
15. Svâtî.
14. Citrâ.

Our calculations, it should be remarked, are founded upon the assumption that, at the time when the observations were made of which our text records the results, the vernal equinox coincided with the initial point of the Hindu sidereal sphere, or with the beginning of the portion of the asterism Açvinî, a point 10′ eastward on the ecliptic from the star ζ Piscium: this was actually the case (see above, under i. 27) about A. D. 560. The question how far this assumption is supported by evidence contained in the data themselves will be considered later. To fill out the table, we have also added the intervals in right ascension and in polar longitude.

The stars of which the text thus accurately defines the positions do not, in most cases, by themselves alone, constitute the asterisms (*nakshatra*); they are only the principal members of the several groups of stars—each, in the calculation of conjunctions (*yoga*) between the planets and the asterisms (see below, vv. 14–15), representing its group, and therefore called (see below, vv. 16–19) the "junction-star" (*yogatârâ*) of the asterism.

It will be at once noticed that while, in a former passage (ii. 64), the ecliptic was divided into twenty-seven equal arcs, as portions for the asterisms, we have here presented to us twenty-eight asterisms, very unequally distributed along the ecliptic, and at greatly varying distances from it. And it is a point of so much consequence, in order to the right understanding of the character and history of the whole system, to apprehend clearly the relation of the groups of stars to the arcs allotted to them, that we have prepared the accompanying diagram (Fig. 31) in illustration of that relation. The figure represents, in two parts, the circle of the ecliptic: along the central lines is marked its division into arcs of ten and five degrees: upon the outside of these lines it is farther divided into equal twenty-sevenths, or arcs of 13° 20′, and upon the inside into equal twenty-eighths, or arcs of 12° 51$\frac{3}{7}$′; these being the portions (*bhoga*) of two systems of asterisms, twenty-seven and twenty-eight in number respectively. The starred lines which run across all the divisions mark the polar longitudes, as stated in the text, of the junction-stars of the asterisms. The names of the latter are set over against them, in the inner columns: the names of the portions in the system of twenty-seven are given in full in the outer columns, and those in the system of twenty-eight are also placed opposite the portions, upon the inside, in an abbreviated form.

The text nowhere expressly states which one of the twenty-eight asterisms which it recognizes is, in its division of the ecliptic into only twenty-seven portions, left without a portion. That Abhijit, the twenty-second of the series, is the one thus omitted, however, is clearly implied in the statements of the fourth and fifth verses. Those statements, which have caused difficulty to more than one expounder of the passage, and have been variously misinterpreted, are made entirely clear by supplying the words "asterism" and "portion" throughout, where they are to be understood, thus: "the asterism Uttara-Ashâḍhâ is at the middle of the portion styled Pûrva-Ashâḍhâ; the asterism Abhijit, likewise, is at the end of the portion Pûrva-Ashâḍhâ; the position of the asterism Çravana is at the end of the portion receiving its name from Uttara-Ashâḍhâ; while the asterism Çravishṭhâ is between the third and fourth quarters of the portion named for Çravaṇa." After this interruption to the regularity of correspondence of the two systems—the asterism Abhijit being left without a portion, and the portion Çravishṭhâ containing no asterism—they go on again harmoniously together to the close. The figure illustrates clearly this condition of things, and shows that, if Abhijit be left out of account, the two systems agree so far as this—that twenty-six asterisms fall within the limits of portions bearing the same name, while all the discordances are confined to one portion of the ecliptic, that comprising the 20th to the 23d portions. If, on the other hand, the ecliptic be divided into twenty-eighths, and if these be assigned as portions to the twenty-eight asterisms, it is seen from the figure that the discordances between the two systems will be very great; that only in twelve instances will a portion be occupied by the asterism bearing its own name, and by that alone; that in sixteen cases asterisms will be found to fall within the limits of portions of different name; that four portions will be left without any asterism at all, while four others will contain two each.

These discordances are enough of themselves to set the whole subject of the asterisms in a new light. Whereas it might have seemed, from what we have seen of it heretofore, that the system was founded upon a division of the ecliptic into twenty-seven equal portions, and the selection of a star or a constellation to mark each portion, and to be, as it were, its ruler, it now appears that the series of twenty-eight asterisms may be something independent of, and anterior to, any division of the ecliptic into equal arcs, and that the one may have been only artificially brought into connection with the other, complete harmony between them being altogether impossible. And this view is fully sustained by evidence derivable from outside the Hindu science of astronomy, and beyond the borders of India. The Pârsîs, the Arabs, and the Chinese, are found also to be in possession of a similar system of division of the heavens into twenty-eight portions, marked or separated by as many single stars or constellations. Of the Pârsî system little or nothing is known excepting the number and names of the divisions, which are given in the second chapter of the Bundehesh (see Anquetil du Perron's Zendavesta, etc., ii. 349). The Arab divisions are styled ***manâzil al-ḳamar***, "lunar mansions, stations of the moon," being brought into special connection with the moon's revolution; they are marked, like the Hindu "portions," by groups of stars. The first extended comparison of the Hindu asterisms and the Arab mansions was made by Sir William Jones, in the second volume of the Asiatic Researches, for 1790: it was, however, only a rude and imperfect sketch, and led its author to no valuable or trustworthy conclusions. The same comparison was taken up later, with vastly more learning and acuteness, by Colebrooke, whose valuable article, published also in the Asiatic Researches, for 1807 (ix. 323, etc.; Essays ii. 321, etc.), has ever since remained the chief source of knowledge respecting the Hindu asterisms and their relation to the lunar mansions of the Arabs. To Anquetil (as above) is due the credit of the first suggestion of a coincidence between the Pârsî, Hindu, and Chinese systems: but he did nothing more than suggest it: the origin, character, and use of the Chinese divisions were first established, and their primitive identity with the Hindu asterisms demonstrated, by Biot, in a series of articles published in the Journal des Savants for 1840: and he has more recently, in the volume of the same Journal for 1859, reviewed and restated his former exposition and conclusions. These we shall present more fully hereafter: at present it will be enough to say that the Chinese divisions are equatorial, not zodiacal; that they are named *sieu*, "mansions"; and that they are the intervals in right ascension between certain single stars, which are also called *sieu*, and have the same title with the divisions which they introduce. We propose to present here a summary comparison of the Hindu, Arab, and Chinese systems, in connection with an identification of the stars and groups of stars forming the Hindu asterisms, and with the statement of such information respecting the latter, beyond that given in our text, as will best contribute to a full understanding of their character.

The identification of the asterisms is founded upon the positions of their principal or junction-stars, as stated in the astronomical text-books, upon the relative places of these stars in the groups of which they form a part, and upon the number of stars composing each group, and the figure by which their arrangement is represented: in a few cases, too, the names themselves of the asterisms are distinctive, and assist the identification. The number and configuration of the stars forming the groups are not stated in our text; we derive them mainly from Colebrooke, although ourselves also having had access to, and compared, most of his authorities, namely the Çâkalya-Sanhitâ, the Muhûrta-Cintâmaṇi, and the Ratnamâlâ (as cited by Jones, As. Res., ii. 294). Sir William Jones, it may be remarked, furnishes (As. Res., ii. 293, plate) an engraved copy of drawings made by a native artist of the figures assigned to the asterisms. For the number of stars in each group we have an additional authority in al-Bîrûnî, the Arab savant of the eleventh century, who travelled in India, and studied with especial care the Hindu astronomy. The information furnished by him with regard to the asterisms we derive from Biot, in the Journal des Savants for 1845 (pp. 39–54); it professes to be founded upon the Khaṇḍa-Kaṭaka* of Brahmagupta. Al-Bîrûnî also gives an identification of the asterisms, so far as the Hindu astronomers of his day were able to furnish it to him, which was only in part: he is obliged to mark seven or eight of the series as unknown or doubtful. He speaks very slightingly of the practical acquaintance with the heavens possessed by the Hindus of his time, and they certainly have not since improved in this respect; the modern investigators of the same subject, as Jones and Colebrooke, also complain of the impossibility of obtaining from the native astronomers of India satisfactory identifications of the asterisms and their junction-stars. The translator, in like manner, spent much time and effort in the attempt to derive such information from his native assistant, but was able to arrive at no results which could constitute any valuable addition to those of Colebrooke. It is evident that for centuries past, as at present, the native

---

* The true form of the name is not altogether certain, it being known only through its Arabic transcription: it seems to designate rather a chapter in a treatise than a complete work of its author.

tradition has been of no decisive authority as regards the position and composition of the groups of stars constituting the asterisms: these must be determined upon the evidence of the more ancient data handed down in the astronomical treatises.

In order to an exact comparison of the positions of the junction-stars as defined by the Hindus with those of stars contained in our catalogues, we have reduced the polar longitudes and latitudes to true longitudes and latitudes, by the following formulas (see Fig. 30):

$$(1 \div \cos \mathrm{A}a) \cot \mathrm{ELC} = \tan \mathrm{S}ab$$
$$\sin \mathrm{S}ab \sin \mathrm{S}a = \sin \mathrm{S}b$$
$$\tan \mathrm{S}b \cot \mathrm{S}ab = \sin ab$$

A *a* being the polar longitude as stated in the text ($= \mathrm{L}a + 180°$), S *a* the polar latitude, E L C the inclination of the ecliptic, S *b* the true latitude, and *a b* a quantity to be added to or subtracted from the polar longitude to give the true longitude. The true positions of the stars compared we take from Flamsteed's Catalogus Brittanicus, subtracting in each case 15° 42′ from the longitudes there given, in order to reduce them to distances from the vernal equinox of A. D. 560, assumed to coincide with the initial point of the Hindu sphere. There is some discordance among the different Hindu authorities, as regards the stated positions of the junction-stars of the asterisms. The Çâkalya-Sanhitâ, indeed, agrees in every point precisely with the Sûrya-Siddhânta. But the Siddhânta-Çiromaṇi often gives a somewhat different value to the polar longitude or latitude, or both. With it, so far as the longitude is concerned, exactly accord the Brahma-Siddhânta, as reported by Colebrooke, and the Khaṇḍa-Kaṭaka, as reported by al-Bîrûnî. The latitudes of the Brahma-Siddhânta also are virtually the same with those of the Siddhânta-Çiromaṇi, their differences never amounting, save in a single instance, to more than 3′: but the latitudes of the Khaṇḍa-Kaṭaka often vary considerably from both. The Graha-Lâghava, the only other authority accessible to us, presents a series of variations of its own, independent of those of either of the other treatises. All these differences are reported by us below, in treating of each separate asterism. The presiding divinities of the asterisms we give upon the authority of the Tâittirîya-Sanhitâ (iv. 4. 10. 1–3), the Tâittirîya-Brâhmaṇa (iii. 1. 1, 2, as cited by Weber, Zeitsch. f. d. K. d. Morg., vii. 266 etc., and Ind. Stud., i. 90 etc.), the Muhûrta-Cintâmaṇi, and Colebrooke: those of about half the asterisms are also indirectly given in our text, in the form of appellations for the asterisms derived from them.

The names and situations of the Arab lunar stations are taken from Ideler's Untersuchungen über die Sternnamen: for the Chinese mansions and their determining stars we rely solely upon the articles of Biot, to which we have already referred.

It has seemed to us advisable, notwithstanding the prior treatment by Colebrooke of the same subject, to enter into a careful re-examination and identification of the Hindu asterisms, because we could not accept in the bulk, and without modification, the conclusions at which he arrived. The identifications by Ideler of the Arab mansions, more thorough and correct than any which had been previously made, and Biot's comparison of the Chinese *sieu*, have placed new and valuable materials in our hands: and these—together with a more exact comparison than was attempted by Colebrooke of the positions given by the Hindus to their junction-stars with the data of the modern catalogues, and a new and independent combination of the various materials which he himself furnishes—while they have led us to accept the greater number of his identifications, often establishing them more confidently than he was able to do, have also enabled us in many cases to alter and amend his results. Such a re-examination was necessary, in order to furnish safe ground for a more detailed comparison of the three systems, which, as will be seen hereafter, leads to important conclusions respecting their historical relations to one another.

1. *Açvinî*; this treatise exhibits the form *açvini*; in the older lists, as also often elsewhere, we have the dual *açvinâu*, *açvayujâu*, "the two horsemen, or Açvins." The Açvins are personages in the ancient Hindu mythology somewhat nearly corresponding to the Castor and Pollux of the Greeks. They are the divinities of the asterism, which is named from them. The group is figured as a horse's head, doubtless in allusion to its presiding deities, and not from any imagined resemblance. The dual name leads us to expect to find it composed of two stars, and that is the number allotted to the asterism by the Çâkalya and Khanḍa-Kaṭaka. The Sûrya-Siddhânta (below, v. 16) designates the northern member of the group as its junction-star: that this is the star β Arietis (magn. 3.2), and not α Arietis (magn. 2), as assumed by Colebrooke, is shown by the following comparison of positions:

Açvinî	long., A. D. 560,	11° 59′	lat.	9° 11′ N.
β Arietis . . .	do.	13° 56′	do.	8° 28′ N.
α Arietis . . .	do.	17° 37′	do.	9° 57′ N.

Colebrooke was misled in this instance by adopting, for the number of stars in the asterism, three, as stated by the later authorities, and then applying to the group as thus composed the designation given by our text of the relative position of the junction-star as the northern, and he accordingly overlooked the very serious error in the determination of the longitude thence resulting. Indeed, throughout his comparison, he gives too great weight to the determination of latitude, and too little to that of longitude: we shall see farther on that the accuracy of the latter is, upon the whole, much more to be depended upon than that of the former.

Considered as a group of two stars, Açvinî is composed of β and γ Arietis (magn. 4.3); as a group of three, it comprises also α in the same constellation.

There is no discordance among the different authorities examined by us as regards the position of the junction-star of Açvinî, either in latitude or in longitude. The case is the same with the 8th, 10th, 12th, and 13th asterisms, and with them alone.

The first Arab *manzil* is likewise composed of β and γ Arietis, to which some add α: it is called ash-Sharaṭân, "the two tokens"—that is to say, of the opening year.

The Chinese series of *sieu* commences, as did anciently the Hindu system of asterisms, with that which is later the third asterism. The twenty-seventh *sieu*, named Leu (M. Biot has omitted to give us the signification of these titles), is β Arietis, the Hindu junction-star.

2. *Bharaṇî;* also, as plural, *bharaṇyas;* from the root *bhar*, "carry": in the Tâittirîya lists the form *apabharaṇî*, "bearer away," in singular and plural, is also found. Its divinity is Yama, the ruler of the world of departed spirits; it is figured as the *yoni*, or pudendum muliebre. All authorities agree in assigning it three stars, and the southernmost is pointed out below (v. 18) as its junction-star. The group is unquestionably to be identified with the triangle of faint stars lying north of the back of the Ram, or 35, 39, and 41 Arietis: they are figured by some as a distinct constellation, under the name of Musca Borealis. The designation of the southern as the junction-star is not altogether unambiguous, as 35 and 41 were, in A. D. 560, very nearly equidistant from the equator; the latter would seem more likely to be the one intended, since it is nearer the ecliptic, and the brightest of the group—being of the third magnitude, while the other two are of the fourth: the defined position, however, agrees better with 35, and the error in longitude, as compared with 41, is greater than that of any other star in the series:

Bharaṇî	24° 35′	11° 6′ N.
35 Arietis (*a* Muscæ) . .	26° 54′	11° 17′ N.
41 Arietis (*c* Muscæ) . .	28° 10′	10° 26′ N.

The Graha-Lâghava gives Bharaṇî 1° more of polar longitude: this would reduce by the same amount the error in the determination of its longitude by the other authorities.

The second Arab *manzil*, al-Buṭain, "the little belly"—i. e., of the Ram—is by most authorities defined as comprising the three stars in the haunch of the Ram, or ε, δ, and ϱ^3 (or else ζ) Arietis. Some, however, have regarded it as the same with Musca; and we cannot but think that al-Bîrûnî, in identifying, as he does, Bharaṇî with al-Buṭain, meant to indicate by the latter name the group of which the Hindu asterism is actually composed.

The last Chinese *sieu*, Oei, is the star 35 Arietis, or *a* Muscæ.

3. *Kṛttikâ;* or, as plural, *kṛttikâs:* the appellative meaning of the word is doubtful. The regent of the asterism is Agni, the god of fire. The group, composed of six stars, is that known to us as the Pleiades. It is figured by some as a flame, doubtless in allusion to its presiding divinity: the more usual representation of it is a razor, and in the choice of this symbol is to be recognized the influence of the etymology of the name, which may be derived from the root *kart*, "cut;" in the configuration of the group, too, may be seen, by a sufficiently prosaic eye, a broad-bladed knife, with a short handle. If the designation given below (v. 18) of the southern member of the group as its junction-star, be strictly true, this is not Alcyone, or η Tauri (magn. 3), the brightest of the six, but either Atlas (27 Tauri: magn. 4) or Merope (23 Tauri: magn. 5): the two latter were very nearly equally distant from the equator of A. D. 560, but Atlas is a little nearer to the ecliptic. The defined position agrees best with Alcyone, nor can we hesitate to regard this as actually the junction-star of the asterism. We compare the positions below:

Kṛttikâ	39° 8′	4° 44′ N.
Alcyone	39° 58′	4° 1′ N.
27 Tauri	40° 20′	3° 53′ N.
23 Tauri	39° 41′	3° 55′ N.

The Siddhânta-Çiromaṇi etc. give Kṛttikâ 2′ less of polar longitude than the Sûrya-Siddhânta, and the Graha-Lâghava, on the other hand, 30′ more: the latter, with the Khaṇḍa-Kaṭaka, agree with our text as regards the polar latitude, which the others reckon at 4° 30′, instead of 5°.

The Pleiades constitute the third *manzil* of the Arabs, which is denominated ath-Thuraiyâ, "the little thick-set group," or an-Najm, "the constellation." Alcyone is likewise the first Chinese *sieu*, which is styled Mao.

4. *Rohiṇî*, "ruddy"; so named from the hue of its principal star. Prajâpati, "the lord of created beings," is the divinity of the asterism. It contains five stars, in the grouping of which Hindu fancy has seen the figure of a wain (compare v. 13, below); some, however, figure it as a temple. The constellation is the well-known one in the face of Taurus to which we give the name of the Hyades, containing ϵ, δ, γ, ϑ, α Tauri; the latter, the most easterly (v. 19) and the brightest of the group—being the brilliant star of the first magnitude known as Aldebaran—is the junction-star, as is shown by the annexed comparison of positions:

Rohiṇî	48° 9′	4° 49′ S.
Aldebaran . . .	49° 45′	5° 30′ S.

The Siddhânta-Çiromaṇi etc. here again present the insignificant variation from the polar longitude of our text, of 2′ less: the former also makes its polar latitude 4° 30′: the Graha-Lâghava reads, for the polar longitude, 49°. All these variations add to the error of defined position.

The fourth Arab *manzil* is composed of the Hyades: its name is ad-Dabarân, "the follower"—i. e., of the Pleiades. We would suggest the inquiry whether this name may not be taken as an indication that the Arab system of mansions once began, like the Chinese, and like the Hindu system originally, with the Pleiades. There is, certainly, no very obvious propriety in naming any but the second of a series the "following" (*sequens* or *secundus*). Modern astronomy has retained the title as that of the principal star in the group, to which alone it was often also applied by the Arabs.

The second Chinese *sieu*, Pi, is the northernmost member of the same group, or ϵ Tauri, a star of the third to fourth magnitude.

5. *Mṛgaçîrsha*, or *mṛgaçiras*, "antelope's head": with this name the figure assigned to the asterism corresponds: the reason for the designation we have not been able to discover. Its divinity is Soma, or the moon. It contains three stars, of which the northern (v. 16) is the determinative. These three can be no other than the faint cluster in the head of Orion, or λ, φ^1, φ^2 Orionis, although the Hindu measurement of the position of the junction-star, λ (magn. 4), is far from accurate, especially as regards its latitude:

Mṛgaçîrsha	61° 3′	9° 49′ S.
λ Orionis	63° 40′	13° 25′ S.

In this erroneous determination of the latitude all authorities agree: the Graha-Lâghava adds 1° to the error in polar longitude, reading 62° instead of 63°.

Here again there is an entire harmony among the three systems compared. The Arab *manzil*, al-Haḳ'ah, is composed of the same stars which make up the Hindu asterism: the third *sieu*, named Tse, is the Hindu junction-star, λ Orionis.

6. *Ârdrâ*, "moist:" the appellation very probably has some meteorological ground, which we have not traced out: this is indicated also by the choice of Rudra, the storm-god, as regent of the asterism. It comprises a single star only, and is figured as a gem. It is impossible not to regard the bright star of the first magnitude in Orion's right shoulder, or α Orionis, as the one here meant to be designated, notwithstanding the very grave errors in the definition of its position given by our text: the only visible star of which the situation at all nearly answers to that definition is 135 Tauri, of the sixth magnitude; we add its position below, with that of α Orionis:

Ârdrâ	65° 50′	8° 53′ S.
α Orionis	68° 43′	16° 4′ S.
135 Tauri . . .	67° 38′	9° 10′ S.

The distance from the sun at which the heliacal rising and setting of Ârdrâ is stated below (ix. 14) to take place would indicate a star of about the third magnitude; this adds to the difficulty of its identification with either of the two stars compared. We confess ourselves unable to account for the confusion existing with regard to this asterism, of which al-Bîrûnî also could obtain no intelligible account from his Indian teachers. But it is to be observed that all the authorities, excepting our text and the Çâkalya-Sanhitâ, give Ârdrâ 11° of polar latitude instead of 9°, which would reduce the error of latitude, as compared with α Orionis, to an amount very little greater than will be met with in one or two other cases below, where the star is situated south of the ecliptic; and it is contrary to all the analogies of the system that a faint star should have been selected to form by itself an asterism. The Siddhânta-Çiromaṇi etc. make the polar longitude of the asterism 20′ less than that given by the Sûrya-Siddhânta, and the Graha-Lâghava 1° 20′ less: these would add so much to the error of longitude.

Here, for the first time, the three systems which we are comparing disagree with one another entirely. The Chinese have adopted for the determinative of their fourth *sieu*, which is styled Tsan, the upper star in Orion's belt, or δ Orionis (2)—a strange and arbitrary selection, for which M. Biot is unable to find any explanation. The Arabs have established their sixth station close to the ecliptic, in the feet of Pollux, naming it al-Han'ah, "the pile": it comprises the two stars γ (2.3) and ξ (4.3) Geminorum: some authorities, however, extend the limits of the mansion so far as to include also the stars in the foot of the other twin, or η, ν, μ Geminorum; of which the latter is the next Chinese *sieu*.

7. *Punarvasu;* in all the more ancient lists the name appears as a dual, *punarvasû*: it is derived from *punar*, "again," and *vasu*, "good, brilliant": the reason of the designation is not apparent. The regent

of the asterism is Aditi, the mother of the Âdityas. Its dual title indicates that it is composed of two stars, of nearly equal brilliancy, and two is the number allotted to it by the Çâkalya and Khaṇḍa-Kaṭaka, the eastern being pointed out below (v. 19) as the junction-star. The pair are the two bright stars in the heads of the Twins, or α and β Geminorum, and the latter (1.2) is the junction-star. The comparison of positions is as follows:

Punarvasu	92° 52′	6° 0′ N.
β Geminorum . . .	93° 14′	6° 39′ N.

The Graha-Lâghava adds 1° to the polar longitude of Punarvasu as stated by the other authorities.

Four stars are by some assigned to this asterism, and with that number corresponds the representation of its arrangement by the figure of a house: it is quite uncertain which of the neighboring stars of the same constellation are to be added to those above mentioned to form the group of four, but we think ι (magn. 4) and υ (5) those most likely to have been chosen: Colebrooke suggests ϑ (3.4) and τ (5.4).

The determinative of the fifth *sieu*, Tsing, is μ Geminorum (3), which, as we have seen, is reckoned among the stars composing the sixth *manzil*: the seventh *manzil* includes, like the Hindu asterism, α and β Geminorum: it is named adh-Dhirâ', "the paw"—i. e., of the Lion; the figure of Leo (see Ideler, p. 152 etc.) being by the Arabs so stretched out as to cover parts of Gemini, Cancer, Canis Minor, and other neighboring constellations.

8. *Pushya;* from the root *push*, "nourish, thrive"; another frequent name, which is the one employed by our treatise, is *tishya*, which is translated "auspicious"; Amara gives also *sidhya*, "prosperous." Its divinity is Bṛhaspati, the priest and teacher of the gods. It comprises three stars—the Khaṇḍa-Kaṭaka alone seems to give it but one—of which the middle one is the junction-star of the asterism. This is shown by the position assigned to it to be δ Cancri (4):

Pushya	106° 0′	0° 0′
δ Cancri	108° 42′	0° 4′ N.

The other two are doubtless γ (4.5) and ϑ (6) of the same constellation: the asterism is figured as a crescent and as an arrow, and the arrangement of the group admits of being regarded as representing a crescent, or the barbed head of an arrow. Were the arrow the only figure given, it might be possible to regard the group as composed of γ, ϑ, and β (4), the latter representing the head of the arrow, and the nebulous cluster, Præsepe, between γ and ϑ, the feathering of its shaft: ϑ (105° 43′—0° 48′ S.) would then be the junction-star.

The Arab *manzil*, an-Nathrah, "the nose-gap"—i. e., of the Lion—comprises γ and δ Cancri, together with Præsepe; or, according to some authorities, Præsepe alone. The sixth *sieu*, Kuei, is ϑ Cancri, a star which is, at present, only with difficulty distinguished by the naked eye. Ptolemy rates it as of the fourth magnitude, like γ and δ: perhaps it is one of the stars of which the brilliancy has sensibly diminished during the past two or three thousand years, or else a variable star of very long period. The possibility of such changes requires to be taken into account, in comparing our heavens with those of so remote a past.

9. *Âçleshâ ;* or, as plural, *âçleshâs ;* the word is also written *âçreshâ:* its appellative meaning is "entwiner, embracer." With the name accord the divinities to whom the regency of the asterism is assigned, which are *sarpâs,* the serpents. The number of stars in the group is stated as five by all the authorities excepting the Khaṇḍa-Kaṭaka, which reads six: their configuration is represented by a wheel. The star α Cancri (4) is pointed out by Colebrooke as the junction-star of Âçleshâ, apparently from the near correspondence of its latitude with that assigned to the latter, for he says nothing in connection with it of his native helpers: but α Cancri is not the eastern (v. 19) member of any group of five stars; nor, indeed, is it a member of any distinct group at all. Now the name, figure, and divinity of Âçleshâ are all distinctive, and point to a constellation of a bent or circular form: and if we go a little farther southward from the ecliptic, we find precisely such a constellation, and one containing, moreover, the corresponding Chinese determinative. The group is that in the head of Hydra, or η, σ, δ, ε, ρ Hydræ, σ and ρ being of the fifth magnitude, and the rest of the fourth: their arrangement is conspicuously circular. There can be no doubt, therefore, that the situation of the asterism is in the head of Hydra, and ε Hydræ, its brightest star (being rated in the Greenw. Cat. as of magnitude 3.4, while δ is 4.5), is the junction-star:

Âçleshâ	109° 59′	6° 56′ S.
ε Hydræ	112° 20′	11° 8′ S.
α Cancri	113° 5′	5° 31′ S.

The error of the Hindu determination of the latitude is, indeed, very considerable, yet not greater than we are compelled to accept in one or two other cases. The Khaṇḍa-Kaṭaka increases it 1°, giving the asterism 6° instead of 7° of polar latitude. The Siddhânta-Çiromaṇi etc. deduct 1° from the polar longitude of the Sûrya-Siddhânta, and the Graha-Lâghava deducts 2°: both variations would add to the error in longitude.

The Arab *manzil* is, in this instance, far removed from the Hindu asterism, being composed of ξ Cancri (5) and λ Leonis (5.4), and called at-Ṭarf, "the look"—i. e., of the Lion. The seventh Chinese *sieu,* Lieu, is, as already noticed, included in the Hindu group, being δ Hydræ.

10. *Maghâ ;* or, as plural, *maghâs ;* "mighty." The *pitaras,* Fathers, or *manes* of the departed, are the regents of the asterism, which is figured as a house. It is, according to most authorities, composed of five stars, of which the southern (v. 18) is the junction-star. Four of these must be the bright stars in the neck and side of the Lion, or ζ, γ, η, and α Leonis, of magnitudes 4.5, 2, 3.4, and 1.2 respectively; but which should be the fifth is not easy to determine, for there is no other single star which seems to form naturally a member of the same group with these: ν (5), π (5), or ρ (4) might be forced into a connection with them. This difficulty would be removed by adopting, with the Khaṇḍa-Kaṭaka, six as the number of stars included in the asterism: it would then be composed of all the stars forming the conspicuous constellation familiarly known as "the Sickle." The star α Leonis, or Regulus, the most brilliant of the group, is the junction-star, and its position is defined with unusual precision:

Maghâ	129° 0′	0° 0′
Regulus	129° 49′	0° 27′ N.

The tenth *manzil*, aj-Jabhah, "the forehead"—i. e., of the Lion—is also composed of ζ, γ, η, α Leonis.

The eighth, ninth, and tenth *sieu* of the Chinese system altogether disagree in position with the groups marking the Hindu and Arab mansions, being situated far to the southward of the ecliptic, in proximity, according to Biot, to the equator of the period when they were established. The eighth, Sing, is α Hydræ (2), having longitude (A. D. 560) 127° 16′, latitude 22° 25′ S.

11, 12. *Phalgunî;* or, as plural, *phalgunyas;* the dual, *phalgunyâu*, is also found: this treatise presents the derivative form *phâlgunî*, which is not infrequently employed elsewhere. The word is likewise used to designate a species of fig-tree: its derivation, and its meaning, as applied to the asterisms, is unknown to us. Here, as in two other instances, later (the 20th and 21st, and the 26th and 27th asterisms), we have two groups called by the same name, and distinguished from one another as *pûrva* and *uttara*, "former" and "latter"—that is to say, coming earlier and later to their meridian-transit. The true original character and composition of these three double asterisms has been, if we are not mistaken, not a little altered and obscured in the description of them furnished to us; owing, apparently, to the ignorance or carelessness of the describers, and especially to their not having clearly distinguished the characteristics of the combined constellation from those of its separate parts. In each case, a couch or bedstead (*çayyâ, mañca, paryanka*) is given as the figure of one or both of the parts, and we recognize in them all the common characteristic of a constellation of four stars, forming together a regular oblong figure, which admits of being represented—not unsuitably, if rather prosaically—by a bed. This figure, in the case of the Phalgunîs, is composed of δ, ϑ, β, and 93 Leonis, a very distinct and well-marked constellation, containing two stars, δ and β, of the second to third magnitude, one, ϑ, of the third, and one, 93, of the fourth. The symbol of a bed, properly belonging to the whole constellation, is given by all the authorities to both the two parts into which it is divided. Each of these latter has two stars assigned to it, and the junction-stars are said (v. 18) to be the northern. The first group is, then, clearly identifiable as δ and ϑ Leonis, the former and brighter being the distinctive star:

Pûrva-Phalgunî . . .	139° 58′	11° 19′ N.
δ Leonis	141° 15′	14° 19′ N.
ϑ Leonis	143° 24′	9° 40′ N.

The Siddhânta-Çiromaṇi etc., and the Graha-Lâghava, give Pûrva-Phalgunî respectively 3° and 4° more of polar longitude than the Sûrya-Siddhânta. These are more notable variations than are found in any other case, and they appear to us to indicate that these treatises intend to designate ϑ, the southern member of the group, as its junction-star: we have accordingly added its position also above.

In the latter group, the junction-star is evidently β Leonis:

Uttara-Phalgunî . . .	150° 10′	12° 5′ N.
β Leonis	151° 37′	12° 17′ N.

This star, however, is not the northern, but the southern, of the two composing the asterism: its description as the southern we cannot but regard as simply an error, founded on a misapprehension of the composition of the double group. To al-Bîrûnî, β Leonis and another star to the northward, in the Arab constellation Coma Berenices, were pointed out as forming the asterism Uttara-Phalgunî. The Çâkalya gives it five stars, probably adding to β Leonis the four small stars in the head of the Virgin, ξ^1, ν, π, and o, of magnitudes four to five and five.

The regents of Pûrva and Uttara-Phalgunî are Bhaga and Aryaman, or Aryaman and Bhaga, two of the Âdityas.

The two corresponding Arab mansions are called az-Zubrah, "the mane"—i. e., of the Lion—and as-Sarfah, "the turn": they agree as nearly as possible with the Hindu asterisms, the former being composed of δ and ϑ Leonis, the latter of β Leonis alone. The Chinese *sieu*, named respectively Chang and Y, are υ^1 Hydræ (5)* and α Crateris (4).

13. *Hasta*, "hand." Savitar, the sun, is regent of the asterism, which, in accordance with its name, is figured as a hand, and contains five stars, corresponding to the five fingers. These are the five principal stars in the constellation Corvus, a well-marked group, which bears, however, no very conspicuous resemblance to a hand. The stars are named—counting from the thumb around to the little finger, according to our apprehension of the figure—β, α, ϵ, γ, and δ Corvi. The text gives below (v. 17) a very special description of the situation of the junction-star in the group, but one which is unfortunately quite hard to understand and apply: we regard it as most probable, however (see note to v. 17), that γ (3) is the star intended: the defined position, in which all the authorities agree, would point rather to δ (3):

Hasta	174° 22′	10° 6′ S.
γ Corvi	170° 44′	14° 29′ S.
δ Corvi	173° 27′	12° 10′ S.

The Hindu and Chinese systems return, in this asterism, to an accordance with one another: the eleventh *sieu*, Chin, is the star γ Corvi. The Arab system holds its own independent course one point farther: its thirteenth mansion comprises the five bright stars β, η, γ, δ, ϵ Virginis, which form two sides, measuring about 15° each, of a great triangle: the mansion is named al-Auwâ', "the barking dog."

14. *Citrâ*, "brilliant." This is the beautiful star of the first magnitude α Virginis, or Spica, constituting an asterism by itself, and figured as a pearl or as a lamp. Its divinity is Tvashtar, "the shaper, artificer." Its longitude is very erroneously defined by the Sûrya-Siddhânta:

Citrâ	180° 48′	1° 50′ S.
Spica	183° 49′	2° 2′ S.

All the other authorities, however, saving the Çâkalya, remove this error, by giving Citrâ 183° of polar longitude, instead of 180°. The only variation from the definition of latitude made by our text is offered by the Siddhânta-Çiromaṇi, which, varying for once from the Brahma-Siddhânta, reads 1° 45′ instead of 2°.

---

* It is, apparently, by an original error of the press, that M. Biot, in all his tables, calls this star ν^1.

Spica is likewise the fourteenth *manzil* of the Arabs, styled by them as-Simâk, and the twelfth *sieu* of the Chinese, who call it Kio.

15. *Svâti*, or *svâti;* the word is said to mean "sword." The Tâittirîya-Brâhmaṇa calls the asterism *nishṭyâ*, "outcast," possibly from its remote northern situation. It is, like the last, an asterism comprising but a single brilliant star, which is figured as a coral bead, gem, or pearl. In the definition of its latitude all authorities agree; the Graha-Lâghava makes its polar longitude 198° only, instead of 199°. The star intended is plainly α Bootis, or Arcturus:

Svâtî	183° 2′	33° 50′ N.
Arcturus. . . .	184° 12′	30° 57′ N.

In this instance, the Hindus have gone far beyond the limits of the zodiac, in order to bring into their series of asterisms a brilliant star from the northern heavens: the other two systems agree in remaining near the ecliptic. The fourteenth Chinese *sieu*, Kang, is κ Virginis (4.5): the Arab *manzil*, al-Ghafr, "the covering," includes the same star, together with ι, and either λ or φ Virginis.

16. *Viçâkhâ*, "having spreading branches": in all the earlier lists the name appears as a dual, *viçâkhe.* The asterism is also placed under the regency of a dual divinity, *indrâgnî*, Indra and Agni. We should expect, then, to find it composed, like the other two dual asterisms, the 1st and 7th, of two stars, nearly equal in brilliancy, and two is actually the number assigned to the group by the Çâkalya and the Khaṇḍa-Kaṭaka. Now the only two stars in this region of the zodiac forming a conspicuous pair are α and β Libræ, both of the second magnitude, and as these two compose the corresponding Arab mansion, while the former of them is the Chinese *sieu*, we have the strongest reasons for supposing them to constitute the Hindu asterism also. There are, however, difficulties in the way of this assumption. The later authorities give Viçâkhâ four stars, and the defined position of the junction-star identifies it neither with α nor β, but with the faint star ι (4.3) in the the same constellation. Colebrooke, overlooking this star, suggests α or κ Libræ (5): the following comparison of positions will show that neither of them can be the one meant to be pointed out:

Viçâkhâ	213° 31′	1° 25′ S.
ι Libræ	211° 0′	1° 48′ S.
α Libræ	205° 5′	0° 23′ N.
κ Libræ	217° 45′	0° 2′ N.

The group is figured as a *toraṇa:* this word Jones and Colebrooke translate "festoon," but its more proper meaning is "an outer door or gate, a decorated gateway." And if we change the designation of situation of the junction-star in its group, given below (v. 16), from "northern" to "southern," we find without difficulty a quadrangle of stars, viz. ι, α, β, γ (4.5) Libræ, which admits very well of being figured as a gateway. Nor is it, in our opinion, taking an unwarrantable liberty to make such an alteration. The whole scheme of designations we regard as of inferior authenticity, and as partaking of the confusion and uncertainty of the later knowledge of the Hindus respecting their system of asterisms. That they were long ago doubtful of the position of Viçâkhâ is shown by the fact that al-Bîrûnî was obliged to mark it in his list as "unknown." Very probably the Sûrya-Siddhânta, in calling ι the northern member of the group, intended to include with it only the star 20 Libræ (3.4), situated about 6° to the south of it. Upon the whole, then, while we regard the identification of Viçâkhâ as in some respects more doubtful than that of any other asterism in the series, we yet believe that it was originally composed of the two stars α and β Libræ, and that later the group was extended to include also ι and γ, and, as so extended, was figured as a gateway. The selection, contrary to general usage, of the faintest star in the group as its junction-star, may have been made in order to insure against the reversion of the asterism to its original dual form.

The variations of the other authorities from the position as stated in our text are of small importance: the Siddhânta-Çiromaṇi etc. give Viçâkhâ 55′ less of polar longitude, and the Graha-Lâghava 1° less; of polar latitude, the Siddhânta-Çiromaṇi gives it 10′, the Graha-Lâghava 30′ less; the Khaṇḍa-Kaṭaka agrees here, as also in the two following asterisms, with the Sûrya-Siddhânta.

The sixteenth Arab *manzil,* comprising, as already noticed, α and β Libræ, is styled az-Zubânân, "the two claws"—i. e., of the Scorpion: the name of the corresponding Chinese mansion, having for its determinative α Libræ, is Ti.

17. *Anurâdhâ;* or, as plural, *anurâdhâs:* the word means "success." The divinity is Mitra, "friend," one of the Âdityas. According to the Çâkalya, the asterism is composed of three stars, and with this our text plainly agrees, by designating (v. 18) the middle as the junction-star: all the other authorities give it four stars. As a group of three, it comprises β, δ, π Scorpionis, δ (2.3) being the junction-star; as the fourth member we are doubtless to add ϱ Scorpionis (5.4). It is figured as a *bali* or *vali;* this Colebrooke translates "a row of oblations"; we do not find, however, that the word, although it means both "oblation, offering," and "a row, fold, ridge," is used to designate the two combined: perhaps it may better be taken as simply "a row;" the stars of the asterism, whether considered as three or four, being disposed in nearly a straight line. The comparison of positions is as follows:

Anurâdhâ	224° 44′	2° 52′ S.
δ Scorpionis . . .	222° 34′	1° 57′ S.

The Siddhânta-Çiromaṇi and Graha-Lâghava estimate the latitude of Anurâdhâ somewhat more accurately, deducting from the polar latitude, as given by our text, 1° 15′ and 1° respectively: the Siddhânta-Çiromaṇi etc. also add the insignificant amount of 5′ to the polar longitude of the Sûrya-Siddhânta.

The corresponding Arab *manzil,* named al-Iklîl, "the crown," contains also the three stars β, δ, π Scorpionis, some authorities adding ϱ to the group. The Chinese *sieu,* Fang, is π (3), the southernmost and the faintest of the three.

18. *Jyeshṭhâ,* "oldest." The Tâittirîya-Sanhitâ, in its list of asterisms, repeats here the name *rohiṇî,* "ruddy," which we have had above as that of the 4th asterism: the appellation has the same ground in this

as in the other case, the junction-star of Jyeshṭhâ being also one of those which shine with a reddish light. The regent is Indra, the god of the clear sky. The group contains, according to all the authorities, three stars, and the central one (v. 18) is the junction-star. This is the brilliant star of the first magnitude α Scorpionis, or Antares; its two companions are σ (3.4) and τ (3.4) in the same constellation:

Jyeshṭhâ	230° 7′	3° 50′ S.
Antares	229° 44′	4° 31′ S.

The constellation is figured as a ring, or ear-ring; by this may be understood, perhaps, a pendent ear-jewel, as the three stars of Jyeshṭhâ form nearly a straight line, with the brightest in the middle.

The Siddhânta-Çiromaṇi and Graha-Lâghava add to the polar longitude of the junction-star of the asterism, as stated in our text, 5′ and 1° respectively, and they deduct from its polar latitude 30′ and 1° respectively, making the definition of its position in both respects less accurate.

Antares forms the eighteenth *manzil*, and is styled al-Ḳalb, "the heart"—i. e., of the Scorpion: σ and τ are called an-Niyât, "the *præcordia*." The Chinese *sieu*, Sin, is the westernmost of the three, or σ.

19. *Mûla*, "root." The presiding divinity of the asterism is *nirṛti*, "calamity," who is also regent of the south-western quarter. It comprises, according to the Çâkalya, nine stars; their configuration is represented by a lion's tail. The stars intended are those in the tail of the Scorpion, or ϵ, μ, ζ, η, ϑ, ι, $\varkappa$, υ, λ Scorpionis, all of them of the third, or third to fourth, magnitude. Other authorities count eleven stars in the group, probably reckoning μ and ζ as four stars; each being, in fact, a group of two closely approximate stars, named in our catalogues μ^1 (3), μ^2 (4), ζ^1 (4.5), ζ^2 (3). The Khaṇḍa-Kaṭaka alone gives Mûla only two stars, which are identified by al-Bîrûnî with the Arab *manzil* ash-Shaulah, or λ and υ Scorpionis. The Tâittirîya-Sanhitâ, too, gives the name of the asterism as *vicṛtâu*, "the two releasers": the Vicṛtâu are several times spoken of in the Atharva-Veda as two stars of which the rising promotes relief from lingering disease (*kshetriya*): it is accordingly probable that these are the two stars in the sting of the Scorpion, and that they alone have been regarded by some as composing the asterism: their healing virtue would doubtless be connected with the meteorological conditions of the time at which their heliacal rising takes place. Our text (v. 19) designates the eastern member of the group as its junction-star: it is uncertain whether the direction is meant to apply to the group of two, or to that of nine stars: if, as seems probable, λ is the star pointed out by the definition of position, it is strictly true only of the pair λ and υ, since ι, $\varkappa$, and ϑ are all farther eastward than λ:

Mûla	242° 52′	8° 48′ S.
λ Scorpionis . . .	244° 53′	13° 44′ S.

The Graha-Lâghava gives a more accurate statement of the longitude, adding 1° to the polar longitude as defined by all the other authorities: but it increases the error in latitude, by deducting 1° from that presented by our text: the Siddhânta-Çiromaṇi, in like manner, deducts 30′, while the Khaṇḍa-Kaṭaka adds the same amount.

The Tâittirîya-Sanhitâ makes *pitaras*, the Fathers, the presiding divinities of this asterism, as well as of the tenth.

Bentley states (Hind. Astr., p. 5) that Mûla was originally reckoned as the first of the asterisms, and was therefore so named, as being their root or origin; also that, at another time, or in a different system, the series was made to begin with Jyeshṭhâ, which thence received its title of "eldest." These statements are put forth with characteristic recklessness, and apparently, like a great many others in his pretended history of Hindu astronomy, upon the unsupported authority of his own conjecture. It is, in many cases, by no means easy to discover reasons for the particular appellations by which the asterisms are designated: but we would suggest that Mûla may perhaps have been so named from its being considerably the lowest, or farthest to the southward, of the whole series of asterisms, and hence capable of being looked upon as the root out of which they had grown up the heavens. It would even be possible to trace the same conception farther, and to regard Jyeshṭhâ as so styled because it was the first, or "oldest," outgrowth from this root, while the Viçâkhe, "the two diverging branches," were the stars in which the series broke into two lines, the one proceeding northward, to Svâtî or Arcturus, the other westward, to Citrâ or Spica. We throw out the conjecture for what it may be worth, not being ourselves at all confident of its accordance with the truth.

The nineteenth Arab *manzil* is styled ash-Shaulah, "the sting"—i. e., of the Scorpion—and comprises, as already noticed, υ and λ Scorpionis. The determinative of the seventeenth *sieu*, Uei, is included in the Hindu asterism, being μ^2 Scorpionis.

20, 21. *Ashâḍhâ;* or, as plural, *ashâḍhâs;* this treatise presents the derivative form *âshâḍhâ*, which is not infrequent elsewhere: the word means "unsubdued." Here, again, we have a double group, divided into two asterisms, which are distinguished as *pûrva* and *uttara*, "former and latter." Their respective divinities are *âpas*, "the waters," and *viçve devâs*, "the collective gods." Two stars are ordinarily allotted to each asterism, and in each case the northern is designated (v. 16) as the junction-star. By some authorities each group is figured as a bed or couch; by others, the one as a bed and the other as an elephant's tusk; and here, again, there is a difference of opinion as to which is the bed and which the tusk. The true solution of this confusion is, as we conceive, that the two asterisms taken together are figured as a bed, while either of them alone is represented by an elephant's tusk. The former group must comprise δ (3.4) and ϵ (3.2) Sagittarii, the former being the junction-star; this is shown by the following comparison of positions:

Pûrva-Ashâḍhâ	254° 39′	5° 28′ S.
δ Sagittarii	254° 32′	6° 25′ S.

The Graha-Lâghava gives Pûrva-Ashâḍhâ 1° more of polar longitude, and 30′ less of polar latitude, than the Sûrya-Siddhânta: the Siddhânta-Çiromaṇi etc. give it 10′ less of the latter.

The latter of the two groups contains, as its southern star, ζ Sagittarii (3.4), and its northern and junction-star can be no other than σ (2.3) in the same constellation, notwithstanding the error in the Hindu determination of its latitude, which led Colebrooke to regard τ (4.3) as the star intended: we subjoin the positions:

Uttara-Ashâḍhâ	260° 23′	4° 59′ S.
σ Sagittarii	262° 21′	3° 24′ S.
τ Sagittarii	264° 48′	5° 1′ S.

The only variation from the position of the junction-star of this asterism as stated in our text is presented by the Graha-Lâghava, which makes its polar longitude 261° instead of 260°.

The Çâkalya (according to Colebrooke: our MS. is defective at this point) and the Khaṇḍa-Kaṭaka assign four stars to each of the Ashâḍhâs, and the former represents each as a bed. It would not be difficult to establish two four-sided figures in this region of the constellation Sagittarius, each including the stars above mentioned, with two others: the one would be composed of γ^2 (4.3), δ, ε, η (4—the star is also called β Telescopii), the other of φ (4.3), σ, τ, and ζ: such is unquestionably the constitution of the two asterisms, considered as groups of four stars; they are thus identified also, it may be remarked, by al-Bîrûnî. The junction-stars would still be δ and σ, which are the northernmost in their respective constellations; nor is there any question as to which four among the eight are selected to make up the double asterism, since δ, ε, ζ, and σ both form the most regular quadrangular figure, and are the brightest stars.

The determinatives of the eighteenth and nineteenth mansions of the Chinese, Ki and Teu, are γ^2 and φ Sagittarii, which are included in the two quadruple groups as stated above. The twentieth *manzil* comprehends all the eight stars which we have mentioned, and is styled an-Na'âim, "the pasturing cattle": some also understand each group of four as representing an ostrich, na'âm. The twenty-first *manzil*, on the other hand, al-Baldah, "the town," is described as a vacant space above the head of Sagittarius, bounded by faint stars, among which the most conspicuous is π Sagittarii (4.5).

22. *Abhijit*, "conquering." The regent of the asterism is Brahma. The position assigned to its junction-star, which is described as the brightest (v. 19) in a group of three, identifies it with α Lyræ, or Vega, a star which is exceeded in brilliancy by only one or two others in the heavens:

Abhijit	264° 10′	59° 58′ N.
Vega	265° 15′	61° 46′ N.

The other authorities compared (excepting the Çâkalya) define the position in latitude of Abhijit more accurately, adding 2° to the polar latitude given by the Sûrya-Siddhânta: the Graha-Lâghava also improves the position in longitude by adding 1° 20′, while the Siddhânta-Çiromaṇi etc. increase the error by deducting 1° 40′.

The Tâittirîya-Sanhitâ (iv. 4. 10) omits Abhijit from its list of the asterisms: the probable reason of its omission in some authorities, or in certain connections, and its retention in others, we shall discuss farther on.

Abhijit is figured as a triangle, or as the triangular nut of the *çṛngâṭa*, an aquatic plant; this very distinctly represents the grouping of α Lyræ with the two other fainter stars of the same constellation, ε and ζ, both of the fifth magnitude.

In this and the two following asterisms—as once before, in the fifteenth of the series—the Hindus have gone far from the zodiac, in order to bring into their system brilliant stars from the northern heavens, while the Chinese and the Arab systems agree in remaining in the immediate neighborhood of the ecliptic. The twentieth *sieu* is named Nieu, and is the star β Capricorni (3), situated in the head of the Goat: the twenty-second *manzil*, Sa'd adh-Dhâbiḥ, "felicity of the sacrificer," contains the same star, the group being α (composed of two stars, each of magnitude 3.4) and β Capricorni.

23. *Çravaṇa,* "hearing, ear"; from the root *çru,* "hear": another name for the asterism, *çroṇâ,* found occurring in the Tâittirîya lists, is perhaps from the same root, but the word means also "lame." Çravaṇa comprises three stars, of which the middle one (v. 18) is the junction-star: they are to be found in the back and neck of the Eagle, namely as γ, α, and β Aquilæ; α, the determinative, is a star of the first to second magnitude, while γ and β are of the third and fourth respectively:

Çravaṇa	282° 29′	29° 54′ N.
α Aquilæ . . .	281° 41′	29° 11′ N.

All the authorities agree as to the polar latitude of Çravaṇa: the Siddhânta-Çiromaṇi etc. give it 2° less of polar longitude than our treatise, and the Graha-Lâghava even as much as 5° less.

The regent of the asterism is Vishṇu, and its figure or symbol corresponds therewith, being three footsteps, representatives of the three steps by which Vishṇu is said, in the early Hindu mythology, to have strode through heaven. The Çâkalya, however, gives a trident as the figure belonging to Çravaṇa. Possibly the name is to be regarded as indicating that it was originally figured as an ear.

The Chinese *sieu* corresponding in rank with Çravaṇa is called Nü, and is the faint star ε Aquarii (4.3). The Arab *manzil* Sa'd Bula', "felicity of a devourer," or al-Bula', "the devourer," etc., includes the same star, being composed of ε, μ (4.5), ν (5) Aquarii, or, according to others, of ε and 7 (6) Aquarii, or of μ and ν.

24. *Çravishṭhâ;* the word is a superlative formation from the same root from which came the name of the preceding asterism, and means, probably, "most famous." Another and hardly less frequent appellation is *dhanishṭhâ*, an irregular superlative from *dhanin,* "wealthy." The class of deities known as the *vasus,* "bright, good," are the regents of the asterism. It comprises four stars, or, according to the Çâkalya and Khaṇḍa-Kaṭaka, five: the former, which is given by so early a list as that of the Tâittirîya-Brâhmaṇa, is doubtless the original number. The group is the conspicuous one in the head of the Dolphin, composed of β, α, γ, δ Delphini, all of them stars of the third, or third to fourth, magnitude, and closely disposed in diamond or lozenge-form: they are figured by the Hindus as a drum or tabor. The junction-star, which is the western (v. 17), is β:

Çravishṭhâ	296° 5′	35° 33′ S.
β Delphini	296° 19′	31° 57′ S.

The only variation from the position assigned in our text to the junction-star of Çravishṭhâ is presented by the Graha-Lâghava, which gives it 286°, instead of 290°, of polar longitude. Perhaps its intention is to point out ζ (5) as the junction-star: this is doubtless the one added to the other four, on account of its close proximity to them, to make up the group of five; it lies only about half a degree westward from β.

The name of the twenty-fourth *manzil*, Sa'd as-Su'ûd, "felicity of felicities"—i. e., "most felicitous"—exhibits an accordance with that of the Hindu asterism which possibly is not accidental. The two are, however, as already noticed, far removed in position from one another, the Arab mansion being composed of the two stars β (3) and ξ (5.4), in the left shoulder of Aquarius, to which some add also 46, or c^1, Capricorni (6). The corresponding *sieu*, Hiü, is the first of them, or β Aquarii.

25. *Çatabhishaj*, "having a hundred physicians": the form *çatabhishâ*, which seems to be merely a corruption of the other, also occurs in later writings. It is, as we should expect from the title, said to be composed of a hundred stars, of which the brightest (v. 19) is the junction-star. This, from its defined position, can only be λ Aquarii (4):

Çatabhishaj	319° 51′	0° 29′ S.
λ Aquarii	321° 33′	0° 23′ S.

The rest of the asterism is to be sought among the yet fainter stars in the knee of Aquarius, and the stream from his jar: of course, the number one hundred is not to be taken as an exact one, nor are we to suppose it possible to trace out with any distinctness the figure assigned to the group, which is a circle. The Khaṇḍa-Kaṭaka, according to al-Bîrûnî, gives Çatabhishaj only a single star, but this is probably an error of the Arab traveller: he is unable to point out which of the stars in Aquarius is to be regarded as constituting the asterism.

The regent of the 25th asterism, according to nearly all the authorities, is Varuṇa, the chief of the Âdityas, but later the god of the waters: the Tâittirîya-Sanhitâ alone gives to it and to the 14th asterism, as well as to the 18th, Indra as presiding divinity: this is perhaps mere blundering.

The Graha-Lâghava places the junction-star of Çatabhishaj precisely on the ecliptic: the Siddhânta-Çiromaṇi etc. give it 20′, instead of 30′, of polar latitude south.

The corresponding lunar mansion of the Arabs, Sa'd al-Akhbiyah, "the felicity of tents," comprises the three stars in the right wrist and hand of the Water-bearer, or γ (3), ζ (4), η (4) Aquarii, together with a fourth, which Ideler supposes to be π (5). Since, however, the twenty-third Chinese determinative, Goei, is α Aquarii (3), a star so near as readily to be brought into the same group with the other three, we are inclined to regard it as altogether probable that the mansion was, at least originally, composed of α, γ, ζ, and η.

26, 27. *Bhâdrapadâ;* as plural, *bhâdrapadâs:* also *bhadrapadâ;* from *bhadra*, "beautiful, happy," and *pada*, "foot." Another frequent appellation is *proshṭhapadâ: proshṭha* is said to mean "carp" and "ox"; the latter signification might perhaps apply here. We have here, once more, a double asterism, divided into two parts, which are distinguished from one another as *pûrva* and *uttara*, "former" and "latter." All authorities agree in assigning two stars to each of the two groups; but there is not the same accordance as regards the figures by which they are represented: by some the one, by others the other, is called a couch or bed, the alternate one, in either case, being pronounced a bi-faced figure: the Muhûrta-Cintâmaṇi calls the first a bed, and the second twins. It admits, we apprehend, of little or no question that the Bhâdrapadâs are properly the four bright stars β, α, γ Pegasi, and α Andromedæ—all of them commonly reckoned as of the second magnitude—which form together a nearly perfect square, with sides measuring about 15°: the constellation, a very conspicuous one, is familiarly known as the "Square of Pegasus." The figure of a couch or bed, then, belongs, as in the case of the other two double asterisms, already explained, to the whole constellation, and not to either of the two separate asterisms into which it is divided, while, on the other hand, either of these latter is properly enough symbolized by a pair of twins, or by a figure with a double face. The appropriateness of the designation "feet," found as a part of both the names of the whole constellation, is also sufficiently evident, if we regard the group as thus composed. The junction-star of the former half-asterism is, by its defined position, clearly shown to be α Pegasi:

Pûrva-Bhâdrapadâ	334° 25′	22° 30′ N.
α Pegasi	333° 27′	19° 25′ N.

The Graha-Lâghava gives the junction-star 1° less of polar longitude, which would bring its position to a yet closer accordance, in respect to longitude, with α Pegasi: the error in latitude, which is common to all the authorities, is not greater than we have met with several times elsewhere. But we are told below (v. 16) that the principal star of each of these asterisms is the northern, and this would exclude β Pegasi altogether, bringing in as the other member of the first pair some more southern star, perhaps ζ Pegasi (3.4). The confusion is not less marked, although of another character, in the case of the second asterism: in the definition of position of its junction-star we find a longitude given which is that of one member of the group, and a latitude which is that of the other, as is shown by the following comparison:

Uttara-Bhâdrapadâ	347° 16′	24° 1′ N.
γ Pegasi	349° 8′	12° 35′ N.
α Andromedæ	354° 17′	25° 41′ N.

If we accept either of these two stars as the one of which the position is meant to be defined, we shall be obliged to admit an error in the determination either of its longitude or of its latitude considerably greater than we have met with elsewhere. Nor is the matter mended by any of the other authorities: the only variation from the data of our text is presented by the Graha-Lâghava, which reads, as the polar latitude of Uttara-Bhâdrapadâ, 27° instead of 26°. There can be no doubt that the two stars recognized as composing the asterism are γ Pegasi and α Andromedæ, but there has evidently been a blundering confusion of the two in making out the definition of position of the junction-star. We would suggest the following as a possible explanation of this confusion: that originally α and γ Pegasi were designated and described as junction-stars of the two half-groups, of which they were respectively the southern members; that afterward, for some reason—perhaps owing to the astrological theory (see above, vii. 21) of the superiority of a northern star—the rank of junction-star was sought to be transferred from the southern to the northern stars of both asterisms; that, in making the transfer, the original constitution of the former group was neglected, while in the latter the attempt was made to define the real position of the northern star, but by simply adding to the polar latitude already stated for γ Pegasi, without altering its polar longitude also. Al-Bîrûnî, it should be remarked, was unable to obtain from his Hindu informants any satisfactory identification of either of these asterisms, and marks both in his catalogue as "unknown."

The view we have taken of the true character of the two Bhâdrapadâs is powerfully supported by their comparison with the corresponding members of the other two systems. The twenty-sixth and twenty-seventh *manzils*, al-Fargh al-Mukdim and al-Fargh al-Mukhir, "the fore and hind spouts of the water-jar," comprise respectively α and β Pegasi, and γ Pegasi and α Andromedæ; the determinatives of the twenty-fourth and twenty-fifth *sieu*, Che and Pi, are α and γ Pegasi.

The regents of these two asterisms are *aja ekapât* and *ahi budhnya*, the "one-footed goat" and the "bottom-snake," two mythical figures, of obscure significance, from the Vedic pantheon.

28. *Revatî*, "wealthy, abundant." Its presiding divinity is Pûshan, "the prosperer," one of the Âdityas. It is said to contain thirty-two stars, which are figured, like those of Çravishṭhâ, by a drum or tabor; but it would be in vain to attempt to point out precisely the thirty-two which are intended, or to discover in their arrangement any resemblance to the figure chosen to represent it. The junction-star of the group is said (v. 18) to be its southernmost member: all authorities agree in placing it upon the ecliptic, and all excepting our treatise and the Çâkalya make its position exactly mark the initial point of the fixed sidereal sphere. The star intended is, as we have already often had occasion to notice, the faint star ζ Piscium, of about the fifth magnitude, situated in the band which connects the two Fishes. It is indeed very near to the ecliptic, having only 13′ of south latitude. It coincided in longitude with the vernal equinox in the year 572 of our era.

At the time of al-Bîrûnî's visit to India, the Hindus seem to have been already unable to point out distinctly and with confidence the situation in the heavens of that most important point from which they held that the motions of the planets commenced at the creation, and at which, at successive intervals, their universal conjunction would again take place; for he is obliged to mark the asterism as not certainly identifiable. He also assigns to it, as to Çatabhishaj, only a single star.

The twenty-sixth Chinese *sieu*, Koei, is marked by ζ Andromedæ (4), which is situated only 35′ east in longitude from ζ Piscium, but which has 17° 36′ of north latitude. The last *manzil*, Baṭn al-Ḥût, "the fish's belly," or ar-Rishâ, "the band," seems intended to include the stars composing the northern Fish, and with them probably the Chinese determinative also: but it is extended so far northward as to take in the bright star β Andromedæ (2), and to this star alone the name of the mansion is sometimes applied, although its situation, so far from the ecliptic (in lat. 25° 56′ N.), renders it by no means suited to become the distinctive star of one of the series of lunar stations.

We present, in the annexed table, a general conspectus of the correspondences of the three systems; and, in order to bring out those correspondences in the fullest manner possible, we have made the comparison in three different ways: noting, in the first place, the cases in which the three agree with one another; then those in which each agrees with one of the others; and finally, those in which each agrees with either the one or the other of the remaining two.

Correspondences of the Hindu, Arab, and Chinese Systems of Asterisms.

No	Hindu Name.	Hindu with Arab and Chinese.	Hindu with Arab	Hindu with Chinese.	Arab with Chinese.	Hindu with Arab or Chinese.	Arab with Hindu or Chinese.	Chinese with Hindu or Arab.
1	Açvinî,	1	1	1	1	1	1	1
2	Bharaṇî,	2*	2*	2	2*	2	2*	2
3	Kṛttikâ,	3	3	3	3	3	3	3
4	Rohiṇî,	4	4	4	4	4	4	4
5	Mṛgaçîrsha,	5	5	5	5	5	5	5
6	Ârdrâ,	..	..	..	†	..	..	..
7	Punarvasu,	..	6	..	†	6	6	†
8	Pushya,	6‡	7	6	..	7	7	6
9	Âçleshâ,	..	..	7	..	8	..	7
10	Maghâ,	..	8	..	..	9	8	..
11	P.-Phalgunî,	..	9	..	..	10	9	..
12	U.-Phalgunî,	..	10	..	..	11	10	..
13	Hasta,	..	..	8	..	12	..	8
14	Citrâ,	7	11	9	6	13	11	9
15	Svâtî,	..	..	..	7	..	12	10
16	Viçâkhâ,	8	12	10	8	14	13	11
17	Anurâdhâ,	9	13	11	9	15	14	12
18	Jyeshṭhâ,	10	14	12	10	16	15	13
19	Mûla,	11‡	15	13	..	17	16	14
20	P.-Ashâḍhâ,	12	16	14	11	18	17	15
21	U.-Ashâḍhâ,	..	..	15	..	19	..	16
22	Abhijit,	..	..	..	12	..	18	17
23	Çravaṇa,	..	..	..	13	..	19	18
24	Çravishṭhâ,	..	..	..	14	..	20	19
25	Çatabhishaj,	..	..	..	15	..	21	20
26	P.-Bhâdrapadâ,	13	17	16	16	20	22	21
27	U.-Bhâdrapadâ,	14	18	17	17	21	23	22
28	Revatî,	..	..	..	18§	..	24§	23§

* This supposes the second *manzil* to be composed of the stars in Musca, as defined by some authorities. † The sixth *manzil* includes, according to many authorities, the fifth *sieu*, but as there is, at any rate, a discordance in the order of succession, we have not reckoned this among the correspondences. ‡ We reckon these two as cases of general coincidence, because, although the Chinese *sieu* is not contained in the Arab mansion, the Hindu asterism includes them both, and the virtual correspondence of the three systems is beyond dispute. § Here we assume the Chinese *sieu* to be comprised among the stars forming the last *manzil*, which is altogether probable, although nowhere distinctly stated.

Owing to the different constitution of the systems, their correspondences are somewhat diverse in character: we account the Hindu asterisms and the Arab mansions to agree, when the groups which mark the two are composed, in whole or in part, of the same stars: we account the Chinese system to agree with the others, when the determinative of a *sieu* is to be found among the stars composing their groups. We have prefixed to the whole the numbers and titles of the Hindu asterisms, for the sake of easy reference back to the preceding detailed identifications and comparisons.

After this exhibition of the concordances existing among the three systems, it can, we apprehend, enter into the mind of no one to doubt that all have a common origin, and are but different forms of one and the same system. The questions next arise—is either of the three the original from which the others have been derived? and if so, which of them is entitled to the honor of being so regarded? and are the other two independent and direct derivatives from it, or does either of them come from the other, or must both acknowledge an intermediate source? In endeavoring to answer these questions, we will first exhibit the views of M. Biot respecting the origin and character of the Chinese *sieu*, as stated in the volumes for 1840 and 1859 of the Journal des Savants.

According to Biot, the *sieu* form an organic and integral part of that system by which the Chinese, from an almost immemorial antiquity, have been accustomed to make their careful and industrious observations of celestial phenomena. Their instruments, and their methods of observation, have been closely analogous with those in use among modern astronomers in the West: they have employed a meridian-circle and a measure of time, the clepsydra, and have observed meridian-transits, obtaining right ascensions and declinations of the bodies observed. To reduce the errors of their imperfect time-keepers, they long ago selected certain stars near the equator, of which they determined with great care the intervals in time, and to these they referred the positions of stars or planets coming to the meridian between them. The stars thus chosen are the *sieu*. Twenty-four of them were fixed upon more than two thousand years before our era (M. Biot says, about B. C. 2357: but it is obviously impossible to fix the date, by internal evidence, within a century or two, nor is the external evidence of a more definite character); the considerations which governed their selection were three: proximity to the equator of that period, distinct visibility—conspicuous brilliancy not being demanded for them—and near agreement in respect to time of transit with the upper and lower meridian-passages of the bright stars near the pole, within the circle of perpetual apparition: M. Biot finds reason to believe that these circumpolar stars had been earlier observed with special care, and made standards of comparison, and that, when it was afterward seen to be desirable to have stations near the equator, such stars were adopted as most nearly agreed with them in right ascension. The other four, being the 8th, 14th, 21st, and 28th, the accession of which completed the system of twenty-eight, were added in the time of Cheu-Kong, about B. C. 1100, because they marked very nearly the positions of the equinoxes and solstices at that epoch: the bright star of the Pleiades, however, which had originally been made the first of the series, from its near approach to the vernal equinox of that remoter era, still maintained, as it has ever since maintained, its rank as the first. Since the time of Cheu-Kong the system has undergone no farther modification, but has been preserved unaltered and unimproved, with the obstinate persistency so characteristic of the Chinese, although many of the determinative stars have, under the influence of the precession, become far removed from the equator, one of them even having retrograded into the preceding mansion.

If the history of the Chinese *sieu*, as thus drawn out, is well-founded and true, the question of origin is already solved: the system of twenty-eight celestial mansions is proved to be of native Chinese institution—just as the system of representation of the planetary movements by epicycles is proved to be Greek by the fact that we can trace in the history of Greek science the successive steps of its gradual elaboration. That history rests, at present, upon the authority of M. Biot alone: we are not aware, at least, that any other investigator has gone independently over the same ground; and he has not himself laid before us, in their original form, the passages from Chinese texts which furnish the basis of his conclusions. But we regard them as entitled to be received, upon his authority, with no slight measure of confidence: his own distinguished eminence as a physicist and astronomer, his familiarity with researches into the history and archæology of science, his access to the abundant material for the history of Chinese astronomy collected and worked up by the French missionaries at Pekin, and the zealous assistance of his son, M. Édouard Biot, the eminent Sinologist, whose premature death, in 1850, has been so deeply deplored as a severe loss to Chinese studies—all these advantages, rarely united in such fullness in the person of any one student of such a subject, give very great weight to views arrived at by him as the results of laborious and long-continued investigation. Nor do we see that any general considerations of importance can be brought forward in opposition to those views. It is, in the first place, by no means inconsistent with what we know in other respects of the age and character of the culture of the Chinese, that they should have devised such a system at so early a date. They have, from the beginning, been as much distinguished by a tendency to observe and record as the Hindus by the lack of such a tendency: they have always attached extreme importance to astronomical labors, and to the construction and rectification of the calendar; and the industry and accuracy of their observations is attested by the use made of them by modern astronomers—thus, to take a single instance, of the cometary orbits which have been calculated, the first twenty-five rest upon Chinese observations alone: and once more, it is altogether in accordance with the clever empiricism and practical shrewdness of the Chinese character that they should have originated at the very start a system of observation exceedingly well adapted to its purpose, stopping with that, working industriously on thenceforth in the same beaten track, and never developing out of so promising a commencement anything deserving the name of a science, never devising a theory of the planetary motions, never even recognizing and defining the true character of the cardinal phenomenon of the precession.

Again, although it might seem beforehand highly improbable that a system of Chinese invention should have found its way into the West, and have been extensively accepted there, many centuries before the Christian era, there are no so insuperable difficulties in the way as should destroy the force of strong presumptive evidence of the truth of such a communication. It is well known that in very ancient times the products of the soil and industry of China were sought as objects of luxury in the West, and mercantile intercourse opened and maintained across the deserts of Central Asia; it even appears that, as early as about B. C. 600 (Isaiah xlix. 12), some knowledge of the Sinim, as a far-off eastern nation, had penetrated to Babylon and Judea. On the other hand, we do not know how much, if at all, earlier than this it may be necessary to acknowledge the system of asterisms to have made its appearance in India. The literary memorials of the earliest period, the Vedic period proper, present no evidence of the existence of the system: indeed, it is remarkable how little notice is taken of the stars by the Vedic poets; even the recognition of some of them as planets does not appear to have taken place until considerably later. In the more recent portions of the Vedic texts—as in the nineteenth book of the Atharva-Veda, a modern appendage to that modern collection, and in parts of the Yajur-Veda, of which there is reason to believe that the canon was not closed until a comparatively late period—full lists of the asterisms are found. The most unequivocal evidence of the early date of the system in India is furnished by the character of the divinities under whose regency the several asterisms are placed: these are all from the Vedic pantheon; the popular divinities of later times are not to be found among them; but, on the other hand, more than one whose consequence is lost, and whose names almost are forgotten, even in the epic period of Hindu history, appear in the list. Neither this, however, nor any other evidence known to us, is sufficient to prove, or even to render strongly probable, the existence of the asterisms in India at so remote a period that the system might not be believed to have been introduced, in its fully developed form, from China.

If, now, we make the attempt to determine, upon internal evidence, which of the three systems is the primitive one, a detailed examination of their correspondences and differences will lead us first to the important negative conclusion that no one among them can be regarded as the immediate source from which either of the other two has been derived. It is evident that the Hindu asterisms and the Arab *manâzil* constitute, in many respects, one and the same system: both present to us constellations or groups of stars, in place of the single determinatives of the Chinese *sieu;* and not only are those groups composed in general of the same stars, but in several cases—as the 7th, 10th, 11th, and 12th members of the series—where they differ widely in situation from the Chinese determinatives, they exhibit an accordance with one another which is too close to be plausibly looked upon as accidental. But if it is thus made to appear that neither can have come independently of the other from a Chinese original, it is no less certain that neither can have come through the other from such an original; for each has its own points of agreement with the *sieu*, which the other does not share—the Hindu in the 9th, 13th, and 21st asterisms, the Arab in the 15th, 22nd, 23rd, 24th, and 25th mansions. The same considerations show, inversely, that the Chinese system cannot be traced to either of the others as its source, since it agrees in several points with each one of them where that one differs from the third. It becomes necessary, then, to introduce an additional term into the comparison; to assume the existence of a fourth system, differing in some particulars from each of the others, in which all shall find their common point of union. Such an assumption is not to be looked upon as either gratuitous or arbitrary. Not only do the mutual relations of the three systems point distinctly toward it, but it is also supported by general considerations, and will, we think, be found to remove many of the difficulties which have embarrassed the history of the general system. It has been urged as a powerful objection to the Chinese origin of the twenty-eight-fold division of the heavens, that we find traces of its existence in so many of the countries of the West, geographically remote from China, and in which Chinese influence can hardly be supposed to have been directly felt. And it is undoubtedly true that neither India nor Arabia has stood in ancient times in such relations to China as should fit it to become the immediate recipient of Chinese learning, and the means of its communication to surrounding peoples. The great route of intercourse between China and the West led over the table-land of Central Asia, and into the north-eastern territory of Iran, the seat of the Zoroastrian religion and culture: thence the roads diverged, the one leading westward, the other south-eastward into India, through the valley of the Cabul, the true gate of the Indian peninsula. Within or upon the limits of this central land of Iran we conceive the system of mansions to have received that form of which the Hindu *nakshatras* and the Arab *manâzil* are the somewhat altered representatives: precisely where, and whether in the hands of Semitic or of Aryan races, we would not at present attempt to say. There are, as has been noticed above, traces of an Iranian system to be found in the Bundehesh; but this is a work which, although probably not later than the times of Persia's independence under her Sassanian rulers, can pretend to no high antiquity, and no like traces have as yet been pointed out in the earliest Iranian memorial, the Zendavesta. Weber (Ind. Literaturgeschichte, p. 221), on the other hand, sees in the *mazzaloth* and *mazzaroth* of the Scriptures (Job xxxviii. 32; II Kings xxiii. 5)—words radically akin with the Arabic *manzil*—indications of the early existence of the system in question among the western Semites, and suspects for it a Chaldaic origin: but the allusions appear to us too obscure and equivocal to be relied upon as proof of this, nor is it easy to believe that such a method of division of the heavens should have prevailed so far to the west, and from so ancient a time, without our hearing of it from the Greeks; and especially, if it formed a part of the Chaldaic astronomy. This point, however, may fairly be passed over, as one to be determined, perhaps, by future investigations, and not of essential importance to the present inquiry. The question of originality is at least definitely settled adversely to the claims of both the Hindu and the Arab systems, and can only lie between the Chinese and that fourth system from which the other two have together descended. And as concerns these, we are willing to accept the solution which is furnished us by the researches of M. Biot, supported as we conceive it to be by the general probabilities of the case. Any one who will trace out, by the help of a celestial globe or map,* the positions of the Chinese determinatives, cannot fail to perceive their general approach to a great circle of the sphere which is independent of the ecliptic, and which accords more nearly with the equator of B. C. 2350 than with any other later one. The full explanations and tables of positions given by Biot (Journ. d. Sav., 1840, pp. 243–254) also furnish evidence, of a kind appreciable by all, that the system may have had the origin which he attributes to it, and that, allowing for the limitations imposed upon it by its history, it is consistent with itself, and well enough adapted to the purposes for which it was designed. With the positions of its determinative stars seem to have agreed those of the constellations adopted by the common parent of the Hindu and Arab systems, excepting in five or six points: those points being where the Chinese make their one unaccountable leap from the head to the belt of Orion, and again, where the *sieu* are drawn off far to the southward, in the constellations Hydra and Crater: and this, in our view, looks much more as if the series of the *sieu* were the original, whose guidance had been closely followed excepting in a few cases, than as if the asterisms composing the other systems had been independently selected from the groups of stars situated along the zodiac, with the intention of forming a zodiacal series. It is easy to see, farther, how the single determinatives of the *sieu* should have become the nuclei for constellations such as are presented by the other systems; but if, on the contrary, the *sieu* had been selected by the Chinese, in each case, from groups previously constituted, there appears no reason why their brightest stars should not have been chosen, as they were chosen later by the Hindus, in the establishment of junction-stars for the asterisms.

We would suggest, then, as the theory best supported by all the evidence thus far elicited, that a knowledge of the Chinese astronomy, and with it the Chinese system of division of the heavens into twenty-eight mansions, was carried into Western Asia at a period not much later than B. C. 1100, and was there adopted by some western people, either Semitic or Iranian. That in their hands it received a new form, such as adapted it to a ruder and less scientific method of observation, the limiting stars of the mansions being converted into zodiacal groups or constellations, and in some instances altered in position, so as to be brought nearer to the general planetary path of the ecliptic. That in this changed form, having become a means of roughly determining and describing the places and movements of the planets, it passed into the keeping of the Hindus—very probably along with the first knowledge of the planets themselves—and entered upon an independent career of history in India. That it still maintained itself in its old seat, leaving its traces later in the Bundehesh; and that it made its way so far westward as finally to become known to, and adopted by, the Arabs. The farther

* We propose to furnish at the close of this publication, in connection with the additional notes, such a map of the zodiacal zone of the heavens as will sufficiently illustrate the character and mutual relations of the three systems compared.

modifications introduced into it by the latter people all have in view a single purpose, that of establishing its stations in the immediate neighborhood of the ecliptic: to this purpose the whole Arab system is not less constantly faithful than is the Chinese to its own guiding principle. The Hindu sustains in this respect but an unfavorable comparison with the others: the arbitrary introduction, in the 15th, 22nd, 23rd, and 24th asterisms, of remote northern stars, greatly impairs its unity, and also furnishes an additional argument of no slight force against its originality; for, on the one hand, the derivation of the others from it becomes thereby vastly more difficult, and, on the other, we can hardly believe that a system of organic Indian growth could have become disfigured in India by such inconsistencies; they wear the aspect, rather, of arbitrary alterations made, at the time of its adoption, in an institution imported from abroad.

It might, at first sight, appear that the adoption by the Arabs of the *manzil* corresponding to Açvinî as the first of their series indicated that they had derived it from India posterior to the transfer by the Hindus of the first rank from Kṛttikâ, the first of the *sieu*, to Açvinî: but the circumstance seems readily to admit of another interpretation. The names of many of the Arab mansions show the influence of the Greek astronomy, being derived from the Greek constellations: the same influence would fully explain an arrangement which made the series begin with the group coinciding most nearly with the beginning of the Greek zodiac. The transfer on the part of the Hindus, likewise, was unquestionably made at the time of the general reconstruction of their astronomical system under the influence of western science. The two series are thus to be regarded as having been brought into accordance in this respect by the separate and independent working of the same cause.

M. Biot insists strongly, as a proof of the non-originality of the system of asterisms among the Hindus, upon its gross and palpable lack of adaptedness to the purpose for which they used it; he compares it to a gimlet out of which they have tried to make a saw. In this view we can by no means agree with him: we would rather liken it to a hatchet, which, with its edge dulled and broken, has been turned and made to do duty as a hammer, and which is not ill suited to its new and coarser office. Indeed, taking the Hindu system in its more perfect and consistent form, as applied by the Arabs, and comparing it with the Chinese *sieu* at any time within the past two thousand years, we are by no means sure that the advantage in respect to adaptation would not be generally pronounced to be upon the side of the former. The distance of many of the *sieu* during that period from the equator, the faintness of some among them, the great irregularity of their intervals, render them anything but a model system for measuring distances in right ascension. On the other hand, to adopt a series of conspicuous constellations along the zodiac, by their proximity to which the movements of the planets shall be marked, is no unmotived proceeding: just such a division of the ecliptic among twelve constellations preceded and led the way to the Greek method of measuring by signs, having exact limits, and independent of the groups of stars which originally gave name to them. M. Biot's error lies in his misapprehension, in two important respects, of the character of the Hindu asterisms: in the first place, he constantly treats them as if they were, like the *sieu*, single stars, the intervals between whose circles of declination constituted the accepted divisions of the zodiac; and in the second place, he assumes them to have been established for the purpose of marking the moon's daily progress from point to point along the ecliptic. Now, as regards the first of these points, we have already shown above that the conversion of the Chinese determinatives into constellations took place, in all probability, before their introduction to the knowledge of the Hindus: there is, indeed, an entire unanimity of evidence to the effect that the Hindu system is from its inception one of groups of stars: this is conclusively shown by the original dual and plural names of the asterisms, or by their otherwise significant titles—compare especially those of the 13th and 25th of the series. The selection of a "junction-star" to represent the asterism appears to be something comparatively modern: we regard it as posterior to the reconstruction of the Hindu astronomy upon a truly scientific basis, and the determination, by calculation, of the precise places of the planets: this would naturally awaken a desire for, and lead to, a similarly exact determination of the position of some star representing each asterism, which might be employed in the calculation of conjunctions, for astrological purposes; the astronomical uses of the system being no longer of much account after the division of the ecliptic into signs. And the choice of the junction-star has fallen, in the majority of cases, not upon the Chinese determinative itself, but upon some other and more conspicuous member of the group originally formed about the latter. Again, there is an entire absence of evidence that the "portions" of the asterisms, or the arcs of the ecliptic named from them, were ever measured from junction-star to junction-star: whatever may be the discordance among the different authorities respecting their extent and limits, they are always freely, and often arbitrarily, taken from parts of the ecliptic adjacent to, or not far removed from, the successive constellations.

As regards the other point noticed, it is, indeed, not at all to be wondered at that M. Biot should treat the Hindu *nakshatras* as a system bearing special relations to the moon, since, by those who have treated of them, they have always been styled "houses of the moon," "moon-stations," "lunar asterisms," and the like. Nevertheless, these designations seem to be founded only in carelessness, or in misapprehension. In the Sûrya-Siddhânta, certainly, there is no hint to be discovered of any particular connection between them and the moon, and for this reason we have been careful never to translate the term *nakshatra* by any other word than simply "asterism." Nor does the case appear to have been otherwise from the beginning. No one of the general names for the asterisms (*nakshatra*, *bha*, *dhishṇya*) means literally anything more than "star" or "constellation": their most ancient and usual appellation, *nakshatra*, is a word of doubtful etymology (it may be radically akin with *nakta*, nox, νύξ, "night"), but it is not infrequently met with in the Vedic writings, with the general signification of "star," or "group of stars": the moon is several times designated as "sovereign of the *nakshatras*," but evidently in no other sense than that in which we style her "queen of night"; for the same title is in other passages given to the sun, and even also to the Milky Way. When the name came to be especially applied to the system of zodiacal asterisms, we have seen above that a single one of the series, the 5th, was placed under the regency of the moon, as another, the 13th, under that of the sun: this, too, by no means looks as if the whole design of the system was to mark the moon's daily motions. Naturally enough, since the moon is the most conspicuous of the nightly luminaries, and her revolutions more rapid and far more important than those of the others, the asterisms would practically be brought into much more frequent use in connection with her movements: their number, likewise, being nearly accordant with the number of days of her sidereal revolution, could not but tempt those who thus employed them to set up an artificial relation between the two. Hence the Arabs distinctly call their divisions of the zodiac, and the constellations which mark them, "houses of the moon," and, until the researches of M. Biot, no one, so far as we are aware, had ever questioned that the number of the asterisms or mansions, wherever found, was derived from and dependent on that of the days in the moon's revolution. It was most natural, then, that Western scholars, having first made acquaintance with the Arab system, should, on finding the same in India, call it by the same name: nor is it very strange, even, that Ideler should have gone a step farther, and applied the familiar title of "lunar stations" to the Chinese *sieu* also; an error for which he is sharply criticised by M. Biot (Journ. d. Sav., 1859, p. 480). The latter cites from al-Bîrûnî (Journ. d. Sav. 1845, p. 49; 1859, pp. 487–8) two passages derived by him from Varâha-mihira and Brahmagupta respectively, in which are recorded attempts to establish a systematic relation between the asterisms and the moon's true and mean daily motions. One of these passages is exceedingly obscure, and both are irreconcilable with one another, and with what we know of the system of asterisms from other sources: two conclusions, however, bearing upon the present matter, are clearly derivable from them: first, that, as the "portions" assigned to the asterisms had no natural and fixed limits, it was possible for any Hindu system-maker so to define them as to bring them into a connection with the moon's daily motions: and secondly, that such a connection was never deemed an essential feature of the system, and hence no one form of it was generally recognized and accepted. The considerations adduced by us above are, we think, fully sufficient to account for any such isolated attempts at the establishment of a connection as al-Bîrûnî, who naturally sought to find in the Hindu *nakshatras* the correlatives of his own *manâzil al-ḳamar*, was able to discover among the works of Hindu astronomers; there is no good reason why we should deprive the former of their true character, which is that of zodiacal constellations, rudely marking out divisions of the ecliptic, and employable for all the purposes for which such a division is demanded.

The reason of the variation in the number of the asterisms, which are reckoned now as twenty-eight and now as twenty-seven, is a point of no small difficulty in the history of the system. M. Biot makes the acute suggestion that the omission of Abhijit from the series took place because the mansion belonging to that asterism was on the point of becoming extinguished, the circle of declination of its junction-star being brought by the precession to a coincidence with that of the junction-star of the preceding asterism about A. D. 972. But it has been shown above that M. Biot's view of the nature of a *nakshatra*—that it is, namely, the arc of the ecliptic intercepted between the circles of declination of two successive junction-stars—is altogether erroneous: however nearly those circles might approach one another, there would still be no difficulty in assigning to each asterism its "portion" from the neighboring region of the ecliptic. Again, this explanation would not account for the early date of the omission of Abhijit, which, as already noticed, is found wanting in one of the most ancient lists, that of the Tâittirîya-Sanhitâ. It is to be observed, moreover, that M. Biot, in calculating the period of Abhijit's disappearance, has adopted τ Sagittarii as the junction-star of Uttara-Ashâḍhâ, while we have shown above that σ, and not τ, is to be so regarded: and this substitution would defer until several centuries later the date of coincidence of the two circles of declination. According to the Hindu measurements, indeed (see the table of positions of the junction-stars, near the beginning of this note), Abhijit is farther removed from the preceding asterism, both in polar longitude and in right ascension, than are five of the other asterisms from their respective predecessors: nor does the Hindu astronomical system acknowledge or make allowance for the alteration of position of the circles of declination under the influence of the precession: their places, as data for the calculation of conjunctions, are ostensibly laid down for all future time. For these various reasons, M. Biot's explanation is to be rejected as insufficient. A more satisfactory one, in our opinion, may be found in the fact, illustrated above (see Fig. 31, beginning of this note), that the asterisms are in general so distributed as to accord quite well with a division of the ecliptic into twenty-seven equal portions, but not with a division into twenty-eight equal portions; that the region where they are too much crowded together is that from the 20th to the 23rd asterism, and that, among those situated in this crowded quarter, Abhijit is farthest removed from the ecliptic, and so is more easily left out than any of the others, in dividing the ecliptic into portions. We cannot consider it at all doubtful that Abhijit is as originally and truly a part of the system of asterisms as any other constellation in the series, which is properly composed of twenty-eight members, and not of twenty-seven: the analogy of the other systems, and the fact that treatises like this Siddhânta, which reckon only twenty-seven divisions of the ecliptic, are yet obliged, in treating of the asterisms as constellations, to regard them as twenty-eight, are conclusive upon this point. The whole difficulty and source of discordance seems to lie in this—how shall there, in any systematic method of division of the ecliptic, be found a place and a portion for a twenty-eighth asterism? The Khaṇḍa-Kaṭaka, as cited by al-Bîrûnî—in making out, by a method which is altogether irrespective of the actual positions of the asterisms with reference to the zodiac, the accordance already referred to between their portions and the moon's daily motions—allots to Abhijit so much of the ecliptic as is equivalent to the mean motion of the moon during the part of a day by which her revolution exceeds twenty-seven days.

Others allow it a share in the proper portions of the two neighboring asterisms: thus the Muhûrta-Mâlâ, a late work, of date unknown to us, says: "the last quarter of Uttara-Ashâḍhâ and the first fifteenth of Çravaṇa together constitute Abhijit: it is so to be accounted when twenty-eight asterisms are reckoned; not otherwise." Ordinarily, however, the division of the ecliptic into twenty-seven equal "portions" is made, and Abhijit is simply passed by in their distribution. After the introduction of the modern method of dividing the circle into degrees and minutes, this last way of settling the difficulty would obviously receive a powerful support, and an increased currency, from the fact that a division by twenty-seven gave each portion an even number of minutes, 800, while a division by twenty-eight yielded the awkward and unmanageable quotient $771\frac{3}{7}$.

Much yet remains to be done, before the history and use of the system of asterisms, as a part of the ancient Hindu astronomy and astrology, shall be fully understood. There is in existence an abundant literature, ancient and modern, upon the subject, which will doubtless at some time provoke laborious investigation, and repay it with interesting results. To us hardly any of that literature is accessible, and only the final results of wide-extended and long-continued studies upon it could be in place here. We have already allotted to the *nakshatras* more space than to some may seem advisable: our excuse must be the interest of the history of the system, as part of the ancient history of the rise and spread of astronomical science; the importance attaching to the researches of M. Biot, the inadequate attention hitherto paid them, and the recent renewal of their discussion in the Journal des Savants; and finally and especially, the fact that in and with the asterisms is bound up the whole history of Hindu astronomy, prior to its transformation under the overpowering influence of western science. In the modern astronomy of India, the *nakshatras* are of subordinate consequence only, and appear as hardly more than reminiscences of a former order of things: from the Sûrya-Siddhânta might be struck out every line referring to them, without serious alteration of the character of the treatise.

Before bringing this note to a close, we present, in the annexed table, a comparison of the true longitudes and latitudes of the junction-stars of the twenty-eight asterisms, as derived by calculation from the positions stated in our text, with the actual longitudes and latitudes of the stars with which they are probably to be identified. In a single case, (the 27th asterism), we compare the longitude of one star and the latitude of another; the reason of this is explained above, in connection with the identification of the asterism. We add columns giving the errors of the Hindu determinations of position: in that for the latitude north direction is regarded as positive, and south direction as negative.

Upon examining the column of errors of latitude presented in this table, it will be seen that they are too considerable, and too irregular, both in amount and in direction, to be plausibly accounted for otherwise than as direct errors of observation and calculation. The grossest of them, as has already been pointed out, are committed in the measurement of southern latitudes, when of considerable amount, and they are

Positions, and Errors of Position, of the Junction-Stars of the Asterisms.

No.	Name.	Longitude, A. D. 560. Hindu.	Longitude, A. D. 560. True.	Longitude, A. D. 560. Hindu error.	Latitude. Hindu.	Latitude. True.	Latitude. Hindu error.	Star compared.
		° ′	° ′	° ′	° ′	° ′	° ′	
1	Açvinî,	11 59	13 56	− 1 57	9 11 N.	8 28 N.	+ 0 43	β Arietis.
2	Bharaṇî,	24 35	26 54	− 2 19	11 6 "	11 17 "	− 0 11	35 Arietis, *a* Muscæ.
3	Kṛttikâ,	39 8	39 58	− 0 50	4 44 "	4 1 "	+ 0 43	η Tauri, Alcyone.
4	Rohiṇî,	48 9	49 45	− 1 36	4 49 S.	5 30 S.	+ 0 41	α Tauri, Aldebaran.
5	Mṛgaçîrsha,	61 3	63 40	− 2 37	9 49 "	13 25 "	+ 3 36	λ Orionis.
6	Ârdrâ,	65 50	68 43	− 2 53	8 53 "	16 4 "	+ 7 11	α Orionis.
7	Punarvasu,	92 52	93 14	− 0 22	6 0 N.	6 39 N.	− 0 39	β Gemin., Pollux.
8	Pushya,	106 0	108 42	− 2 42	0 0	0 4 "	− 0 4	δ Cancri.
9	Âçleshâ,	109 59	112 20	− 2 21	6 56 S.	11 8 S.	+ 4 12	ε Hydræ.
10	Maghâ,	129 0	129 49	− 0 49	0 0	0 27 N.	− 0 27	α Leonis, Regulus.
11	P.-Phalgunî,	139 58	141 15	− 1 17	11 19 N.	14 19 "	− 3 0	δ Leonis.
12	U.-Phalgunî,	150 10	151 37	− 1 27	12 5 "	12 17 "	− 0 12	β Leonis.
13	Hasta,	174 22	173 27	+ 0 55	10 6 S.	12 10 S.	+ 2 4	δ Corvi.
14	Citrâ,	180 48	183 49	− 3 1	1 50 "	2 2 "	+ 0 12	α Virginis, Spica.
15	Svâtî,	183 2	184 12	− 1 10	33 50 N.	30 57 N.	+ 2 53	α Bootis, Arcturus.
16	Viçâkhâ,	213 31	211 0	+ 2 31	1 25 S.	1 48 S.	+ 0 23	ι Libræ.
17	Anurâdhâ,	224 44	222 34	+ 2 10	2 52 "	1 57 "	− 0 55	δ Scorpionis.
18	Jyeshṭhâ,	230 7	229 44	+ 0 23	3 50 "	4 31 "	+ 0 41	α Scorp., Antares.
19	Mûla,	242 52	244 33	− 1 41	8 48 "	13 44 "	+ 4 56	λ Scorpionis.
20	P.-Ashâḍhâ,	254 39	254 32	+ 0 7	5 28 "	6 25 "	+ 0 57	δ Sagittarii.
21	U.-Ashâḍhâ,	260 23	262 21	− 1 58	4 59 "	3 24 "	− 1 25	σ Sagittarii.
22	Abhijit,	264 10	265 15	− 1 5	59 58 N.	61 46 N.	− 1 48	α Lyræ, Vega.
23	Çravaṇa,	282 29	281 41	+ 0 48	29 54 "	29 19 "	+ 0 35	α Aquilæ, Atair.
24	Çravishṭhâ,	296 5	296 19	− 0 14	35 33 "	31 57 "	+ 3 36	β Delphini.
25	Çatabhishaj,	319 50	321 33	− 1 43	0 28 S.	0 23 S.	− 0 5	λ Aquarii.
26	P.-Bhâdrapadâ,	334 25	333 27	+ 0 58	22 30 N.	19 25 N.	+ 3 5	α Pegasi.
27	U.-Bhâdrapadâ,	347 16	349 8	− 1 52	24 1 "	25 41 "	− 1 40	γ Peg. & α Androm.
28	Revatî,	359 50	359 50	0 0	0 0	0 13 S.	+ 0 13	ζ Piscium.

all in the same direction, giving the star a place too far to the north. The column of errors in longitude, on the other hand, shows a very marked preponderance of *minus* errors, their sum being 33° 54′, while the sum of *plus* errors is only 7° 52′. Upon taking the difference of these sums, and dividing it by twenty-eight, we find the average error of longitude to be − 56′, the greatest deviation from it in either direction being − 2° 4′ and + 3° 27′.* So far as this goes, it would indicate that the Hindu measurements of position were made from a vernal equinox situated about 1° to the eastward of that of A. D. 560, and so at a time seventy years previous to the date we have assumed for them, or about A. D. 490. In our present ignorance of the methods of observation

* In a comparison in which a high degree of exactness was desired, and was not, in the nature of the case, unattainable, it would of course be necessary to take into account the proper motions of the stars compared. This we have not thought it worth while, in the present instance, to do. We may remark, however, that the junction-star of the 15th asterism, Arcturus, has a much greater proper motion than any other in the series; and that, if this were allowed for, according to its value as determined by Main (Mem. Roy. Astr. Soc., vol. xix, 4to, 1851), the Hindu error of longitude would be diminished about 22′, but that of latitude increased about 35′.

employed by the Hindus for this purpose, such a determination of date cannot, indeed, be relied upon as exact or conclusive, yet it is the best and surest that we can attain. The general conclusion, at any rate, stands fast, that the positions of the junction-stars of the asterisms were fixed not far from the time when the vernal equinox coincided with the initial point of the Hindu sidereal sphere, or during the sixth century of our era.

Since, according to the Hindu theory, the initial point of the sidereal sphere is also, for all time, the mean place of the vernal equinox, which always reverts to it after a libration of 27° in either direction (see above, iii. 9–12), we are not surprised to find the positions of the asterisms primarily defined upon the supposition of their coincidence. But it is not a little strange that the effect of the precession in altering the direction of the circles of declination drawn through the junction-stars, and so the polar longitudes and latitudes of the latter, should be made no account of (see, however, the latter half of v. 12, below, and the note upon it), and that directions for calculating the conjunctions of the planets with the asterisms according to their positions as thus stated should be given (vv. 14–15), unaccompanied by any hint that a modification of the data of the process would ever be found necessary. This carelessness is perhaps to be regarded as an additional evidence of the small importance attached, after the reconstruction of the Hindu astronomy, to calculations in which the asterisms were concerned; although it also tends strongly to prove what we have suggested above (note to iii. 9–12), that in the construction of the Hindu astronomical system the precession was ignored altogether. It is to be noticed that the two systems of *yogas* (see above, ii. 65, and additional note upon that passage), originally founded upon actual conjunctions with the asterisms, have been divorced from any real connection with them. A like consideration might restrain us from accepting the determinations of position here presented as the best results which Hindu observers and instruments were capable of attaining; yet, in the absence of other tests of their powers, we cannot well help drawing the conclusion that the accuracy of a Hindu observation is not to be relied on within a degree or two.

10. Agastya is at the end of Gemini, and eighty degrees south; and Mṛgavyâdha is situated in the twentieth degree of Gemini;

11. His latitude (*vikshepa*), reckoned from his point of declination (*apakrama*), is forty degrees south: Agni (*hutabhuj*) and Brahmahṛdaya are in Taurus, the twenty-second degree;

12. And they are removed in latitude (*vikshipta*), northward, eight and thirty degrees respectively. . . .

In connection with the more proper subject of this chapter we also have laid before us, here and in a subsequent passage (vv. 20–21), the defined positions of a few fixed stars which are not included in the system of zodiacal asterisms. The definition is made in the same manner as before, by polar longitudes and latitudes. It is not at all difficult to identify the stars referred to in these verses; they were correctly pointed out by Colebrooke, in his article already cited (As. Res., vol. ix). Agastya is *α* Navis, or Canopus, a star of the first magnitude, and one of the most brilliant in the southern heavens. Its remote southern position, only 37° from the pole, renders it invisible to an observer stationed much to the northward of the Tropic of Cancer. Its Hindu name is that of one of the old Vedic *ṛshis*, or inspired sages. The comparison of its true position with that assigned it by our text—which, in this instance, does not require to be reduced to true longitude and latitude—is as follows:

Agastya	90° 0′	80° 0′ S.
Canopus	85° 4′	75° 50′ S.

The error of position is here very considerable, and the variations of the other authorities from the data of our text are correspondingly great. The Siddhânta-Çiromaṇi and (according to Colebrooke) the Brahma-Siddhânta give Agastya 87° of polar longitude, and 77° of latitude, which is a fair approximation to the truth: the Graha-Lâghava also places it correctly in lat. 76° S., but makes its longitude only 80°, which is as gross an error as that of the Sûrya-Siddhânta, but in the opposite direction. The Çâkalya-Sanhitâ agrees precisely with our treatise as respects the positions of these four stars, as it does generally in the numerical data of its astronomical system.

Mṛgavyâdha, "deer-hunter"—it is also called Lubdhaka, "hunter"—is α Canis Majoris, or Sirius, the brightest of the fixed stars:

Mṛgavyâdha	76° 23′	39° 52′ S.
Sirius	84° 7′	39° 32′ S.

Here, while all authorities agree with the correct determination of the latitude of Sirius presented by our text, the Siddhânta-Çiromaṇi etc. greatly reduce its error of longitude, by giving the star 86°, instead of 80°, of polar longitude: the Graha-Lâghava reads 81°.

The star named after the god of fire, Agni, and called in the text by one of his frequent epithets, *hutabhuj*, "devourer of the sacrifice," is the one which is situated at the extremity of the northern horn of the Bull, or β Tauri: it alone of the four is of the second magnitude only:

Agni	54° 5′	7° 44′ N.
β Tauri	62° 32′	5° 22′ N.

The very gross error in the determination of the longitude of this star is but slightly reduced by the Graha-Lâghava, which gives it 53°, instead of 52°, of polar longitude. The Siddhânta-Çiromaṇi and Brahma-Siddhânta omit all notice of any of the fixed stars excepting Canopus and Sirius.

Brahmahṛdaya, "Brahma's heart," is α Aurigæ, or Capella:

Brahmahṛdaya	60° 29′	28° 53′ N.
Capella	61° 50′	22° 52′ N.

The Graha-Lâghava, leaving this erroneous determination of latitude unamended, adds a greater error of longitude, in the opposite direction to that of our text, by giving the star 4° more of polar longitude.

We shall present these comparisons in a tabular form at the end of the chapter, in connection with the other passage of similar import.

12. . . . Having constructed a sphere, one may examine the corrected (*sphuṭa*) latitude and polar longitude (*dhruvaka*).

What is the true meaning and scope of this passage, is a question with regard to which there may be some difference of opinion. The commentator explains it as intended to satisfy the inquiry whether the polar longitudes and latitudes, as stated in the text, are constant, or whether they are subject to variation. Now although, he says, owing to the precession, the values of these quantities are not unalterably fixed, yet they are given by the text as they were at its period, and as if they were constant, while the astronomer is directed to determine them for his own time by actual observation. For this purpose he is to take such a sphere as is described below (chap. xiii)—of which the principal parts, and the only ones which would be brought into use in this process, are hoops or circles representing the colures, the equator, and the ecliptic—and is to suspend upon its poles an additional movable circle, graduated to degrees: this would be, of course, a revolving circle of declination. The sphere is next to be adjusted in such manner that its axis shall point to the pole, and that its horizon shall be water-level. Then, in the night, the junction-star of Revatî (ζ Piscium) is to be looked at through a hole in the centre of the instrument, and the corresponding point of the ecliptic, which is 10′ east of the end of the constellation Pisces, is to be brought over it; after that, it will be necessary only to bring the revolving circle of declination, as observed through the hole in the centre of the instrument, over any other star of which it is desired to determine the position, and its polar longitude and latitude may be read off directly upon the ecliptic and the movable circle respectively.

Colebrooke (As. Res., ix. 326; Essays, ii. 324) found this passage similarly explained in other commentaries upon the Sûrya-Siddhânta to which he had access, and also met with like directions in the commentaries on the Siddhânta-Çiromaṇi.

There are, however, very serious objections to such an interpretation of the brief direction contained in the text. It is altogether inconsistent with the whole plan and method of a Hindu astronomical treatise to leave, and even to order, matters of this character to be determined by observation. Observation has no such important place assigned to it in the astronomical system: with the exception of terrestrial longitude and latitude, which, in the nature of things, are beyond the reach of a treatise, it is intended that the astronomer should find in his text-book everything which he needs for the determination of celestial phenomena, and should resort to instruments and observation only by way of illustration. The sphere of which the construction is prescribed in the thirteenth chapter is not an instrument for observation: it is expressly stated to be "for the instruction of the pupil," and it is encumbered with such a number and variety of different circles, including parallels of declination for all the asterisms and for the observed fixed stars, that it could not be used for any other purpose: it will be noticed, too, that the commentary is itself obliged to order here the addition of the only appliances—the revolving circle of declination and the hole through the centre—which make of it an instrument for observation. The simple and original meaning of the passage seems to be that, having constructed a sphere in the manner to be hereafter described, one may examine the places of the asterisms as marked upon it, and note their coincidence with the actual positions of the stars in the heavens. And we would regard the other interpretation as forced upon the passage by the commentators, in order to avoid the difficulty pointed out by us above (near the end of the note on the last passage but one) and to free the Siddhânta from the imputation of having neglected the precessional variation of the circles of declination. M. Biot pronounces the method of observation explained by the commentators "almost impracticable," and it can, accordingly, hardly be that by which the positions of the asterisms were at first laid down, or by which they could be made to undergo the necessary corrections. Another method, more in accordance with the rules and processes of the third chapter, and which appears to us to be more authentic and of higher value, is described by Colebrooke (as above) from the Siddhânta-Sarvabhâuma, being there cited from the Siddhânta-Sundara; it is as follows:

"A tube, adapted to the summit of the gnomon, is directed toward the star on the meridian: and the line of the tube, pointed to the star, is prolonged by a thread to the ground. The line from the summit of the gnomon to the base is the hypothenuse; the height of the gnomon is the perpendicular; and its distance from the extremity of the thread is the base of the triangle. Therefore, as the hypothenuse is to its base, so is the radius to a base, from which the sine of the angle, and consequently the angle itself, are known. If it exceed the latitude [of the place of observation], the declination is south; or, if the contrary, it is north. The right ascension of the star is calculated from the hour of night, and from the right ascension of the sun for that time. The declination of the corresponding point of the ecliptic being found, the sum or difference of the declinations, according as they are of the same or of different denominations, is the distance of the star from the ecliptic. The longitude of the same point is computed; and from these elements, with the actual precession of the equinox, may be calculated the true longitude of the star; as also its latitude on a circle passing through the poles of the ecliptic."

The Siddhânta-Sarvabhâuma also gives the true longitudes and latitudes of the asterisms, professedly as thus obtained by observation and calculation, and they are reported by Colebrooke in his general table of data respecting the asterisms.

If we are not mistaken, the amount and character of the errors in the stated latitudes of the asterisms tend to prove that this, or some kindred process, was that by which their positions were actually determined.

13. In Taurus, the seventeenth degree, a planet of which the latitude is a little more than two degrees, south, will split the wain of Rohiṇî.

The asterism Rohiṇî, as has been seen above, is composed of the five principal stars in the head of Taurus, in the constellation of which is seen the figure of a wain. The divinity is Prajâpati. The distances of its stars in longitude from the initial point of the sphere vary from 45° 46′ (γ) to 49° 45′ (α): hence the seventeenth degree of the second sign—the reckoning commencing at the initial point of the sphere, taken as coinciding also with the vernal equinox—is very nearly the middle of the wain. The latitude of its stars, again, varies from 2° 36′ (ϵ) to 5° 47′ (ϑ) S.; hence, to come into collision with, or to enter, the wain, a planet must have more than two degrees of south latitude. The Siddhânta does not inform us what would be the consequences of such an occurrence; that belongs rather to the domain of astrology than of astronomy. We cite from the Pañcatantra (vv. 238–241) the following description of these consequences, derived from the astrological writings of Varâha-mihira:*

"When Saturn splits the wain of Rohiṇî here in the world, then Mâdhava rains not upon the earth for twelve years.

"When the wain of Prajâpati's asterism is split, the earth, having as it were committed a sin, performs, in a manner, her surface being strewn with ashes and bones, the *kâpâlika* penance.

"If Saturn, Mars, or the descending node splits the wain of Rohiṇî, why need I say that, in a sea of misfortune, destruction befalls the world?

"When the moon is stationed in the midst of Rohiṇî's wain, then men wander recklessly about, deprived of shelter, eating the cooked flesh of children, drinking water from vessels burnt by the sun."

Upon what conception this curious feature of the ancient Hindu astrology is founded, we are entirely ignorant.

14. Calculate, as in the case of the planets, the day and night of the asterisms, and perform the operation for apparent longitude (*dṛkkarman*), as before: the rest is by the rules for the conjunction (*melaka*) of planets, using the daily motion of the planet as a divisor: the same is the case as regards the time.

15. When the longitude of the planet is less than the polar longitude (*dhruvaka*) of the asterism, the conjunction (*yoga*) is to come; when greater, it is past: when the planet is retrograding (*vakragati*), the contrary is to be recognized as true of the conjunction (*samâgama*).

The rules given in the preceding chapter for calculating the conjunction of two planets with one another apply, of course, with certain modifications, to the calculation of the conjunctions of the planets with the asterisms. The text, however, omits to specify the most important of these modifications—that, namely, in determining the apparent longitude of an asterism, one part of the process prescribed in the case of a planet, the *ayanadṛkkarman*, or correction for ecliptic deviation, is to be omitted altogether; since the polar longitude of the asterism, which is given, corresponds in character with the *âyana graha*, or longitude of the planet as affected by ecliptic deviation, which must be ascertained by the *ayanadṛkkarman*. The commentary notices the omission, but offers neither explanation nor excuse for it. The other essential modification—that, the asterism being fixed, the motion of the planet alone is

* Our translation represents the verses as amended in their readings by Benfey (Pantschatantra etc., 2r Theil, nn. 234–237). In the third of the verses, however, the reading of the published text, *çaçi*, "moon," would seem decidedly preferable to *çikhi*, "descending node"; since the node, being always necessarily in the ecliptic, can never come into collision with Rohiṇî's wain.

to be used as divisor in determining the place and time of the conjunction—is duly noticed.

The inaccuracies in the Hindu process for determining apparent longitudes, which, as above noticed, are kept within bounds, where the planets alone are concerned, by the small amount of their latitudes, would be liable in the case of many of the asterisms to lead to grave errors of result.

16. Of the two Phalgunîs, the two Bhâdrapadâs, and likewise the two Ashâḍhâs, of Viçâkhâ, Açvinî, and Mṛgaçîrsha (*sâumya*), the junction-star (*yogatârâ*) is stated to be the northern (*uttara*):
17. That which is the western northern star, being the second situated westward, that is the junction-star of Hasta; of Çravishṭhâ it is the western:
18. Of Jyeshṭhâ, Çravaṇa, Anurâdhâ (*maitra*), and Pushya (*bârhaspatya*), it is the middle star: of Bharaṇî, Kṛttikâ (*âgneya*), and Maghâ (*pitrya*), and likewise of Revatî, it is the southern:
19. Of Rohiṇî, Punarvasu (*âditya*), and Mûla, it is the eastern, and so also of Âçleshâ (*sârpa*): in the case of each of the others, the junction-star (*yogatârakâ*) is the great (*sthûla*) one.

We have had occasion above, in treating of the identification of the asterisms, to question the accuracy of some of these designations of the relative position of the junction-stars in the groups containing them. We do not regard the passage as having the same authenticity and authority with that in which the determinations of the polar longitudes and latitudes are given; and indeed, we are inclined to suspect that all which follows the fifteenth verse in the chapter may be a later addition to its original content. It is difficult to see otherwise why the statements given in verses 20 and 21 of the positions of certain stars should be separated from those presented above, in verses 10–12. A designation of the relative position of the junction-star in each group ought also properly to be connected with a definition of the number of stars composing each, and a description of its configuration—such as are presented along with it by other treatises, as the Çâkalya-Sanhitâ. The first is even in some points ambiguous unless accompanied by the others, since there are cases in which the same star has a different position in its asterism according as the latter is to be regarded as including a less or a greater number of stars. In this respect also, then, the passage looks like a disconnected fragment. Nor is the method of designation so clear and systematic as to inspire us with confidence in its accuracy. Upon a consideration of the whole series of asterisms, it is obvious that the brightest member of each group is generally selected as its junction-star. Hence we should expect to find a general rule to that effect laid down, and then the exceptions to it specially noted, together with the cases in which such a designation would be equivocal. Instead of this, we have the junction-stars of only two asterisms containing more than one star, namely Abhijit and Çatabhishaj, described by their superior brilliancy, while that of the former is not less capable of being pointed out by its position than are any of the others in the series. Again, there are cases in which it is questionable which star is meant to be pointed out in a group of which the constitution is not doubtful, owing to the very near correspondence of more than one star with the position as defined. And once more, where, in a single instance, a special effort has apparently been made to fix the position of the junction-star beyond all doubt or cavil, the result is a failure; for it still remains a matter of dispute how the description is to be understood, and which member of the group is intended. The case referred to is that of Hasta, which occupies nearly all of verse 17. That Colebrooke was not satisfied as to the meaning of the description is clear from the fact that he specifies, as the star referred to, "γ or δ Corvi." His translation of the verse, "2nd W. of 1st N. W.", conveys to us no intelligible meaning whatever, as applied to the actual group. He evidently understood *paçcimottaratârâyâ* as a single word, standing by euphony for *-târâyâs*, ablative of *-târâ*. Our own rendering supposes it divided into the two independent words *paçcimottaratârâ yâ*, or the three *paçcimâ uttaratârâ yâ*. This interpretation is, in the first place, supported by the corresponding passage in the Çâkalya-Sanhitâ, which reads, "of Hasta, the north-western (*vâyavî*): it is also the second western." Again, it applies without difficulty to one of the stars in the group, namely to γ, which we think most likely to be the one pointed out—and mainly, because either of the others would admit of being more simply and briefly designated, δ as the northern, β as the eastern, α as the southern, and ϵ as the western star. We should, then, regard the description as unambiguous, were it not for what is farther added, "being the second situated westward;" for γ is the first or most westerly of the five in longitude, and the third in right ascension, while the second in longitude and in right ascension respectively are the two faint stars ϵ and α. We confess that we do not see how the difficulty is to be solved without some emendation of the text.

We conceive ourselves to be justified, then, in regarding this passage as of doubtful authenticity and inferior authority: as already partaking, in short, of that ignorance and carelessness which has rendered the Hindu astronomers unable, at any time during the past thousand years, to point out in the heavens the complete series of the groups of stars composing their system of asterisms. None of the other authorities accessible to us gives a description of the relative places of the junction-stars, excepting the Çâkalya-Sanhitâ, and our manuscript of its text is so defective and corrupt at this point that we are able to derive from it with confidence the positions of only about a third of the stars. So far, it accords with the Sûrya-Siddhânta, save that it points out as the junction-star of Pûrva-Ashâḍhâ the brightest, instead of the northernmost, member of the group; and here there is a difference in the mode of designation only, and not a disagreement as regards the star designated.

20. Situated five degrees eastward from Brahmahṛdaya is Prajâpati: it is at the end of Taurus, and thirty-eight degrees north.

21. Apâmvatsa is five degrees north from Citrâ: somewhat greater than it, as also six degrees to the north of it, is Âpas.

The three stars whose positions are defined in this passage are not mentioned in the Çâkalya-Sanhitâ, nor in the Siddhânta-Çiromaṇi and (according to Colebrooke) the Brahma-Siddhânta; only the latter of them, Âpas, is omitted by the Graha-Lâghava, being noticed in the Sûrya-Siddhânta alone. It may fairly be questioned, for the reason remarked above, whether the original text of our treatise itself contained the last two verses of this chapter: moreover, at the end of the next chapter (ix. 18), where those stars are spoken of which never set heliacally, on account of their high northern situation, Prajâpati is not mentioned among them, as it ought to be, if its position had been previously stated in the treatise. Still farther on (xiii. 9), in the description of the armillary sphere, it is referred to by the name of Brahma, which, according to the commentary on this passage, and to Colebrooke, it also customarily bears. Perhaps another evidence of the unauthenticity of the passage is to be seen in the fact that the two definitions of the polar longitude of Prajâpati do not, if taken in connection with verse 11, appear to agree with one another: a star which is 5° east from the position of Brahmahṛdaya, as there stated, is not "at the end of Taurus," but at its twenty-seventh degree: this may, however, be merely an inaccurate expression, intended to mean that the star is in the latter part, or near the end, of Taurus. The Graha-Lâghava, which defines the positions of all these stars directly, by degrees of polar longitude and latitude, and not by reference either to the signs or to other stars, gives Prajâpati 61° of polar longitude, or 5° more than it assigned to Brahmahṛdaya: it also adds 1° to the polar latitude as stated in our text. The star referred to can hardly be any other than that in the head of the Wagoner, or δ Aurigæ (4):

Prajâpati	67° 11′	36° 49′ N.
δ Aurigæ	69° 54′	30° 49′ N.

The error of latitude is about the same with that which was committed with reference to Brahmahṛdaya, or Capella. Why so faint and inconspicuous a star should be found among the few of which the Hindu astronomers have taken particular notice is not easy to discover.

The position of the star named Apâmvatsa, "Waters' Child," is described in our text by reference to Citrâ, or Spica Virginis: it is said to be in the same longitude, 180°, and 5° farther north; and this, since Citrâ itself is in lat. 2° S., would make the latitude of Apâmvatsa 3° N. The Graha-Lâghava gives it this latitude directly, and also makes its longitude agree with that of Spica, which, as already noticed, it places at the distance of 183° from the origin of the sphere. Âpas, "Waters" (the commentary, however, treats the word as a singular masculine, Âpa), is put 6° north of Apâmvatsa, or in lat. 9° N. It is identified by Colebrooke with δ Virginis (3), and doubtless correctly:

Âpas	176° 23′	8° 15′ N.
δ Virginis	171° 28′	8° 38′ N.

Colebrooke pronounces Apâmvatsa to comprise "the nebulous stars marked b 1, 2, 3" in Virgo. We can find, however, no such stars upon any map, or in any catalogue, accessible to us, and hence presume that Colebrooke must have been misled here by some error of the authority on which he relied. There is, on the other hand, a star, ϑ Virginis (4), situated directly between Spica and δ, and at such a distance from each as shows almost beyond question that it is the star intended:

Apâmvatsa	178° 48′	2° 45′ N.
ϑ Virginis	178° 12′	1° 45′ N.

It is not less difficult in this than in the former case to account for the selection of these stars, among the hundreds equalling or excelling them in brilliancy, as objects of special attention to the astronomical observers of ancient India. Perhaps we have here only the scattered and disconnected fragments of a more complete and shapely system of stellar astronomy, which flourished in India before the scientific reconstruction of the Hindu astronomy transferred the field of labor of the astronomer from the skies to his text-books and his tables of calculation.

The annexed table gives a comparative view of the positions of the seven stars spoken of in this and a preceding passage (vv. 10–12) as defined by our text and as determined by modern observers:

Positions of certain Fixed Stars.

Name.	Hindu position:				True position:		Star compared.
	pol. long.	pol. lat.	long.	lat.	long.	lat.	
	° ′	° ′	° ′	° ′	° ′	° ′	
Agastya,	90 0	80 0 S.	90 0	80 0 S.	85 4	75 50 S.	α Argûs, Canopus.
Mṛgavyâdha,	80 0	40 0 S.	76 23	39 52 S.	84 7	39 32 S.	α Canis Maj., Sirius.
Agni,	52 0	8 0 N.	54 5	7 44 N.	62 32	5 22 N.	β Tauri.
Brahmahṛdaya,	52 0	30 0 N.	60 29	28 53 N.	61 50	22 52 N.	α Aurigæ, Capella.
Prajâpati,	57 0	38 0 N.	67 11	36 49 N.	69 54	30 49 N.	δ Aurigæ.
Apâmvatsa,	180 0	3 0 N.	178 48	2 45 N.	178 12	1 45 N.	ϑ Virginis.
Âpas,	180 0	9 0 N.	176 23	8 15 N.	171 28	8 38 N.	δ Virginis.

The gross errors in the determinations of position of these stars give us a yet lower idea of the character of Hindu observations than we derived from our examination of the junction-stars of the asterisms.

The essay of Colebrooke in the ninth volume of the Asiatic Researches, to which we have already so often referred, gives farther information of much interest respecting such matters connected with the Hindu astronomy of the fixed stars as are passed without notice in our treatise. He states the rules laid down by different authorities for calculating the time of heliacal rising of Agastya, or Canopus, upon which depends the performance of certain religious ceremonies. He also presents a view of the Hindu doctrine of the Seven Sages, or *ṛshis*, by which name are known the bright stars in Ursa Major forming the well-known constellation of the Wain, or Dipper. To these stars the ancient astronomers of India, and many of the modern upon their authority, have attributed an independent motion about the pole of the heavens, at the rate of 8′ yearly, or of a complete revolution in 2700 years. The Sûrya-Siddhânta alludes in a later passage (xiii. 9) to the Seven Sages, but it evidently is to be understood as rejecting the theory of their proper motion, which is also ignored by the Siddhânta-Çiromaṇi. That so absurd a dogma should have originated and gained a general currency in India, and that it should still maintain itself in many of the astronomical text-books, is, however, too striking and significant a circumstance to be left out of sight in estimating the character of the ancient and native Hindu astronomy.

CHAPTER IX.

OF HELIACAL RISINGS AND SETTINGS.

CONTENTS:—1, subject of the chapter; 2–3, under what circumstances, and at which horizon, the planets rise and set heliacally; 4–5, method of calculating their distances in oblique ascension from the sun; 6–9, distances from the sun at which they disappear and re-appear; 10–11, how to find the time of heliacal setting or rising, past or to come; 12–15, distances from the sun at which the asterisms and fixed stars disappear and re-appear; 16–17, mode of determining their times of rising and setting; 18, what asterisms and stars never set heliacally.

1. Now is set forth the knowledge of the risings (*udaya*) and settings (*astamaya*) of the heavenly bodies of inferior brilliancy, whose orbs are overwhelmed by the rays of the sun.

The terms used for the heliacal settings and risings of the heavenly bodies, or their disappearance in the sun's neighborhood and their return to visibility, are precisely the same with those employed to denote their rising (*udaya*) and setting (*asta*, *astamaya*, *astamana*) above and below the horizon. The title of the chapter, *udayâstâdhikâra*, is literally translated in our heading.

2. Jupiter, Mars, and Saturn, when their longitude is greater than that of the sun, go to their setting in the west; when it is less, to their rising in the east: so likewise Venus and Mercury, when retrograding.

3. The moon, Mercury, and Venus, having a swifter motion, go to their setting in the east when of less longitude than the sun; when of greater, to their rising in the west.

These specifications are of obvious meaning and evident correctness. The planets which have a slower motion than the sun, and so are overtaken by him, make their last appearance in the west, after sunset, and emerge again into visibility in the east, before sunrise: of those which move more rapidly than the sun, the contrary is true: Venus and Mercury belong to either class, according as their apparent motion is retrograde or direct.

4. Calculate the longitudes of the sun and of the planet—in the west, for the time of sunset; in the east, for that of sunrise—and then make also the calculation of apparent longitude (*dṛkkarman*) of the planet.

5. Then the ascensional equivalent, in respirations, of the interval between the two (*lagnântaraprâṇâs*) will give, when divided by sixty, the degrees of time (*kâlânçâs*); or, in the west, the ascensional equivalent, in respirations, of the interval between the two when increased each by six signs.

Whether a planet will or will not be visible in the west after sunset, or in the east before sunrise, is in this treatise made to depend solely upon the interval of time by which its setting follows, or its rising precedes, that of the sun, or upon its distance from the sun in oblique ascension; to the neglect of those other circumstances—as the declination of the two bodies, and the distance and direction of the planet from the ecliptic—which variously modify the limit of visibility as thus defined. The ascertainment of the distance in oblique ascension, then, is the object of the rules given in these verses. In explaining the method of the process, we will consider first the case of a calculation made for the eastern horizon. The time of sunrise having been determined, the true longitudes and rates of motion of the sun and the planet in question are found for that moment, as also the latitude of the planet. Owing to the latter's removal in latitude from the ecliptic, it will not pass the horizon at the same moment with the point of the ecliptic which determines its longitude, and the point with which it does actually rise must be found by a separate process. This is accomplished by calculating the apparent longitude of the planet, according to the method taught in the seventh chapter. There is nothing in the language of the text which indicates that the calculation is not to be made in full, as there prescribed, and for the given moment of sunrise: as so conducted, however, it would evidently yield an erroneous result; for, the planet being above the horizon, the point of the ecliptic to which it is then referred by a circle through the north and south points of the horizon is not the one to which it was referred by the horizon itself at the moment of its own rising. The commentary removes this difficulty, by specifying that the *akshadṛkkarman*, or that part of the process which gives the correction for latitude, is to be performed "only as taught in the first half-verse"—that is, according to the former part of vii. 8, which contains the rule for determining the amount of the correction at the horizon—omitting the after process, by which its value is made to correspond to the altitude of the planet at the given time. Having thus ascertained the points of the ecliptic which rise with the sun and with the planet respectively, the corresponding equatorial interval, or the distance of the planets in oblique ascension, is found by a rule already given (iii. 50). The result is expressed in respirations of sidereal time, which are equivalent to minutes of the equator (see above, i. 11–12); they are reduced to degrees by dividing by sixty: and the degrees thus found receive the technical name of "time-degrees" (*kâlânçâs, kâlabhâgâs*); they are also called below "degrees of setting" (*astânçâs*), and "degrees of visibility" (*dṛçyânçâs*).

If the planet for which the calculation is made has greater longitude than the sun, the process, being adapted to the time of sunset, and to the western horizon, requires a slight modification, owing to the fact that the equivalents of the signs in oblique ascension (iii. 42–45) are given only as measured at the eastern horizon. Since 180 degrees of the ecliptic are always above the horizon, any given point of the ecliptic will set at the same moment that another 180° distant from it rises; by adding, then, six signs to the calculated positions of the sun and the planet, and ascertaining, by iii. 50, the ascensional difference of the two points so found, the interval between the setting of the sun and that of the planet will be determined.

Before going on to explain how, from the result thus obtained, the time of the planet's disappearance or re-appearance may be derived, the text defines the distances from the sun, in oblique ascension or "degrees of time," at which each planet is visible.

6. The degrees of setting (*astânçâs*) are, for Jupiter, eleven; for Saturn, fifteen; for Mars, moreover, they are seventeen:

7. Of Venus, the setting in the west and the rising in the east take place, by reason of her greatness, at eight degrees; the setting in the east and the rising in the west occur, owing to her inferior size, at ten degrees:

8. So also Mercury makes his setting and rising at a distance from the sun of twelve or fourteen degrees, according as he is retrograding or rapidly advancing.

9. At distances, in degrees of time (*kâlabhâgâs*), greater than these, the planets become visible to men; at less distances they become invisible, their forms being swallowed up (*grasta*) by the brightness of the sun.

The moon, it will be noticed, is omitted here; her heliacal rising and setting are treated of at the beginning of the next following chapter.

In the case of Mercury and Venus, the limit of visibility is at a greater or less distance from the sun according as the planet is approaching its inferior or superior conjunction, the diminution of the illuminated portion of the disk being more than compensated by the enlargement of the disk itself when seen so much nearer to the earth.

Ptolemy treats, in the last three chapters (xiii. 7–9) of his work, of the disappearance and reappearance of the planets in the neighborhood of the sun, and defines the limits of visibility of each planet when in the sign Cancer, or where the equator and ecliptic are nearly parallel. His limits are considerably different from those defined in our text, being, for Saturn, 14°; for Jupiter, 12° 45′; for Mars, 14° 30′; for Venus and Mercury, in the west, 5° 40′ and 11° 30′ respectively.

10. The difference, in minutes, between the numbers thus stated and the planet's degrees of time (*kâlânçâs*), when divided by the difference of daily motions—or, if the planet be retrograding, by the sum of daily motions—gives a result which is the time, in days etc.

11. The daily motions, multiplied by the corresponding ascensional equivalents (*tallagnâsavas*), and divided by eighteen hundred, give the daily motions in time (*kâlagati*); by means of these is found the distance, in days etc., of the time past or to come.

Of these two verses, the second prescribes so essential a modification of the process taught in the first, that their arrangement might have been more properly reversed. If we have ascertained, by the previous rules, the distance of a planet in oblique ascension from the sun, and if we know the distance in oblique ascension at which it will disappear or re-appear, the interval between the given moment and that at which disappearance or re-appearance will take place may be readily found by dividing by the rate of approach or separation of the two bodies the difference between their actual distance and that of apparition and disparition : but the divisor must, of course, be the rate of approach in oblique ascension, and not in longitude. The former is derived from the latter by the following proportion : as a sign of the ecliptic, or 1800′, is to its equivalent in oblique ascension, as found by iii. 42–45, so is the arc of the ecliptic traversed by each planet in a day to the equatorial equivalent of that arc. The daily rates of motion in oblique ascension thus ascertained are styled the "time-motions" (*kâlagati*), as being commensurate with the "time-degrees" (*kâlânçâs*).

12. Svâtî, Agastya, Mṛgavyâdha, Citrâ, Jyeshṭhâ, Punarvasu, Abhijit, and Brahmahṛdaya rise and set at thirteen degrees.

13. Hasta, Çravana, the Phalgunîs, Çravishṭhâ, Rohiṇî, and Maghâ become visible at fourteen degrees; also Viçâkhâ and Açvinî.

14. Kṛttikâ, Anurâdhâ (*mâitra*), and Mûla, and likewise Âçleshâ and Ârdrâ (*râudrarksha*), are seen at fifteen degrees; so, too, the pair of Ashâḍhâs.

15. Bharaṇî, Pushya, and Mṛgaçîrsha, owing to their faintness, are seen at twenty-one degrees; the rest of the asterisms become visible and invisible at seventeen degrees.

These are specifications of the distances from the sun in oblique ascension (*kâlânçâs*) at which the asterisms, and those other of the fixed stars whose positions were defined in the preceding chapter, make their heliacal risings and settings. The asterisms we are doubtless to regard as represented by their junction-stars (*yogatârâ*). The classification here made of the stars in question, according to their comparative magnitude and brilliancy, is in many points a very strange and unaccountable one, and by no means calculated to give us a high idea of the intelligence and care of those by whom it was drawn up. The first class, comprising such as are visible at a distance of 13° from the sun, is, indeed, almost wholly composed of stars of the first magnitude; one only, Punarvasu (β Geminorum), being of the first to second, and having for its fellow one of the first (α Geminorum). But the second class, that of the stars visible at 14°, also contains four which are of the first magnitude, or the first to second; namely, Aldebaran (Rohiṇî), Regulus (Maghâ), Deneb or β Leonis (Uttara-Phalgunî), and Atair or α Aquilæ (Çravana); and, along with these, one of the second to third magnitude, δ Leonis (Pûrva-Phalgunî), three of the third, and one, ι Libræ (Viçâkhâ), of the fourth. In this last case, however, it might be possible to regard α Libræ, of the second magnitude, as the star which is made to determine the visibility of the asterism. Among the stars of the third class, again, which are visible at 15°, is one, α Orionis (Ârdrâ), which, though a variable star, does not fall below the first to second magnitude; while with it are found ranked six stars of the third magnitude, or of the third to fourth. The class of those which are visible at 17°, and which are left unspecified, contains two stars of the fourth magnitude, but also two of the second, one of which, *α* Andromedæ or *γ* Pegasi (Uttara-Bhâdrapadâ), is mentioned below (v. 18) among those which are never obscured by the too near approach of the sun. The stars forming the class which are not to be seen within 21° of the sun are all of the fourth magnitude, but they are no less distinctly visible than two of those in the preceding class; and indeed, Bharaṇî is palpably more so, since it contains a star of the third magnitude, which is perhaps (see above) to be regarded as its junction-star. Since Agni, Brahma, Apâmvatsa, and Âpas are not specially mentioned, it is to be assumed that they all belong in the class of those visible at 17°, and they are so treated by the commentator: the first of them (*β* Tauri) is a star of the second magnitude; for the rest, see the last note to the preceding chapter.

Some of the apparent anomalies of this classification are mitigated or removed by making due allowance for the various circumstances by which, apart from its absolute brilliancy, the visibility of a star in the sun's neighborhood is favored or the contrary—such as its distance and direction from the equator and ecliptic, and the part of the ecliptic in which the sun is situated during its disappearance. Many of them, however, do not admit of such explanation, and we cannot avoid regarding the whole scheme of classification as one not founded on careful and long-continued observation, but hastily and roughly drawn up in the beginning, and perhaps corrupted later by unintelligent imitators and copyists.

16. The degrees of visibility (*dṛçyânçâs*), if multiplied by eighteen hundred and divided by the corresponding ascensional equivalent (*udayâsavas*), give, as a result, the corresponding degrees on the ecliptic (*kshetrânçâs*); by means of them, likewise, the time of visibility and of invisibility may be ascertained.

This verse belongs, in the natural order of sequence, not after the passage next preceding, with which it has no special connection, but after verse 11. Instead of reducing, as taught in that verse, the motions upon the ecliptic to motions in oblique ascension, the "degrees of time" (*kâlânçâs*) may themselves be reduced to their equivalent upon the corresponding part of the ecliptic, and then the time of disappearance or of re-appearance calculated as before, using as a divisor the sum or difference of daily motions along the ecliptic. The proportion by which the reduction is made is the converse of that before given; namely, as the ascensional equivalent of the sign in which are the sun and the planet is to that sign itself, or 1800′, so are the "degrees of visibility" (*dṛçyânçâs*, or *kâlânçâs*) of the planet to the equivalent distance upon that part of the ecliptic in which it is then situated. The technical name given to the result of the proportion is *kshetrânçâs*: *kshetra* is literally "field, territory," and the meaning of the compound may be thus paraphrased: "the limit of visibility, in degrees, measured upon that part of the ecliptic which is, at the time, the territory occupied by the planets in question, or their proper sphere."

17. Their rising takes place in the east, and their setting in the west; the calculation of their apparent longitude (*dṛkkarman*)

is to be made according to previous rules; the ascertainment of the time, in days etc., is always by the daily motion of the sun alone.

This verse should follow immediately after verse 15, to which it attaches itself in the closest manner. The dislocation of arrangement in the latter part of this chapter is quite striking, and is calculated to suggest a suspicion of interpolations.

The directions given in the verse require no explanation: they are just such an adaptation of the processes already prescribed to the case of the fixed stars as that made in verse 14 of the last chapter. The commentary points out again that the calculation of the correction for latitude (*akshadṛkkarman*) is to be made only for the horizon, or as stated in the first half-verse of the rule.

18. Abhijit, Brahmahṛdaya, Svâtî, Çravaṇa (*vâishṇava*), Çravishṭhâ (*vâsava*), and Uttara-Bhâdrapadâ (*ahirbudhnya*), owing to their northern situation, are not extinguished by the sun's rays.

It may seem that it would have been a more orderly proceeding to omit the stars here mentioned from the specifications of verses 12–15 above; but there is, at least, no inconsistency or inaccuracy in the double statement of the text, since some of the stars may never attain that distance in oblique ascension from the sun which is there pointed out as their limit of visibility. We have not thought it worth the trouble to go through with the calculations, and ascertain whether, according to the data and methods of this treatise, these six stars, and these alone, of those which the treatise notices, would never become invisible at Ujjayinî. It is evident, however, as has already been noticed above (viii. 20–21), that the star called Brahma or Prajâpati (δ Aurigæ) is not here taken into account, since it is 8° north of Brahmahṛdaya, and consequently can not become invisible where the latter does not.

CHAPTER X.

OF THE MOON'S RISING AND SETTING, AND OF THE ELEVATION OF HER CUSPS.

CONTENTS:—1, of the heliacal rising and setting of the moon; 2–5, how to find the interval from sunset to the setting or rising of the moon; 6–8, method of determining the moon's relative altitude and distance from the sun at sunset; 9, to ascertain the measure of the illuminated part of her disk; 10–14, method of delineating the moon's appearance at sunset; 15, how to make the same calculation and delineation for sunrise.

1. The calculation of the heliacal rising (*udaya*) and setting (*asta*) of the moon, too, is to be made by the rules already given. At twelve degrees' distance from the sun she becomes visible in the west, or invisible in the east.

In determining the time of the moon's disappearance in the neighborhood of the sun, or of her emergence into visibility again beyond the sphere of his rays, no new rules are required; the same methods being employed as were made use of in ascertaining the time of heliacal setting and rising of the other planets: they were stated in the preceding chapter. The definition of the moon's limit of visibility would have been equally in order in the other chapter, but is deferred to this in order that the several processes in which the moon is concerned may be brought together. The title of the chapter, *çṛngonnatyadhikâra*, "chapter of the elevation of the moon's cusps" (*çṛnga*, literally "horn"), properly applies only to that part of it which follows the fifth verse.

The degrees spoken of in this verse are, of course, "degrees of time" (*kâlânçâs*), or in oblique ascension.

2. Add six signs to the longitudes of the sun and moon respectively, and find, as in former processes, the ascensional equivalent, in respirations, of their interval (*lagnântarâsavas*): if the sun and moon be in the same sign, ascertain their interval in minutes.

3. Multiply the daily motions of the sun and moon by the result, in nâḍîs, and divide by sixty; add to the longitude of each the correction for its motion, thus found, and find anew their interval, in respirations;

4. And so on, until the interval, in respirations, of the sun and moon is fixed: by so many respirations does the moon, in the light half-month (*çukla*), go to her setting after the sun.

5. Add half a revolution to the sun's longitude, and calculate the corresponding interval, in respirations: by so many respirations does the moon, in the dark half-month (*kṛshṇapaksha*), come to her rising after sunset.

The question here sought to be solved is, how long after sunset upon any given day will take place the setting of the moon in the crescent half-month, or from new to full moon, and the rising of the moon in the waning half-month, or from full to new moon. The general process is the same with that taught in the last chapter, for obtaining a like result as regards the other planets or fixed stars: we ascertain, by the rules of the seventh chapter—applying the correction for the latitude according to its value at the horizon, as determined by the first part of vii. 8—the point of the ecliptic which sets with the moon; and then the distance in oblique ascension between this and the point at which the sun set will measure the required interval of time. An additional correction, however, needs to be applied to the result of this process in the case of the moon, owing to her rapid motion, and her consequent perceptible change of place between the time of sunset and that of her own setting or rising: this is done by calculating the amount of her motion during the interval as first determined, and adding its equivalent in oblique ascension to that interval; then calculating her motion anew for the increased interval and adding its ascensional equivalent—and so on, until the desired degree of accuracy is attained.

The process thus explained, however, is not precisely that which is prescribed in the text. We are there directed to calculate the amount of motion both of the sun and moon during the interval between the setting of the sun and that of the moon, and, having applied them to the longitudes of the two bodies, to take the ascensional equivalent of the distance between them in longitude, as thus doubly corrected, for the precise time of the setting of the moon after sunset. In one point of view this is false and absurd; for when the sun has once passed the horizon, the interval to the setting of the moon will be affected only by her motion, and not at all by his. In another light, the process does not lack reason: the allowance for the sun's motion is equivalent to a reduction of the interval from sidereal (*nâkshatra*) time to civil, or true solar (*sâvana*) time, or from respirations which are thirty-six-hundreths of the earth's revolution on its axis to such as are like parts of the time from actual sunrise to actual sunrise. But such a mode of measuring time is unknown elsewhere in this treatise, which defines (i. 11–12) and employs sidereal time alone, adding (ii. 59) to the sixty nâḍis which constitute a sidereal day so much sidereal time as is needed to make out the length of a day that is reckoned by any other method. It seems necessary, then, either to suppose a notable blunder in this passage, or to recognize in it such a departure from the usual methods of the treatise as would show it to be an interpolation. Probably the latter is the alternative to be chosen: it is, at any rate, that which the commentator prefers: he pronounces the two verses beginning with the second half of verse 2, and ending at the middle of verse 4, to be spurious, and the true text of the Siddhânta to comprise only the first half of verse 2 and the second of verse 4; these would form together a verse closely analogous in its method and expression with verse 5, which teaches the like process for moon-rise, in the waning half-month. Fortified by the authority of the commentator, we are justified in assuming that the Sûrya-Siddhânta originally neglected, in its process for calculating the time of the moon's setting, her motion during the interval between that time and sunset, and that the omission was later supplied by another hand, from some other treatise, which reckoned by solar time instead of sidereal. This does not, however, explain and account for the second half of the second verse; which, if it has any meaning at all, different from that conveyed in the former part of the same verse, seems to signify that when the sun and moon are so near one another as to be in the same sign, the discordance between distances on the ecliptic and their equivalents upon the equator may be neglected, and the difference of longitude in minutes taken for the interval of time in respirations.

If the time is between new and full moon, the object of the process is to obtain the interval from sunset to the setting of the moon; as both take place at the western horizon, the two planets are transferred to the eastern horizon, in order to the measurement of their distance in ascension: if, on the other hand, the moon has passed her full, the time of moonrise is sought; here the sun alone is transferred, by the addition of 180° to his longitude, to the eastern horizon, as taught in verse 5. The equation to be applied to the longitude of both planets is found by the familiar proportion—as sixty nâḍis are to the given interval in nâḍis, so is the true daily motion of the planet to its actual motion during that interval.

6. Of the declinations of the sun and moon, if their direction be the same, take the difference; in the contrary case, take the sum: the corresponding sine is to be regarded as south or north, according to the direction of the moon from the sun.

7. Multiply this by the hypothenuse of the moon's mid-day shadow, and, when it is north, subtract it from the sine of latitude (*aksha*) multiplied by twelve; when it is south, add it to the same.

8. The result, divided by the sine of co-latitude (*lamba*), gives the base (*bhuja*), in its own direction; the gnomon is the perpendicular (*koṭi*); the square root of the sum of their squares is the hypothenuse.

In explaining the method of this process, we shall follow the guidance of the commentator, pointing out afterwards wherein he varies from the strict letter of the text: for illustration we refer to the accompanying figure (Fig. 32).

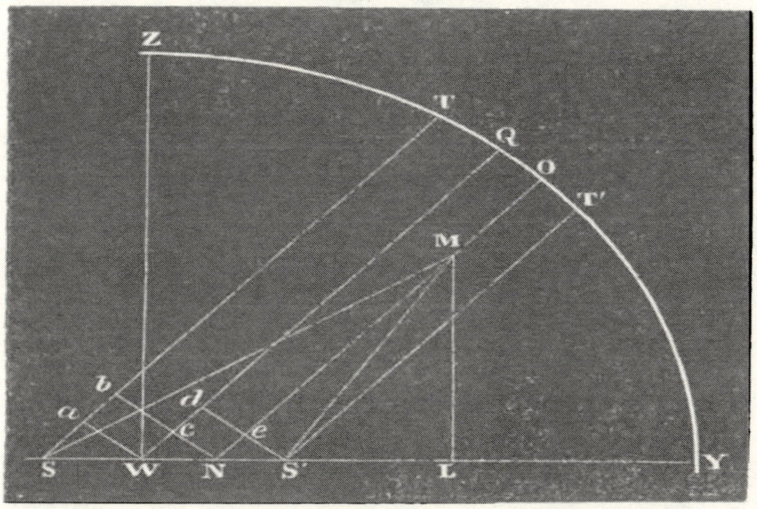

Fig. 32.

The figure represents the south-western quarter of the visible sphere, seen as projected upon the plane of the meridian; Z being the zenith, Y the south point, W Y the intersection of the horizontal and meridian planes, and W the projection of the west point. Let Z Q equal the latitude of the place of observation, and let Q T and Q O be the declinations of the sun and moon respectively, at the given time: then W Q, S T, and N O will be the projections of the equator and of the diurnal circles of the sun and moon. Suppose, now, the sun to be upon the horizon, at S, and the moon to have a certain altitude, being at M: draw from M the perpendicular to the plane of the horizon M L, and join M S: it is required to know the relation to one another of the three sides of the triangle S L M, in order to the delineation of the moon's appearance when at M, or at the moment of sunset.

Now M L is evidently the sine of the moon's altitude at the given time, which may be found by methods already more than once described and illustrated. And S L is composed of the two parts S N and N L, of which the former depends upon the distance of the moon in declination from the sun, and the latter upon the moon's altitude. But S N is one of the sides of a right-angled triangle, in which the angle N S *b* is equal to the observer's co-latitude, and N *b* to the sum of the sine of declination of the sun, *c b* or W *a*, and that of the moon, N *c*. Hence sin *b* S N : *b* N :: R : S N

or sin co-lat. : sum of sines of decl. :: R : S N

and S N = (R × sum of sines of decl.) ÷ sin co-lat.

In like manner, since, in the triangle M N L, the angles at M and N are respectively equal to the observer's latitude and co-latitude,

sin M N L : sin L M N :: M L : N L

or sin co-lat. : sin. lat. :: sin alt. : N L

and N L = (sin alt. × sin lat.) ÷ sin co-lat.

We have thus found the values of M L and the two parts of S L in terms of the general sphere, or of a circle whose radius is tabular radius: it is desired farther to reduce them to terms of a circle in which M L shall equal the gnomon, or twelve digits. And since the gnomon is equal to the sine of altitude in a circle of which the hypothenuse of the corresponding shadow is radius (compare above, iii. 25–27 etc.), this reduction may be effected by multiplying the quantities in question by the hypothenuse of the shadow and dividing by radius. That is to say, representing the reduced values of S N and N L by *s n* and *n l* respectively,

R : hyp. shad. :: M L : gnom.
R : hyp. shad. :: S N : *s n*
R : hyp. shad. :: N L : *n l*

Substituting, now, in the second and third of these proportions the values of S N and N L found for them above, and substituting also in the third the value of the hypothenuse of the shadow derived from the first, we have

$$\text{R : hyp. shad.} :: \frac{\text{R} \times \text{sum sin decl.}}{\text{sin co-lat.}} : sn, \text{ and R} : \frac{\text{R} \times \text{gnom.}}{\text{sin alt.}} :: \frac{\text{sin alt.} \times \text{sin lat.}}{\text{sin co-lat.}} : nl$$

which reduce to

$$sn = \frac{\text{hyp. shad.} \times \text{sum sin decl.}}{\text{sin co-lat.}}, \text{ and } nl = \frac{\text{sin lat.} \times \text{gnom.}}{\text{sin co-lat.}}$$

Hence, if the perpendicular M L be assumed of the constant value of the gnomon, or twelve digits, we have

$$\text{SL} = \frac{(\text{hyp. shad.} \times \text{sum sin decl.}) + (\text{sin lat.} \times \text{gnom.})}{\text{sin co-lat.}}$$

In the case thus far considered the sun and moon have been supposed upon opposite sides of the equator. If they are upon the same side, the sun setting at S′, or if their sines of declination, S′*d* and N *c*, are of the same direction, the value of S′ N, the corresponding part of the base S′ L, will be found by treating in the same manner as before the difference of the sines, S′*e*, instead of their sum. In this case, too, the value of S′*e* being north, S′ N will have to be subtracted from N L to give the base S′ L. Other positions of the two luminaries with respect to one another are supposable, but those which we have taken are sufficient to illustrate all the conditions of the problem, and the method of its solution.

It is evident that, in two points, the process as thus explained by the commentator is discordant with that which the text prescribes. The latter, in the first place, tells us to take, not the sum or difference of the sines of declination, but the sine of the sum or difference of declinations, as the side *b* N of the triangle S N *b*. This seems to be a mere inaccuracy on the part of the text, the difference between the two quantities, which could never be of any great amount, being neglected : it is, however, very hard to see why the less accurate of the two valuations of the quantity in question should have been selected by the text; for it is, if anything, rather less easy of determination than the other. The other discordance is one of much more magnitude and importance : the text speaks of the "hypothenuse of the moon's mid-day shadow" (*madhyâhnenduprabhâkarṇa*), for which the commentary substitutes that of the shadow cast by the moon at the given moment of sunset. The commentator attempts to reconcile the discrepancy by saying that the text means here the moon's shadow as calculated after the method of a noon-shadow ; or again, that the time of sunset is, in effect, the middle of the day, since the civil day is reckoned from sunrise to sunrise : but neither of these explanations can be regarded as satisfactory. The commentator farther urges in support of his understanding of the term, that we are expressly taught above (vii. 11) that the calculation of apparent longitude (*dṛkkarman*) is to be made in the process for finding the elevation of the moon's cusps; while, if the hypothenuse of the moon's meridian-shadow be the one found, there arises no occasion for making that calculation. It seems clear that, unless the commentator's understanding of the true scope and method of the whole process be erroneous, the substitution which he makes must necessarily be admitted. This is a point to which we shall recur later.

9. The number of minutes in the longitude of the moon diminished by that of the sun gives, when divided by nine hundred, her illuminated part (*çukla*) : this, multiplied by the number of digits (*angula*) of the moon's disk, and divided by twelve, gives the same corrected (*sphuṭa*).

The rule laid down in this verse, for determining the measure of the illuminated part of the moon, applies only to the time between new moon and full moon, when the moon is less than 180° from the sun: when her excess of longitude is more than 180°, the rule is to be applied as stated below, in verse 15. As the whole diameter of the moon is illuminated when she is half a revolution from the sun, one half her diameter at a quarter of a revolution's distance, and no part of it at the time of conjunction, it is assumed that the illuminated portion of her diameter will vary as the part of 180° by which she is distant from the sun; and hence that, assuming the measure of the diameter of her disk to be twelve digits, the number of digits illuminated may be found by the following proportion: as half a revolution, or 10,800′, is to twelve digits, so is the moon's distance from the sun in minutes to the corresponding part of the diameter illuminated: the substitution, in the first ratio, of 900 : 1 for 10,800 : 12, gives the rule as stated in the text. Here, it will be noticed, we have for the first and only time the Greek method of measuring the moon's diameter, by equal twelfths, or digits: from this scale a farther reduction is made to the proper Hindu scale, as determined by the methods of the fourth chapter (see above, iv. 2–3, 26), by another proportion: as twelve is to the true diameter in digits, so is the result already found to the true measure of the part of the diameter illuminated.

It is not to be wondered at that the Hindus did not recognize the ellipticity of the line forming the inner boundary of the moon's illuminated part: it is more strange that they ignored the obvious fact that, while the illuminated portion of the moon's spherical surface visible from the earth varies very nearly as her distance from the sun, the apparent breadth of the bright part of her disk, in which that surface is seen projected, must vary rather as the versed sine of her distance.

10. Fix a point, calling it the sun: from that lay off the base, in its own proper direction; then the perpendicular, toward the west; and also the hypothenuse, passing through the extremity of the perpendicular and the central point.

11. From the point of intersection of the perpendicular and the hypothenuse describe the moon's disk, according to its dimensions at the given time. Then, by means of the hypothenuse, first make a determination of directions;

12. And lay off upon the hypothenuse, from the point of its intersection with the disk, in an inward direction, the measure of the illuminated part: between the limit of the illuminated part and the north and south points draw two fish-figures (*matsya*);

13. From the point of intersection of the lines passing through their midst describe an arc touching the three points: as the disk already drawn appears, such is the moon upon that day.

14. After making a determination of directions by means of the perpendicular, point out the elevated (*unnata*) cusp at the extremity of the cross-line: having made the perpendicular (*koṭi*) to be erect (*unnata*), that is the appearance of the moon.

15. In the dark half-month subtract the longitude of the sun increased by six signs from that of the moon, and calculate, in the same manner as before, her dark part. In this case lay off the base in a reverse direction, and the circle of the moon on the west.

Having made the calculations prescribed in the preceding passages, we are now to project their results, and to exhibit a representation of the moon as she will appear at the given time. The annexed figure (Fig. 33) will illustrate the method of the projection.

We first fix upon a point, as S, which shall represent the position of the sun's centre upon the western horizon at the moment of sunset, and we determine, in the manner taught at the beginning of the third chapter, the lines of cardinal direction of which it is the centre. From this point we then lay off the base (*bhuja*) S L, according to its value in digits as ascertained by the previous process, and northward or southward, according to its true direction as determined by the same process. From L, its extremity, is laid off the perpendicular (*koṭi*), which has the fixed value of twelve digits. This, being a line perpendicular to the plane of the horizon, may be regarded as having no proper direction of its own upon the surface of projection: but the text directs us to lay it off westward from L, apparently in order that the observer, standing upon the eastern side of his base S L, and looking westward toward the setting sun, may have his figure duly before him. The western extremity of the perpendicular, M, represents the moon's place, and from that as a centre, and with a radius equal to the semi-diameter of the moon in digits, as ascertained by calculation for the given moment, a circle is described, representing the moon's disk. Next we are to prolong the hypothenuse, S M, to *e*, and to draw, by the usual means, the line *s n* at right angles to it: the directions upon the disk thus determined by the hypothenuse, as the text phrases it, are called by the commentary "moon-directions" (*candradiças*). The sun being at S, the illuminated half of the moon's circumference will be *s w n*, the cusps will be at *s* and *n*, and *w* will be the extremity of the diameter of greatest illumination. From *w*, then, lay off upon the hypothenuse an amount, *w x*, equal to the measure in digits of the illuminated part of the diameter, and through *s*, *x*, and *n* describe an arc of a circle, in the manner already more than once explained (see above, vi. 14–16); the crescent *s w n x* will represent the amount and direction of the moon's illuminated part at the given time. Now we once more make a determination of directions upon the disk according to the perpendicular L M; that is to say, we prolong L M to *e'*, and draw *s' n'* at right angles to it: the directions thus established are styled in the commentary "sun-directions" (*sûryadiças*), although without obvious propriety: they might rather be called "apparent directions," or "directions on the sphere," since *s' n'* should represent a line parallel with the horizon, and *w' e'* one perpendicular to it. The line *s' n'* is called in the text the "cross-line" (*tiryaksûtra*), and whichever of the moon's cusps is found upon that line is, we are told, to be regarded as the elevated (*unnata*) cusp, the other being the depressed one (*nata*). Whenever there is any base (*bhuja*), as S L, or whenever the moon and sun are not upon the same vertical line M L, there will take place, of course, a tilting of the moon's disk, by which one of her cusps will be raised higher above the horizon than the other; the relative value of the base to the perpendicular will determine the amount of the tilting, and of the deflection of the points of direction *n e s w* from *n' e' s' w'*; and the elevated cusp will always be that upon the same side of the perpendicular on which the base lies. What is meant by the latter half of verse 14 is not altogether clear. The commentator explains it in quite a different manner from that in which we have translated it: he understands *koṭi* as meaning in this instance "cusp," which signification it is by derivation well adapted to bear, and does actually receive, although not in any other passage of this treatise: and he explains the verb *kṛtvâ*, "having made," by *dṛshṭvâ*, "having seen": the phrase would then read "beholding the elevated cusp." We cannot accept

Fig. 33.

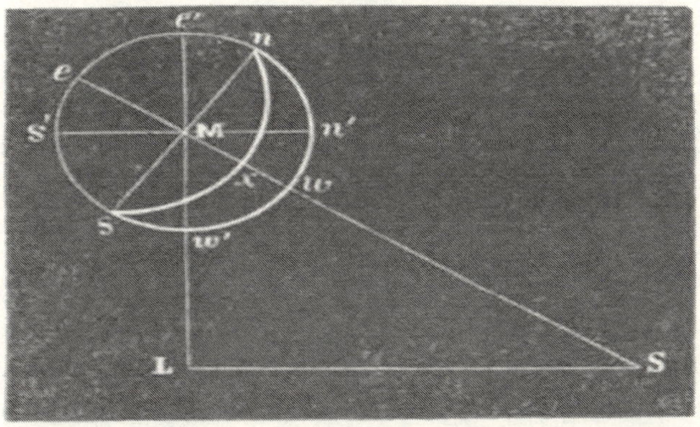

this explanation as a plausible one: to us the meaning seems rather to be that whereas, in the projection, the perpendicular (*koṭi*) L M is drawn on a horizontal surface, we are, in judging of the projection as an actual representation of the moon's position, to conceive of that line as erected, set up perpendicularly.

We have thus far only supposed a case in which the calculations are made for the moment of sunset, the situation of the moon being in the western hemisphere of the heavens. In the text, however, there is nothing whatever to limit or determine the time of calculation, and it is evident that the process of finding the base and perpendicular will be precisely the same, if S (Fig. 32) be taken upon the eastern horizon, and the triangle S L M in the eastern hemisphere. The last verse supposes these to be the conditions of the problem, and lays down rules for determining in such a case the amount of illumination, and for drawing the projection. As regards the measure of the illuminated part, we are to follow the same general method as before, only substituting for the moon's distance in longitude from the sun her distance from the point of opposition, and regarding the result obtained as the measure of that part of the diameter which is obscured (*asita*, "black"); since, during the waning half-month, darkness grows gradually over the moon's face in the same manner as illumination had done during the crescent half-month. But why the base (*bhuja*) is now to be laid off in the opposite to its calculated direction, we find it very hard to see. The commentator says it is because all the conditions of the problem are reversed by our having to calculate and lay off the obscured, instead of the illuminated, part of the moon's disk: but the force of this reason is not apparent. The establishment in the projection of a point representing the position of the sun is, in effect, the one condition which sufficiently determines all the rest: if we are to make a projection corresponding to that drawn in illustration of the other case, we ought, it should seem, to draw the base in its true direction, and, stationing the observer upon the western side of it, looking eastward, to lay off the perpendicular away from him, toward the east; and then to proceed as before, only measuring the obscured part of the diameter from its remoter extremity, instead of from that next the sun. This latter direction is regarded by the commentator as actually conveyed in the final clause of verse 15: he interprets "the circle (*maṇḍala*) of the moon" to mean the dark part of the moon's disk, or that which is to be pointed out as increasing during the waning half-month, and "on the west" to mean on the western side of the complete disk, which is the side now turned away from the sun. It seems to us exceedingly questionable whether the passage fairly admits of this interpretation, but we have no other explanation of it to offer—unless, indeed, it is to be looked upon as a virtual repetition of the former direction to lay off the perpendicular, which determines the position of the moon's disk, towards the west.

We must confess that we feel less satisfied with our comprehension of the scope and methods of this chapter than of any that precedes it. We are disappointed at finding the result arrived at one of so indefinite a character, and of so little significance. The whole laborious calculation seems to be made simply for the sake of delineating the appearance of the moon at a given moment, and pointing out which of her two horns has the greater altitude. No determination is made of the amount of angular deflection, upon which any consequences, meteorological, astrological, or of any other character, could be founded; nor is any hint given of the way in which the results of the process are to be turned to account. Moreover, while the object aimed at seems thus to be merely a projection, a time is selected at which the moon is not ordinarily visible, so that she can not be seen to exhibit an accordance with her delineated appearance! Once more, the whole process is an extremely faulty one: it is, in fact, only when the moon is herself at the horizon that her visible disk can be regarded as in the same plane with lines parallel with and perpendicular to the horizon, or that *e′ w′* and *n′ s′* (Fig. 33) represent actual directions upon her face: anywhere else, the relations of the moon's disk at M in the first figure (Fig. 32) and at M in the other figure (Fig. 33) are so different that the latter cannot fairly represent the former. It would seem, indeed, as if the moment of the moon's own setting or rising were the one for which such a calculation and projection as this would have most significance: at that time, the disappearance or appearance of one of her horns before the other would be such a phenomenon as might seem to a Hindu astronomer worth the trouble of delineating, as a decisive proof of the accuracy of his scientific knowledge. We have not found it possible, however, to make the rules of the text apply to such a case, and the commentary is explicit in its definition of the time of the calculation, as sunset or sunrise alone, to the exclusion of any other moment. But the discordance existing at more than one point in the chapter between the text and the commentary suggests the conjecture that the original design of the one and the traditional interpretation of it represented by the other may be at variance, and we are not without suspicions that the text may have been altered, so as not now fairly and accurately to represent any one consistent process. A better understanding of the general object of the calculation and the use made of its results, and an acquaintance with the solutions of the problem presented by other astronomical treatises, might throw additional light upon these points; but we are not able at present fully to avail ourselves of such assistance, nor is the importance of the subject such as to render incumbent upon us its fuller elucidation.

CHAPTER XI.

OF CERTAIN MALIGNANT ASPECTS OF THE SUN AND MOON.

CONTENTS:—1–5, definition and description of the malignant aspects of the sun and moon, when of equal declination; 6–11, to find the longitude of the sun and moon when their declinations are equal; 12–13, to ascertain the corresponding time; 14–15, to determine the duration of the aspect, and the moment of its beginning and end; 16–18, its continuance and its influences; 19, when such an aspect may occur more than once, or not at all; 20, occurrence of the yoga of like name and character; 21, of unlucky points in the circle of asterisms; 22, caution as to these unlucky aspects and points; 23, introductory to the following chapters.

1. When the sun and moon are upon the same side of either solstice, and when, the sum of their longitudes being a circle, they are of equal declination, it is styled *vâidhṛta.*

2. When the moon and sun are upon opposite sides of either solstice, and their minutes of declination are the same, it is *vyatîpâta*, the sum of their longitudes being a half-circle.

3. Owing to the mingling of the nets of their equal rays, the fire arising from the wrathfulness of their gaze, being driven on by the provector (*pravaha*), is originated unto the calamity of mortals.

4. Since a fault (*pâta*) at this time often causes the destruction of mortals, it is known as *vyatîpâta*, or, by a difference of title, *vâidhṛti.*

5. Being black, of frightful shape, bloody-eyed, big-bellied, the source of misfortune to all, it is produced again and again.

Of all the chapters in the treatise, this is the one which has least interest and value. It is styled *pâtâdhikâra*, "chapter of the *pâtas*," and concerns itself with giving a description of the malignant character of the times when the sun and moon have equal declination, upon the same or opposite sides of the equator, and with laying down rules by which the time of occurrence of those malignant aspects may be calculated. The latter part alone properly falls within the province of an astronomical treatise like the present: the other would better have been left to works of a professedly astrological character. The term *pâta*, applied to the aspects in question, means literally "fall," and hence also either "fault, transgression," or "calamity." We have often met with it above, in the sense of "node of a planet's orbit"; as so used, it was probably first applied to the moon's nodes, because they were the points of danger in her revolution, near which the sun or herself was liable to fall into the jaws of Râhu (see above, iv. 6); and it was then transferred also, though without the same reason, to the nodes of the other planets. As it is employed in this chapter, we translate it simply "aspect." Why the time when the sun and moon are equally distant from the equator should be looked upon as so especially unfortunate is not easy to discover, notwithstanding the lucid explanation furnished in the third verse. For the "provector" (*pravaha*), the wind which carries the planets forward in their orbits, see above, ii. 3. When the equal declinations are of opposite direction, the aspect is denominated *vâidhṛta*, or *vâidhṛti.* This word is a secondary derivative from *vidhṛti*, "holding apart, withholding," or from *vidhṛta:* it has been noted above (under ii. 65) as the name of the last *yoga;* and its use here is not discordant with that, since the twenty-seventh yoga also occurs when the sum of the longitudes of the sun and moon is 360°. The title of the other aspect (*pâta*), which occurs when the sun and moon are equally removed from the equator upon the same side of it, is *vyatîpâta*, which may be rendered "very excessive sin or calamity." This, too, is the name of one of the yogas, but not of that one which occurs when the sum of longitudes of the sun and moon is 180°: the discordance gives occasion for the explanation contained in verse 20, below. The specification of the text, that the aspects take place when the sum of longitudes equals a circle or a half-circle respectively, or when the two luminaries are equally distant from either solstice, or either equinox, is not to be understood as exact: this would be the case if the moon had no motion in latitude; but owing to that motion, the equality of declinations, which is the main thing, occurs at a time somewhat removed from that of equality of distance from the equinoxes: the latter is called in the commentary *madhyapâta*, "the mean occurrence of the aspect." The terms translated by us "upon the same and upon the opposite sides of either solstice" are *ekâyanagata* and *viparîtâyanagata*, literally "situated in the same and in contrary *ayanas*"; *ayana* being, as already pointed out (end of note to iii. 9–12), the name of the halves into which the ecliptic is divided by the solstices.

6. When the longitudes of the sun and moon, being increased by the degrees etc. found for the coincidence of the solstice with its observed place, are together nearly a circle or nearly a half-circle, calculate the corresponding declinations.

7. Then, if the declination of the moon, she being in an odd quadrant, is, when corrected by her latitude (*vikshepa*), greater than the declination of the sun, the aspect (*pâta*) is already past;

8. If less, it is still to come: in an even quadrant, the contrary is the case. If the moon's declination is to be subtracted from her latitude, the rules as to the quadrant are to be reversed.

As in other processes of a similar character (see above, iv. 7–8; vii. 2–6), we are supposed to have found by trial, for the starting-point of the present calculation, the midnight next preceding or following the occurrence of the aspect in question, and to have determined for that moment the longitudes and rates of motion of both bodies, and the moon's latitude. In finding the longitudes, we are to apply the correction for precession; this is the meaning of the expression in verse 6, *dṛktulyasâdhitânçâdi*, which may be literally translated "degrees etc. calculated for accordance with observed place"; the reference is to the similar expression for the precession contained in iii. 11. Next the declinations are to be found, and that of the moon as corrected for her latitude. And since, in the odd quadrants—that is to say, the first and third, counting from the actual vernal equinox—declination is increasing, while in the others it is decreasing, if the declination in an odd quadrant of the moon, the swifter moving body, is already greater than that of the sun, the time of equality of declination is evidently already past, and the converse. But if, on the other hand, the moon's declination (using that term in its Hindu sense) is so small, and her latitude so great, being of opposite directions, that her actual distance from the equator is measured by the excess of the latter above the former, and so is of direction contrary to that of her declination, then, as declination increases, distance from the equator diminishes, or the contrary, and the conditions as formerly stated are reversed throughout.

9. Multiply the sines of the two declinations by radius, and divide by the sine of greatest declination: the difference of the arcs corresponding to the results, or half that difference, is to be added to the moon's longitude when the aspect (*pâta*) is to come;

10. And is to be subtracted from the moon's longitude when the aspect is past. If the same quantity be multiplied by the sun's motion and divided by the moon's motion, the result is an equation, in minutes, which is to be applied to the sun's place, in the same direction as the other to the moon's.

11. So also is to be applied, in the contrary direction, a like equation to the place of the moon's node. This operation is to be repeated, until the declinations of the two bodies come to be the same.

By this process are ascertained the longitudes of the sun and moon at the time when their declinations are equal. Its method may be briefly explained as follows. At the midnight assumed as the starting-point of the whole calculation there is found to be a certain difference in the two declinations: we desire to determine how far the paths of the two luminaries must be traced forward or backward, in order that that difference may be removed; and this must be effected by means of a series of approximations. We commence our calculation with the moon, as being the body of more rapid motion. By a proportion the inverse of that upon which the rule for deriving the declination from the longitude (ii. 28) is founded, we ascertain at what longitude the moon would have the sun's actual declination, and at what longitude she would have her own actual declination, as corrected by her latitude: the difference between the two results is a measure of the amount of motion in longitude, forward or backward, by which she would gain or lose the difference of declination, if the sun remained stationary and her own latitude unchanged. Since, however, that is not the case, we are compelled to calculate the corresponding motion of the sun, and also the moon's latitude in her new position; and in order to the latter, we must correct the place of the node also for its retrograde motion during the interval. The motions of the sun and node are found by the following proportion: as the moon's daily motion is to that of the sun, or to that of the node, so is the correction applied to the moon's place to that which must be applied to the place of the sun, or to that of the node. A new set of positions in longitude having thus been found, the declinations are again to be calculated, and the same approximative process repeated—and so on, until the desired degree of accuracy is attained.

The text permits us to apply, as the correction for the place of the moon, either the whole or the half of the difference of longitude found as the result of the first proportion: it is unessential, of course, in a process of this tentative character, what amount we assume as that of the first correction, provided those which we apply to the places of the sun and node be made to correspond with it: and there may be cases in which we should be conducted more directly to the final result of the process by taking only half of the difference.

12. The aspect (*pâta*) is at the time of equality of declinations; if, then, the moon's longitude, as thus increased or diminished, be less than her longitude at midnight, the aspect is past; if greater, it is to come.

13. The minutes of interval between the moon's longitude as finally established and that at midnight give, when multiplied by sixty and divided by the moon's daily motion, the time of the aspect, in nâdîs.

We had thus far found only the longitudes of the sun and moon at the time of equality of declination, and not that time itself: the latter is now derived from the former by this proportion: as the moon's daily motion is to a day, or sixty nâḍîs, so is the difference between the moon's longitude at midnight and at the time of the aspect to the interval between the latter time and midnight.

14. Multiply the half-sum of the dimensions (*mâna*) of the sun and moon by sixty, and divide by the difference of their daily motions: the result is half the duration (*sthiti*), in nâḍîs etc.

15. The corrected (*sphuṭa*) time of the aspect (*pâta*) is the middle: if that be diminished by the half-duration, the result is the time of the commencement; if increased by the same, it is the time of the end.

16. The time intervening between the moments of the beginning and end is to be looked upon as exceedingly terrible, having the likeness of a consuming fire, forbidden for all works.

The continuance of the centres of the sun and moon at the point of equality of declination is, of course, only momentary; but the aspect and its malignant influences are to be regarded as lasting as long as there is virtual contact of the two disks at that point, or as long as a central eclipse of the sun would last if it took place there. Its half-duration, then, or the interval from its middle to its beginning or end respectively, is found by a proportion, as follows: if in a day, or sixty nâdîs, the two centres of the sun and moon become separated by a distance which is equal to the difference of their daily motions, in how many nâḍîs will they become separated by a distance which is equal to the sum of their semi-diameters? or

diff. d. motions : 60 : : sum semi-diam. : half-duration

And if this amount be subtracted from and added to the time of equality of declination, the results will be the moments at which the aspect will begin and end respectively.

Such is the plain and obvious meaning of the text in this passage. The commentator, however, in accordance with his interpretation of the next following verse (see below), declares that the aspect actually lasts as long as any portion of the moon's disk has the same declination with any portion of that of the sun; and that, accordingly, it commences—the moon's declination being supposed to be increasing—whenever her remoter limb comes to have the same declination with the nearer limb of the sun, and ends when her nearer limb comes to have the same declination with the remoter limb of the sun—the contrary being the case when her declination is decreasing. He acknowledges that the text does not seem to teach this, but puts in the plea which is usual with him when excusing a palpable inaccuracy in the statements or processes of the treatise; namely, that the blessed author of the work, moved by pity for mankind, permitted here the substitution of difference of longitude for difference of declination, in view of the greater ease of its calculation, and the insignificance of the error involved. That error, however, is quite the reverse of insignificant; it is, indeed, so very gross and palpable that we cannot possibly suppose it to have been committed intentionally by the text; we regard it as the easier assumption that the conditions of the continuance of the aspect are differently estimated in the text and in the commentary, being by the former taken to be as we have stated them above, in our explanation of the process. The view of the matter taken by the commentator, it is true, is decidedly the more natural and plausible one: there seems no good reason why an aspect which depends upon equality of declination should be determined as to continuance by motion in longitude, or why the aspect should only occur at all when the two centres are equally distant from the equator; why, in short, there should not be partial aspects, like partial eclipses of the sun. If the doctrine of the commentary is a later development, or an independent form, of that which the text appears to represent, it is a naturally suggested one, and such as might have been expected to arise.

17. While any parts of the disks of the sun and moon have the same declination, so long is there a continuance of this aspect, causing the destruction of all works.

18. So, from a knowledge of the time of its occurrence, very great advantage is obtained, by means of bathing, giving, prayer, ancestral offerings, vows, oblations, and other like acts.

We have translated verse 17 in strict accordance with the interpretation of it presented in the commentary, although we must acknowledge that we do not see how that interpretation is to be reconciled with the actual form of the text. The term *ekâyanagata,* which the commentator renders "having equal declination," is the same with that which in the first verse signified "situated in the same *ayana*"; *maṇḍala,* although it is sometimes used with the meaning "disk," here attributed to it by him, is the word employed in that same verse for a "circle," or "360°"; and *antara,* which he explains by *ekadeça,* "any part," never, so far as we know, is properly used in that sense, while it is of frequent occurrence elsewhere in this treatise with the meaning "interval." The natural rendering of the line would seem to be "when there is between the sun and moon the interval of a circle, situated in the same *ayana.*" This, however, yields no useful meaning, since such a description could only apply to an actual conjunction of the sun and moon. We do not see how the difficulty is to be solved, unless it be allowed us, in view of the discordance already pointed out as existing between the plain meaning of the previous passage and that attributed to it by the commentator, to assume that the text has been tampered with in this verse, and made to furnish a different sense from that it originally had, partly by a forced interpretation, but partly also by such an alteration of its readings as disables it from yielding any other intelligible meaning.

19. When the equality of declinations of the sun and moon takes place in the neighborhood of the equator, the aspect may then again occur a second time: in the contrary case, it may fail to occur.

Near the equinox, where declination changes rapidly, the moon, as the swifter moving body, may come to have twice, in rapid succession, the same declination with the sun, and upon the opposite sides of the equator. Near the solstice, on the other hand, where the ecliptic and equator are nearly parallel, the moon—if she happens to be nearer the equator than the sun is, owing to her latitude—may pass the region in which the aspect would otherwise be liable to occur, without having had a declination equal in amount to that of the sun.

20. If the sum of the longitudes of the sun and moon, in minutes, on being divided by the portion (*bhoga*) of an asterism (*bha*), yields a quotient between sixteen and seventeen, there is another, a third, *vyatîpâta*.

This is simply a special application of the rule formerly given (ii. 65), for finding, for any given time, the current period named *yoga*. The seventeenth of the series, as is shown by the list there given, has the same name, *vyatîpâta*, with one of the aspects treated of in this chapter: judging from verse 22, below, it is also regarded as possessing a like portentous and malignant character.

21. Of the asterisms (*dhishṇya*) Âçleshâ (*sârpa*), Jyeshṭhâ (*âindra*), and Revatî (*paushṇya*), the last quarters are junctions of the asterisms (*bhasandhi*); the first quarter in the asterisms following these respectively is styled *gaṇḍânta*.

22. In all works, one must avoid the terrible trio of *vyatîpâtas*, as also the trio of *gaṇḍântas*, and this trio of junctions of asterisms.

The division of the ecliptic into twenty-sevenths, or asterisms, coincides with its division into twelfths, or signs, at the ends of the ninth, eighteenth, and twenty-seventh asterisms, which are also those of the fourth, eighth, and twelfth signs respectively. To this innocent circumstance it seems to be owing that those points, and the quarters of portions, or arcs of 200′, on either side of them, are regarded and stigmatized as unlucky and ominous. Hence the title *bhasandhi; sandhi* is literally "putting together, joint," and *bha* is, as has been noticed elsewhere (note to iii. 9–12), a name both of the asterisms and of the signs. In which of its various senses the word *gaṇḍa* is used in the compound *gaṇḍânta,* we do not know.

23. Thus hath been related that supreme, pure, excellent, mysterious, and grand system of the heavenly bodies: what else dost thou desire to know?

In this verse re-appears the personality of the revealer of the treatise, the incarnation of a portion of the sun, which has been lost sight of since near the beginning of the work (i. 7). The questions addressed to him, in answer to this appeal, by Maya, the recipient of the revelation, introduce the next chapter, which, with the two that follow it, contains the additional explanations and instructions vouchsafed in reply. The last three chapters confessedly constitute a separate portion of the work, which is here divided into a *pûrva khaṇḍa* and an *uttara khaṇḍa*, or a "former Part" and a "latter Part." It is by no means impossible that the whole second Part is an appendix to the text of the Siddhânta as originally constituted.

The title of the next following chapter is *bhûgolâdhyâya*, "chapter of the earth-globe": in the second part of the treatise the chapters are styled *adhyâya*, "lection," instead of, as hitherto, *adhikâra*, "heading."

CHAPTER XII.

COSMOGONY, GEOGRAPHY, DIMENSIONS OF THE CREATION.

Contents:—1–9, inquiries; 10–28, development of the creative agencies, of the elements, and of the existing creation; 29–31, form and disposition of the stellar and planetary systems; 32–44, situation, form, structure, and divisions of the earth; 45–72, varying phenomena of night and day in different latitudes and zones; 73–77, revolutions of the stars and planets; 78–79, regents of the different divisions of time; 80–90, dimensions of the planetary, stellar, and ethereal orbits.

1. Then the demon Maya, prostrating himself with hands suppliantly joined before him who derived his being from the part of the Sun, and revering him with exceeding devotion, inquired as follows:

2. O blessed one! of what measure is the earth? of what form? how supported? how divided? and how are there in it seven interterranean (*pâtâla*) earths?

3. And how does the sun cause the varying distinction of day and night? how does he revolve about the earth, enlightening all creatures?

4. For what reason are the day and night of the gods and of the demons opposed to one another? or how does that take place by means of the sun's completion of his revolution?

5. Why does the day of the Fathers consist of a month, but that of mortals of sixty nâḍîs? for what reason is not this latter everywhere the case?

6. Whence is it that the regents of the days, years, months, and hours (*horâ*) are not the same? How does the circle of asterisms (*bhagaṇa*) revolve? what is the support of it with the planets?

7. The orbits of the planets and stars, uplifted from the earth one above another—what are their heights? what their intervals? what their dimensions? and what the order in which they are fixed?

8. Why are the rays of the sun hot in the summer, and not so in the winter? how far do his rays penetrate? How many modes of measuring time (*mâna*) are there? and how are they employed?

9. Resolve these my difficulties, O blessed one, creator of creatures! for there is not found besides thee another resolver, who beholdeth all things.

The proper answers to these inquiries commence at about the twenty-seventh verse of the chapter, the preceding philosophical history of the development of the existing creation being apparently volunteered by the revelator. All the questions then find their answers in this chapter, excepting that as to the methods of measuring time, which is disposed of in the fourteenth and concluding chapter. The subject of the thirteenth chapter also seems not to be contemplated in the laying out, in this passage, of the scheme of subjects to be treated of in the remainder of the treatise.

10. Having heard the words thus uttered with devotion by Maya, he then again promulgated this mysterious and supreme Book (*adhyâya*):

11. Listen with concentrated attention: I will proclaim the secret doctrine called the transcendental (*adhyâtma*): there is nothing which may not be bestowed on those who are exceedingly devoted to me.

12. Vâsudeva, the supreme principle of divinity (*brahman*), whose form is all that is (*tat*), the supreme Person (*purusha*), unmanifested, free from qualities, superior to the twenty-five principles, imperishable,

13. Contained within matter (*prakṛti*), divine, pervading everything, without and within, the attractor—he, having in the first place created the waters, deposited in them energy.

14. That became a golden egg, on all sides enveloped in darkness: in it first became manifested the unrestrained, the everlasting one.

15. He in the scripture (*chandas*) is denominated the golden-wombed (*hiraṇyagarbha*), the blessed; as being the first (*âdi*) existence, he is called Âditya; as being generator, the sun.

16. This sun, likewise named Savitar, the supreme source of light (*jyotis*) upon the border of darkness—he revolves, bringing beings into being, the creator of creatures.

17. He is extolled as natural illuminator, destroyer of darkness, great. The Hymns (*ṛcas*) are his disk, the Songs (*sâmâni*) his beams, the Liturgy (*yajûnshi*) his form.

18. He, the blessed one, is composed of the trio of sacred scriptures, the soul of time, the producer of time, mighty, the soul of the universe, all-penetrating, subtle: in him is the universe established.

19. Having made for his chariot, which is composed of the universe, a wheel consisting of the year, and having yoked the seven metres as his steeds, he revolves continually.

20. Three quarters are immortal, secret; this one quarter hath become manifest. In order to the production of the animated creation, he, the mighty one, produced Brahma, the principle of consciousness (*ahânkara*).

21. Bestowing upon him the Scriptures (*veda*) as gifts, and establishing him within the egg as grandfather of all worlds, he himself then revolves, causing existence.

22. Then Brahma, wearing the form of the principle of consciousness (*ahankâra*), produced mind in the creation: from mind was born the moon; from the eyes, the sun, the repository of light;

23. From mind, the ether; thence, in succession, wind, fire, waters, earth—these five elements (*mahâbhûta*) were produced by the successive addition of one quality.

24. Agni and Soma, the sun and moon: then Mars etc. were produced, in succession, from light, earth, ether, water, wind.

25. Again, dividing himself twelve-fold, he, the mighty one, produced what is known as the signs; and yet farther, what has the form of the asterisms (*nakshatra*), twenty-seven-fold.

26. Then he wrought out the whole animate and inanimate creation, from the gods downward, producing forms of matter (*prakṛti*) from the upper, middle, and lower currents (*srotas*).

27. Having produced them in succession, as stated, by a difference of quality and function, he fashioned the distinctive character of each, according to the showing of the Scripture (*veda*)—

28. That is, of the planets, asterisms, and stars, of the earth, and of the universe, he the mighty one; of gods, demons, and mortals, and of the Perfected (*siddha*), in their order.

We do not regard ourselves as called upon to enter into any detailed examination of this metaphysical scheme of development of the creation, or to compare it critically with the similar schemes presented in other Hindu works, as Manu (chap. i), the Purâṇas (see Wilson's Vishṇu Purâṇa, Book I), etc. We will merely explain a few of its expressions, and of the allusions it contains. Vâsudeva is an ordinary epithet of Vishṇu, and its use in the signification here given it seems indicative of Vaishṇava tendencies on the part of the author of the scheme. The twenty-five principles referred to in verse 12 are those established by the Sânkhya philosophy. The reference in verse 15, first half, is to Ṛig-Veda x. 121. In the second half of the same verse we have a couple of false etymologies: *âditya* comes, not from *âdi*, "first," but from *aditi*, "eternity"; and to derive *sûrya*, "sun," from the root *sû*, "generate" (from which *savitar* actually comes), is beyond the usual measure of Hindu theologico-philosophical etymologizing. The Hymns, Songs, and Liturgy are the three bodies of scripture commonly known as the Ṛig-Veda, Sâma-Veda, and Yajur-Veda. The "seven metres" (v. 19) are those which are most often employed in the construction of the Vedic hymns: in parts of the Veda itself they are personified, and marvellous qualities and powers are ascribed to them. The obscure statement contained in the first half of verse 20 comes from verses 3 and 4 of the *purusha*-hymn (Ṛig-Veda x. 90: the hymn is also found in others of the Vedic texts). The second half of verse 22 also nearly coincides with a passage (v. 13) in the same hymn. Of the five elements assumed by the Hindu philosophers, the first, ether, is said to be endowed only with the quality of audibleness; the second, air, has that of tangibility also; the third, fire, has both, along with color; to these qualities the fourth element, water, adds that of savor; the last, earth, possesses audibility, tangibility, color, savor, and odor: this is according to the doctrines of the Sânkhya philosophy. In verses 24 and 25 we have specifications introduced out of consideration for the general character and object of this treatise: as also, in the part assigned to the sun in the history of development, we may perhaps recognize homage paid to its asserted author. For the beings called in verse 28 the "perfected" (*siddha*), see below, verses 31 and 40.

29. This Brahma-egg is hollow; within it is the universe, consisting of earth, sky, etc.; it has the form of a sphere, like a receptacle made of a pair of caldrons.

30. A circle within the Brahma-egg is styled the orbit of the ether (*vyoman*): within that is the revolution of the asterisms (*bha*); and likewise, in order, one below the other,

31. Revolve Saturn, Jupiter, Mars, the sun, Venus, Mercury, and the moon; below, in succession, the Perfected (*siddha*), the Possessors of Knowledge (*vidyâdhara*), and the clouds.

The order of proximity to the earth in which the seven planets are here arranged is, as noticed above (i. 51–52), that upon which depends the succession of their regency over the days of the week, and so also the names of the latter. So far as the first three and the last are concerned, it is a naturally suggested arrangement, which could hardly fail to be hit upon by any nation having sufficient skill to form an order of succession at all: the order in which the sun, Mercury, and Venus are made to follow one another is, on the other hand, a matter of more arbitrary determination, and might have been with equal propriety, for aught we can see, reversed or otherwise varied. Of the supernatural beings called the "possessors of knowledge" (*vidyâdhara*) we read only in this verse: the "perfected" we find again below, in verse 40, as inhabitants of a city on the earth's surface.

32. Quite in the middle of the egg, the earth-globe (*bhûgola*) stands in the ether, bearing the supreme might of Brahma, which is of the nature of self-supporting force.

33. Seven cavities within it, the abodes of serpents (*nâga*) and demons (*asura*), endowed with the savor of heavenly plants, delightful, are the interterranean (*pâtâla*) earths.

34. A collection of manifold jewels, a mountain of gold, is Meru, passing through the middle of the earth-globe, and protruding on either side.

35. At its upper end are stationed, along with Indra, the gods, and the Great Sages (*maharshi*); at its lower end, in like manner, the demons (*asura*) have their place—each the enemy of the other.

36. Surrounding it on every side is fixed next this great ocean, like a girdle about the earth, dividing the two hemispheres of the gods and of the demons.

37. And on all sides of the midst of Meru, in equal divisions of the ocean, upon islands (*dvîpa*), in the different directions, are the eastern and other cities, fashioned by the gods.

38. At a quadrant of the earth's circumference eastward, in the clime (*varsha*) Bhadrâçva, is the city famed as Yamakoṭi, having walls and gateways of gold.

39. To the southward, in the clime Bhârata, is in like manner the great city Lankâ: to the west, in the clime called Ketumâla, is declared to be the city named Romaka.

40. Northward, in the clime Kuru, is declared to be the city called that of the Perfected (*siddha*); in it dwell the magnanimous Perfected, free from trouble.

41. These are situated also at a distance from one another of a quadrant of the earth's circumference; to the north of them, at the same distance, is Meru, the abode of the gods (*sura*).

42. Above them goes the sun when situated at the equinoxes; they have neither equinoctial shadow nor elevation of the pole (*akshonnati*).

43. In both directions from Meru are two pole-stars (*dhruvatârâ*), fixed in the midst of the sky: to those who are situated in places of no latitude (*niraksha*), both these have their place in the horizon.

44. Hence there is in those cities no elevation of the pole, the two pole-stars being situated in their horizon; but their degrees of co-latitude (*lambaka*) are ninety: at Meru the degrees of latitude (*aksha*) are of the same number.

In these verses we have so much of geography as the author of the chapter has seen fit to connect with his astronomical explanations. For a Hindu account of the earth, it is wonderfully moderate, and free from falsehood. The absurd fictions which the Purâṇas put forth as geography are here for the most part ignored, only two or three of the features of their descriptions being retained, and those in an altered form. To the Purâṇas (see especially Wilson's Vishṇu Purâṇa, Book II., chap. ii–vi), the earth is a plain, of immense dimensions. Precisely in the middle of it rises Mount Meru, itself of a size compared with which the earth, as measured by the astronomers, is as nothing: it is said to be 84,000 yojanas high, and buried at the base 16,000 yojanas; it has the shape of an inverted cone, being 32,000 yojanas in diameter at its upper extremity, and only 16,000 at the earth's surface. Out of this mountain the astronomical system makes the axis of the earth, protruding at either extremity, indeed, but of dimensions wholly undefined. As the Purâṇas declare the summit of Meru, and the mountains immediately supporting it, to be the site of the cities inhabited by the different divinities, so also we have here the gods placed upon the northern extremity of the earth's axis, while their foes, the spirits of darkness, have their seat at the southern. The central circular continent, more than 100,000 yojanas in diameter, in the midst of which Meru lies, is named Jambûdvîpa, "the island of the rose-apple tree": it is intersected by six parallel ranges of mountains, running east and west, and connected together by short cross-ranges: the countries lying between these ranges are styled *varshas*, "climes," and are all fully named and described in the Purâṇas, as are the mountain-ranges themselves. The half-moon-shaped strips lying at the bases of the mountains on the eastern, southern, western, and northern edges of the continent, are called by the same names that are given by our text to the four insular climes which it sets up. Bhârata is a real historical name, appearing variously in the early Hindu traditions; Kuru, or Uttara-Kuru, is a title applied in Hindu geography of a less fictitious character to the country or people situated beyond the range of the Himâlaya; the other two names appear to be altogether imaginary. The Purâṇas say nothing of cities in these four climes. Lankâ, as noticed above (i. 62), is properly an appellation of the island Ceylon; and Romaka undoubtedly comes from the name of the great city which was the mistress of the western world at the period of lively commercial intercourse between India and the Mediterranean: the other two cities are pure figments of the imagination. Our treatise, it will be observed, ignores the system of continents, or *dvîpas*, and simply surrounds the earth with an ocean in the midst, like a girdle: the Purâṇas encompass Jambûdvîpa about with six other *dvîpas*, or insular ring-shaped continents, each twice as vast as that which it encloses, and each separated from the next by an ocean of the same extent with itself. Of these seven oceans, the first, which washes the shores of Jambûdvîpa, is naturally enough acknowledged to be composed of salt water: but the second is of syrup, the third of wine, the fourth of clarified butter, the fifth of whey, the sixth of milk, and the last of sweet water. Outside the latter is an uninhabited land of gold, and on its border, as the outmost verge of creation, is the monstrous wall of the Lokâloka mountains, beyond which is only nothingness and darkness.

The author of the Siddhânta-Çiromaṇi, more submissive than the writer of our chapter to the authority of tradition, accepts (Golâdhy., chap. ii) the series of concentric continents and oceans, but gives them all a place in the unknown southern hemisphere, while he regards Jambûdvîpa as occupying the whole of the northern.

The *pâtâlas*, or interterranean cavities, spoken of in verse 33, are also an important feature of the Puranic geography. If our author has not had the good sense to reject them, along with the insular continents, he at least passes them by with the briefest possible notice. In the Purâṇas they are declared to be each of them 10,000 yojanas in depth, and their divisions, inhabitants, and productions are described with the same ridiculous detail as those of the continents on the earth's surface.

It will be observed that the text, although exhibiting in verse 41 a distinct apprehension of the fact that the pole is situated to the northward of all points of the equator alike, yet, in describing the position of the four great cities, speaks as if there were a north direction from Meru, in the continuation of the line drawn to the latter from Lankâ, and an east and west direction at right angles with this.

For the terrestrial equator, considered as a line or circle upon the earth's surface, there is no distinctive name; it is referred to simply as the place "of no latitude" (*niraksha, vyaksha*).

45. In the half-revolution beginning with Aries, the sun, being in the hemisphere of the gods, is visible to the gods: but while in that beginning with Libra, he is visible to the demons, moving in their hemisphere.

46. Hence, owing to his exceeding nearness, the rays of the sun are hot in the hemisphere of the gods in summer, but in that of the demons in winter: in the contrary season, they are sluggish.

47. At the equinox, both gods and demons see the sun in the horizon; their day and night are mutually opposed to each other.

48. The sun, rising at the first of Aries, while moving on northward for three signs, completes the former half-day of the dwellers upon Meru;

49. In like manner, while moving through the three signs beginning with Cancer, he completes the latter half of their day: he accomplishes the same for the enemies of the gods while moving through the three signs beginning with Libra and the three beginning with Capricorn, respectively.

50. Hence are their night and day mutually opposed to one another; and the measure of the day and night is by the completion of the sun's revolution.

51. Their mid-day and midnight, which are opposed to one another, are at the end of each half-revolution from solstice to solstice (*ayana*). The gods and demons each suppose themselves to be uppermost.

52. Others, too, who are situated upon the same diameter (*samasûtrastha*), think one another underneath—as the dwellers in Bhadrâçva and in Ketumâla, and the inhabitants of Lankâ and of the city of the Perfected, respectively.

53. And everywhere upon the globe of the earth, men think their own place to be uppermost: but since it is a globe in the ether, where should there be an upper, or where an under side of it?

54. Owing to the littleness of their own bodies, men, looking in every direction from the position they occupy, behold this earth, although it is globular, as having the form of a wheel.

55. To the gods, this sphere of asterisms revolves toward the right; to the enemies of the gods, toward the left; in a situation of no latitude, directly overhead—always in a westerly direction.

55. Hence, in the latter situation, the day is of thirty ṇâḍîs, and the night likewise: in the two hemispheres of the gods and demons there take place a deficiency and an excess, always opposed to one another.

57. During the half-revolution beginning with Aries, there is always an excess of the day to the north, in the hemisphere of the gods—greater according to distance north—and a corresponding deficiency of the night; in the hemisphere of the demons, the reverse.

58. In the half-revolution beginning with Libra, both the deficiency and excess of day and night in the two hemispheres are the opposite of this: the method of determining them, which is always dependent upon situation (*deça*) and declination, has been before explained.

59. Multiply the earth's circumference by the sun's declination in degrees, and divide by the number of degrees in a circle: the result, in yojanas, is the distance from the place of no latitude where the sun is passing overhead.

60. Subtract from a quarter of the earth's circumference the number of yojanas thus derived from the greatest declination: at the distance of the remaining number of yojanas

61. There occurs once, at the end of the sun's half-revolution from solstice to solstice, a day of sixty nâḍîs, and a night of the same length, mutually opposed to one another, in the two hemispheres of the gods and of the demons.

62. In the intermediate region, the deficiency and excess of day and night are within the limit of sixty nâḍîs; beyond, this sphere of asterisms (*bha*) revolves perversely.

63. Subtract from a quarter of the earth's circumference the number of yojanas derived from the declination found by the sine of two signs: at that distance from the equator the sun is not seen, in the hemisphere of the gods, when in Sagittarius and Capricorn;

64. So also, in the hemisphere of the demons, when in Gemini and Cancer: in the quarter of the earth's circumference where her shadow is lost, the sun may be shown to be visible.

65. Subtract from the fourth part of the earth's periphery (*kakshâ*) the number of yojanas derived from the declination found by the sine of one sign: at the distance from the place of no latitude of the remaining number of yojanas,

66. The sun, when situated in Sagittarius, Capricorn, Scorpio, and Aquarius, is not seen in the hemisphere of the gods; in that of the demons, on the other hand, when in the four signs commencing with Taurus.

67. At Meru, the gods behold the sun, after but a single rising, during the half of his revolution beginning with Aries; the demons, in like manner, during that beginning with Libra.

68. The sun, during his northern and southern progresses (*ayana*) revolves directly over a fifteenth part of the earth's circumference, on the side both of the gods and of the demons.

69. Between those limits, the shadow is cast both southward and northward; beyond them, it falls toward the Meru of either hemisphere respectively.

70. When passing overhead at Bhadrâçva, the sun is rising in Bhârata; it is, moreover, at that time, midnight in Ketumâla, and sunset in Kuru.

71. In like manner also he produces, by his revolution, in Bhârata and the other climes, noon, sunrise, midnight, and sunset, reckoning from east to west.

72. To one going toward Meru, there take place an elevation of the pole (*dhruva*) and a depression of the circle of asterisms; to one going toward the place of no latitude, on the contrary, a depression of the former and an elevation of the latter.

This detailed exposition of the varying relations of day and night in different parts of the globe is quite creditable to the ingenuity, and the distinctness of apprehension, of those by whom it was drawn out. It is for the most part so clearly expressed as to need no additional explanations: we shall append to it only a few brief remarks.

How far, in verse 46, a true statement is given of the cause of the heat of summer and the cold of winter, may be made a matter of some question: the word which we have translated "nearness" (*âsannatâ*) has no right to mean "directness, perpendicularity," and yet, when taken in connection with the preceding verse, it may perhaps admit that signification. The second chapter shows that the Hindus knew very well that the sun is actually nearer to the whole earth in winter, or when near his perigee, than in summer.

The expression *ayanânta*, "at the end of an *ayana*," employed in verses 51 and 61, and which we have rendered by a paraphrase, might perhaps have been as well translated, briefly and simply, "at either solstice." Probably *ayana*, as used in the sense of "solstice" (see above, end of note to iii. 9–12), is an abbreviated form of *ayanânta*, like *jyâ* for *jyârdha* (ii. 15–27), and *aksha* for *akshonnati* (i. 60).

In verse 55, we have translated by "toward the right" and "toward the left" the adverbs *savyam* and *apasavyam*, which mean literally "left-wise" and "right-wise"; that is to say, in such a manner that the left side or the right side respectively of the thing making the revolution is turned toward that about which the revolution is made, this being the Hindu mode of describing the passing of one person about another person or thing, especially in respectful salutation and in religious ceremonial.

The natural measure of the day and of the night is assumed in verse 56 etc. to be the half of a whole day, or thirty nâḍîs, and any deviation from that norm is regarded as an excess (*dhana, vṛddhi*) or a deficiency (*ṛṇa, hâni, kshaya*). The former processes referred to at the end of verse 58 are those taught in ii. 60–62.

We have already above (note to i. 63–65) called attention to the fact that all the Hindu measurements of longitude and latitude upon the earth's surface are made in yojanas, and not in degrees.

The expression "perversely" (*viparîta*) in verse 62 is explained by the commentator to mean "in such manner that the rules as already given cannot be applied"; since the sine of the ascensional difference (*cara*—see ii. 61) as found by them would be greater than radius.

The latter half of verse 64 is obscure: its meaning seems to be, as explained by the commentator, that over a corresponding portion of the earth's surface in the contrary hemisphere the sun is continuously visible during the same period, the shadow of the earth, which is the cause of night, not covering that portion.

73. The circle of asterisms, bound at the two poles, impelled by the provector (*pravaha*) winds, revolves eternally: attached to that are the orbits of the planets, in their order.

74. The gods and demons behold the sun, after it is once risen, for half a year; the Fathers (*pitaras*), who have their station in the moon, for a half-month (*paksha*); and men upon the earth, during their own day.

75. The orbit (*kakshâ*) of one that is situated higher up is large; that of one situated lower down is small. Upon a great orbit the degrees are great; so also, upon a small one, they are small.

76. A planet situated upon a small circuit (*bhramaṇa*) traverses the circle of constellations (*bhagaṇa*) in a little time; one revolving on a large circle (*maṇḍala*), in a long time.

77. The moon, upon a very small orbit, makes many revolutions: Saturn, moving upon a great orbit, makes, as compared with her, a much less number of revolutions.

The connection and orderly succession of subjects is by no means strictly maintained in this part of the chapter. The seventy-fourth verse is palpably out of place, and is, moreover, in great part superfluous; for the statement contained in its first half has already twice been made, in verses 45 and 67, and in the latter passage in nearly the same terms as here: its last specification, too, is of a matter too obvious to call for notice. Nevertheless, the verse cannot well be spared from the chapter, since it contains the only answer which is vouchsafed to the question of verse 5, above, respecting the day and night of the Fathers. In the assignment of the different divisions of time, as single days, to different orders of beings, the month has been given to the *pitaras*, "Fathers," or manes of the departed, and they are accordingly located in the moon, each portion of whose surface enjoys a recurrence of day and night once in each lunar month. The next following verses, 75 to 77, are a rather unnecessary amplification of the idea already expressed in i. 26-27; but they answer well enough here as special introduction to the detailed exhibition of the measurements of the planetary orbits which is to follow. Before that is brought in, however, we have the connection again broken, by the intrusion of the two following verses, respecting the regents of years, months, days, and hours.

78. Counting downward from Saturn, the fourth successively is regent of the day; and the third, in like manner, is declared to be the regent of the year;

79. Reckoning upward from the moon are found, in succession, the regents of the months; the regents of the hours (*horâ*), also, occur in downward order from Saturn.

This passage appears to be introduced here as answer to the inquiry propounded in verse 6, above. Instead, however, of explaining why the different divisions of time are placed under the superintendence and protection of different planets, the text contents itself with reiterating, in a different form, what had already been said before (i. 51-52) respecting the order of succession of the regents of the successive periods; but adding also the important and significant specification respecting the hours, or twenty-fourths of the day. We have sufficiently illustrated the subject, in connection with the other passage; we will only repeat here that, the planets being regarded as standing in the order in which they are mentioned in verse 31, above, their successive regency over the hours is the one fundamental fact upon which all the rest depend, each planet being constituted lord also of the day whose first hour is placed under his charge, and so likewise of the month and of the year over whose first hour and day he is regent—neither the month nor the year, any more than the hour itself, being divisions of time which are known to the Hindus in any other uses, and the name of the hour, *horâ*, which is the Greek ὥρα, betraying the source whence the whole system was introduced into India.

80. The orbit (*kakshâ*) of the asterisms (*bha*) is the circuit (*bhramaṇa*) of the sun multiplied by sixty: by so many yojanas does the circle of the asterisms revolve above all.

81. If the stated number of revolutions of the moon in an Æon (*kalpa*) be multiplied by the moon's orbit, the result is to be known as the orbit of the ether: so far do the rays of the sun penetrate.

82. If this be divided by the number of revolutions of any planet in an Æon (*kalpa*), the result will be the orbit of that planet: divide this by the number of terrestrial days, and the result is the daily eastward motion of them all.

83. Multiply this number of yojanas of daily motion by the orbit of the moon, and divide by a planet's own orbit; the result is, when divided by fifteen, its daily motion in minutes.

84. Any orbit, multiplied by the earth's diameter and divided by the earth's circumference, gives the diameter of that orbit; and this, being diminished by the earth's diameter and halved, gives the distance of the planet.

85. The orbit of the moon is three hundred and twenty-four thousand yojanas: that of Mercury's conjunction (*çîghra*) is one million and forty-three thousand, two hundred and nine:

86. That of Venus's conjunction (*çîghra*) is two million, six hundred and sixty-four thousand, six hundred and thirty-seven: next, that of the sun, Mercury, and Venus is four million, three hundred and thirty-one thousand, five hundred:

87. That of Mars, too, is eight million, one hundred and forty-six thousand, nine hundred and nine; that of the moon's apsis (*ucca*) is thirty-eight million, three hundred and twenty-eight thousand, four hundred and eighty-four:

88. That of Jupiter, fifty-one million, three hundred and seventy-five thousand, seven hundred and sixty-four: of the moon's node, eighty million, five hundred and seventy-two thousand, eight hundred and sixty-four:

89. Next, of Saturn, one hundred and twenty-seven million, six hundred and sixty-eight thousand, two hundred and fifty-five: of the asterisms, two hundred and fifty-nine million, eight hundred and ninety thousand, and twelve:

90. The entire circumference of the sphere of the Brahma-egg is eighteen quadrillion, seven hundred and twelve trillion, eighty billion, eight hundred and sixty-four million: within this is the pervasion of the sun's rays.

We present below the numerical data given in these verses, in a form easier of reference and of comparison with the like data of other treatises:

Planet etc.	Orbit, in yojanas.
Moon,	324,000
" apsis,	38,328,484
" node,	80,572,864
Mercury (conjunction),	1,043,209
Venus (conjunction),	2,664,637
Sun,	4,331,500
Mars,	8,146,909
Jupiter,	51,375,764
Saturn,	127,668,255
Asterisms,	259,890,012
Universe,	18,712,080,864,000,000

We have already more than once (see above, notes to i. 25–27, and iv. 1) had occasion to notice upon what principles the orbits of the planets, as here stated, were constructed by the Hindus. That of the moon (see note to iv. 1) was obtained by a true process of calculation, from genuine data, and is a tolerable approximation to the truth: all the others are manufactured out of this, upon the arbitrary and false assumption that the mean motion of all the planets, each upon its own orbit, is of equal absolute amount, and hence, that its apparent value in each case, as seen by us, is inversely as the planet's distance, or that the dimensions of the orbit are directly as the time employed in traversing it, or as the period of sidereal revolution. These dimensions, then, may be found by various methods: upon dividing the circumference of the moon's orbit by her time of sidereal revolution, we obtain as the amount of her daily motion in yojanas 11,858.717 nearly (more exactly 11,858.71693+); and multiplying this by the time of sidereal revolution of any planet, we obtain that planet's orbit. This is equivalent to making the proportion

moon's sid. rev.: planet's sid. rev. : : moon's orbit: planet's orbit

And since the times of sidereal revolution of the planets are inversely as the number of revolutions made by them in any given period, this proportion, again, is equivalent to

planet's no. of rev. in an Æon : moon's do. : : moon's orbit : planet's orbit

This is the form of the proportion from which is derived the rule as stated in the text, only the latter designates the product of the multiplication of the moon's orbit by her number of revolutions as the orbit of the ether (*âkâça*), or the circumference of the Brahma-egg, within which the whole creation, as above taught, is enclosed. This is the same thing with attributing to the outermost shell of the universe one complete revolution in an Æon (*kalpa*), of 4,320,000,000 years.

There is one feature of the system exposed in this passage which to us is hitherto quite inexplicable: it is the assignment to the asterisms of an orbit sixty times as great as that of the sun. This, according to all the analogies of the system, should imply a revolution of the asterisms eastward about the earth once in each period of sixty sidereal years. The same orbit is found allotted to them in the Siddhânta-Çiromaṇi (Gaṇitâdhy., iv. 5), and it is to be looked upon, accordingly, as an essential part of the general Hindu astronomical system. We do not see how it is to be brought into connection with the other doctrines of the system, or what can be its origin and import—unless, indeed, it be merely an application to the asterisms, in an entirely arbitrary way, of the general law that everything must be made to revolve about the earth as a centre. We have noticed above (note to iii. 9–12) its inconsistency with the doctrine of the precession adopted in this treatise.

The dimensions of the several orbits stated in the text are for the most part correct, being such as are derived by the processes above explained from the numbers of sidereal revolutions given in a former passage (i. 29–34). There is, however, one exception: the orbit of Mercury, as so derived, is 1,043,207.8, and the number adopted by the text—which rejects fractions throughout, taking the nearest whole number—should be, accordingly, —208, and not —209. If we took as divisor the number of Mercury's revolutions in an Æon as corrected by the *bija* (see note to i. 29–34), we should actually obtain for his orbit the value given it by the text; the exact quotient being 1,043,208.73. But as none of the other orbits given are such as would be found by admitting the several corrections of the *bîja*, it seems preferable to assume that the text has at this point become corrupt, or else that the author of the chapter made a blunder in one of his calculations.*

The value of a minute of arc upon the moon's orbit being fifteen yojanas (see note to iv. 2–3), the value, in minutes, of any planet's mean daily motion may be readily found from its orbit by the proportion of which the rule given in verse 83 is a statement, as follows: as the distance, or the orbit, of the planet in question is to that of the moon, so is the moon's mean motion in minutes, or 11,858.717 ÷ 15, to that of the planet.

In verse 84 we are taught to calculate the distance of any planet from the earth's surface: in order to this, we are first to find the diameter of the planet's orbit, adopting, as the ratio of the diameter to the circumference, that of the diameter to the circumference of the earth—the former, of course, as calculated (i. 59) by the false ratio of 1 : $\sqrt{10}$. After being guilty of so gross an inaccuracy, it is quite superfluous, and a mere affectation of exactness, to take into account so trivial a quantity as the radius of the earth, in estimating the planet's distance from the earth.

In the doctrine of the orbits of the planets, as here laid down, we have once more a total negation of the reality of their epicyclical motions, and of their consequently varying distances from the earth in different parts of their revolutions.

CHAPTER XIII.

OF THE ARMILLARY SPHERE, AND OTHER INSTRUMENTS.

CONTENTS:—1–13, construction and equipment of the armillary sphere; 13–15, position of certain points and sines upon it; 15–16, its adjustment and revolution; 17–25, other instruments, especially for the determination of time.

1. Then, having bathed in a secret and pure place, being pure, adorned, having worshipped with devotion the sun, the planets, the asterisms (*bha*), and the elves (*guhyaka*),

2. Let the teacher, in order to the instruction of the pupil—himself beholding everything clearly, in accordance with the knowledge handed down by successive communication, and learned from the mouth of the master (*guru*)—

3. Prepare the wonder-working fabric of the terrestrial and stellar sphere (*bhûbhagola*) . . . .

---

* The last six verses of the chapter, which contain the numerical data, may very possibly be a later addition to its original content: the Âyîn-Akbari (as translated by Gladwin), in its account of the astronomy of the Hindus, which it professedly bases upon the Sûrya-Siddhânta, gives these orbits (8vo. edition, London, 1800, ii. 306), but with the fractional parts of yojanas, as if independently derived from the data and by the rules of the text: the orbit of Mercury it states correctly, as 1,043,207⅔ yojanas.

We have already remarked above (note to xii. 1–9) that the subject of this chapter is one respecting which no inquiries were addressed at the beginning of the preceding chapter by the recipient to the communicator of the revelation, and that the chapter accordingly wears in some measure the aspect of an interpolation. It comes in here as furnishing a means of illustrating to the pupil the mutual relations of the earth and the heavens as explained in the last chapter—and yet not precisely as there explained; for it gives a representation only of the earth and of the one starry concave upon which the apparent movements of all the heavenly bodies are to be traced, and not of the concentric spheres and orbits out of which the universe has been declared to be constructed. The chapter has a peculiar title, unlike that of any other in the treatise: it is styled *jyotishopanishadadhyâya*, "lection of the astronomical Upanishad." Upanishad is the name ordinarily given to such brief treatises, of the later Vedic period, or of times yet more modern, as are regarded as inspired sources of philosophical and theological knowledge, and are looked upon with peculiar reverence: its application to this chapter is equivalent to an assumption for it of especial sanctity and authority. It may possibly also indicate that the chapter is originally an independent treatise, incorporated into the text of the Sûrya-Siddhânta.

The word *bha*, in verse 1, may mean either the asterisms proper (*nakshatra*), or the signs (*râçi*), and is explained by the commentator as intended to include both. The *guhyakas*, "secret ones," are a class of demigods who attend upon Kuvera, the god of wealth, and are the keepers of his treasures: why they are mentioned here, as objects of especial reverence to the astronomical teacher, is not obvious. The commentator explains the word by "Yakshas etc., lesser divinities." In our translation of verse 3 we have followed the reading of the published text, which Colebrooke also appears to have had before him: our own manuscripts read, instead of *bhûbhagola*, *bhûmigola* and *bhûmer gola*, "sphere of the earth" simply.

Colebrooke, in his essay On the Indian and Arabian Divisions of the Zodiac (As. Res., ix. 323 etc.; Essays, ii. 321 etc.) to which we have already so often had occasion to refer, gives a translation of part of this chapter, from the beginning of the third to the middle of the thirteenth verse, as also a brief sketch of the armillary sphere of which the construction is taught in the Siddhânta-Çiromaṇi. He farther furnishes a description, and a comparison with these, of the somewhat similar instruments employed by the Greeks, the Arabs, and the early European astronomers. It has not seemed to us worth while to extract these descriptions and comparisons, or to draw up others from independent and original sources: the object of the Hindu instrument is altogether different from that of the others, since it is intended merely as an illustration of the positions and motions of the heavenly bodies, while those are meant to subserve the purposes of astronomical observation; and its relation to them is determined by this circumstance: while it, of course, possesses some of the circles which enter into the construction of the others, it is, upon the whole, a very different and much more complicated and cumbersome structure. There is nothing in the way of supposing that the first hint of its construction may have been borrowed from the instruments of western nations: but, on the other hand, it may possibly admit also of being regarded as an independent Hindu device.

3. . . . Having fashioned an earth-globe of wood, of the desired size,

4. Fix a staff, passing through the midst of it and protruding at either side, for Meru; and likewise a couple of sustaining hoops (*kakshâ*), and the equinoctial hoop;

5. These are to be made with graduated divisions (*angula*) of degrees of the circle (*bhagaṇa*). . . .

The fixing of a solid globe of wood, representing the earth, in the midst of this instrument, is of itself enough to render impracticable its application to purposes of astronomical observation. For Meru, the axis and poles of the earth, see verse 34 of the preceding chapter. We are not informed of what relative size the globe and the encompassing hoops are to be made; probably their relation is to be such that the globe will be a small one, contained within an ample sphere. The two "supporting hoops," to which are to be attached all the numerous parallels of declination hereafter described, are, of course, to be fastened to the axis at right angles to one another, and to represent the equinoctial and solstitial colures. The commentary directly prescribes this, and the text also assumes it in a later passage (v. 10).

Colebrooke, following the guidance of the commentators, treats the former half of verse 5 as belonging to the following passage, instead of the preceding. It can, however, admit of no reasonable question that the connection as established in our translation is the true one: it is demanded by the natural construction of the verses, and also yields a decidedly preferable sense.

5. . . . Farther—by means of the several day-radii, as adapted to the scale established for those other circles,

6. And by means of the degrees of declination and latitude (*vikshepa*) marked off upon the latter—at their own respective distances in declination, according to the declination of Aries etc., three

7. Hoops are to be prepared and fastened: these answer also inversely for Cancer etc. In the same manner, three for Libra etc., answering also inversely for Capricorn etc.,

8. And situated in the southern hemisphere, are to be made and fastened to the two hoop-supporters. . . .

The grammatical construction of this passage is excessively cumbrous and intricate, and we can hardly hope that the version which we have given of it will be clearly understood without farther explanations. Its meaning, however, is free from ambiguity. We have thus far only three of the circles out of which our instrument is to be constructed, namely those intended to represent the two colures and the equator: we are next to add hoops for the diurnal circles described by the sun when at the points of connection between the different signs of the zodiac. Of these there will be, of course, three north of the equator, one for the sun at the end of Aries and at the beginning of Virgo, one for the sun at the end of Taurus and at the beginning of Leo, and one for the sun at the end of Gemini and the beginning of Cancer, or at the solstice: also, in the southern hemisphere, three others corresponding to these. The dimensions of which they must be made are to be determined by their several radii (which are called day-radii—see above, ii. 60), as ascertained by calculation and reduced to the same scale upon which the colures and equator were constructed. They are then to be attached to the two general supporting hoops, or colures, each at its proper distance from the equator; this distance is ascertained by calculating the declination of the sun when at the points in question, and is determined upon the instrument by the graduation of the two supporting hoops. This graduation is in the text called that for declination (*krânti*) and latitude (*vikshepa*): it will be remembered that, according to Hindu usage, the latter means distance from the ecliptic as measured upon a circle of declination.

8. . . . Those likewise of the asterisms (*bha*) situated in the southern and northern hemispheres, of Abhijit,

9. Of the Seven Sages (*saptarshayas*), of Agastya, of Brahma etc., are to be fixed . . . .

If the orders given in these verses are to be strictly followed, our instrument must now be burdened with forty-two additional circles of diurnal revolution, namely those of the twenty-seven junction-stars (*yogatârâ*) of the asterisms and of that of Abhijit—which is here especially mentioned, as not being always ranked among the asterisms (see above, p. 208 etc.)—those of the seven other fixed stars of which the positions were stated in the eighth chapter (vv. 10–12 and 20–21), and also those of the Seven Sages, or the conspicuous stars in Ursa Major (see end of the last note to the eighth chapter). Such impracticable directions, however, cannot but inspire the suspicion that the instrument may never have been constructed except upon paper.

9. . . . Just in the midst of all, the equinoctial (*vâishuvatî*) hoop is fixed.

10. Above the points of intersection of that and the supporting hoops are the two solstices (*ayana*) and the two equinoxes (*vishuvat*) . . . .

We have already noticed (note to iii. 6) that the celestial equator derives its name from the equinoxes through which it passes. It seems a little strange that the adjustment of the hoop representing it to the two supporting hoops, which we should naturally regard as the first step in the construction of the instrument, is here assumed to be deferred until after all the other circles of declination are fixed in their places.

The word translated "above" (*ûrdhvam*) in verse 10 requires to be understood in two very different senses, as is pointed out by the commentator, to make the definitions of position of the solstices and of the equinoxes both correct: the latter are situated precisely at the intersection of the equinoctial colure with the equator; the former at a distance of 24° above and below the intersection of the equator with the other colure, or at the intersection of the colure with the third parallel of the sun's declination, on either side of the equator.

We are next taught how to fix in its proper position the hoop which is to represent the ecliptic.

10. . . . From the place of the equinox, with the exact number of degrees, as proportioned to the whole circle,

11. Fix, by oblique chords, the spaces (*kshetra*) of Aries and the rest; and so likewise another hoop, running obliquely from solstice (*ayana*) to solstice,

12. And called the circle of declination (*krânti*): upon that the sun constantly revolves, giving light: the moon and the other planets also, by their own nodes, which are situated in the ecliptic (*apamaṇḍala*),

13. Being drawn away from it, are beheld at the limit of their removal in latitude (*vikshepa*) from the corresponding point of declination. . . .

Instead of simply directing that a circle or hoop, of the same dimensions as those of the equator and colures, be constructed to represent the ecliptic, and then attached to the others at the equinoxes and solstices, the text regards it as necessary to fix, upon the six diurnal circles of the sun of which the construction and adjustment were taught above, in verses 5–8, the points of division of all the twelve signs, before the ecliptic hoop can be added to the instrument. In the compound *tiryagjyâ*, in verse 11, which we have rendered "oblique chords," we conceive *jyâ* to have its own more proper meaning of "chord," instead of that of "sine," which, by substitution for *jyârdha* (see note to ii. 15–27, near the end), it has hitherto uniformly borne. We are to ascertain by calculation the measure of the chord of 30°, to reduce it to the scale of dimensions adopted for the other great circles of the instrument, and then, commencing from either equinox, to lay it off, in an oblique direction, to the successive diurnal circles, northward and southward, thus fixing the positions upon them of the initial and final points of the twelve signs; and through all these points the ecliptic hoop is to be made to pass.

It does not appear that separate hoops for the orbits of the other planets, attached to the ecliptic at their respective nodes, are to be added to the instrument.

In verse 12 we have a name for the ecliptic, *apamaṇḍala*, which does not occur elsewhere in the treatise. The word might be literally translated "off-circle," and regarded as designating the circle which deviates in direction from the neighboring equator; but it is more probably an abbreviation for *apakramamaṇḍala*, which would mean, like the ordinary terms *krântimaṇḍala*, *krântivṛtta*, "circle of declination."

13. . . . The orient ecliptic-point (*lagna*) is that at the orient horizon; the occident point (*astamgachat*) is similarly determined.

14. The meridian ecliptic-point (*madhyama*) is as calculated by the equivalents in right ascension (*lankodayâs*), for mid-heaven (*khamadhya*) above. The sine which is between the meridian

(*madhya*) and the horizon (*kshitija*) is styled the day-measure (*antyâ*).

15. And the sine of the sun's ascensional difference (*caradala*) is to be recognized as the interval between the equator (*vishuvat*) and the horizon. . . .

These verses contain an unnecessary and fragmentary, as also a confused and blundering, definition of the positions upon the sphere of a few among the points and lines which have been used in the calculations of the earlier parts of the treatise. We are unwilling to believe that the passage is anything but a late interpolation, made by an awkward hand. For the point of the ecliptic termed *lagna*, or that one which is at any given moment passing the eastern horizon, or rising, see iii. 46–48, and note upon that passage. The like point at the western horizon, which the commentator here calls *astalagna*, "*lagna* of setting," and which the text directs us to find "in a corresponding manner," has never been named or taken into account anywhere in the treatise: we have seen above (as for instance, in ix. 4–5) that all its processes into which distance in ascension enters as an element are transferred for calculation from the occident to the orient horizon. For *madhyalagna*, the point of the ecliptic situated upon the meridian, see above, iii. 49 and note. Although we have ordinarily translated the term by "meridian ecliptic-point," this being a convenient and exact definition of the point actually referred to, we do not regard the word *madhya*, occurring in it, as meaning "meridian" in the sense in which it is used in modern astronomy, namely the great circle passing through the observer's zenith and the north and south points of his horizon. For it deserves to be noted that the text has no distinctive name for the meridian, and nowhere makes any reference to it as a circle on the sphere: it will be seen just below that, while the position of the horizon is defined, the meridian is not contemplated as a circle of sufficient consequence to require to be represented upon the illustrative armillary sphere. The commentator not very infrequently has occasion to speak of the meridian, and styles it *yâmyottaravṛtta*, "south and north circle," or *ûrdhvayâmyottaravṛtta*, "uppermost south and north circle." In the latter half of verse 14, where we have translated *madhya* by "meridian," it would have been more exact to say "mid-heaven," or "the sun at the middle of his visible revolution," or "the sun when at the point called *madhyalagna*." For the "day-measure" (*antyâ*), see above, iii. 34–36. Its definition given here is as bad as it could well be: for, passing over the fact that the line in question is not properly a sine, and moreover that the text does not tell us in which of the numberless possible directions it is to be drawn from the meridian to the horizon, the line which it is attempted to describe is not the one which the treatise regards as the *antyâ*, but the correspondent of the latter in the small circle described by the sun. That is to say, the text here substitutes the line D A in Fig. 8, above (p. 88), for the line E G. A similar blunder is made in defining the sine of the sun's ascensional difference (*carajyâ*): the line A B in the same figure, which is the "earth-sine" (*kujyâ*, *kshitijyâ*), is taken, instead of its equivalent in terms of a great circle, C G. Moreover, the text reads "equator" (*vishuvat*—E C in the figure) here for "east and west hour-circle" (*unmanḍala*—C P): the commentator restores the latter, and excuses the substitution by a false translation of the latter half of iii. 6, making it mean "the east and west hour-circle is likewise denominated the equinoctial circle."

In verse 14, *lankodayâs* is substituted for the more usual term *lankodayâsavas* (see above, iii. 49, and note), in the sense of "equivalents of the signs in right ascension," literally, "at Lankâ."

15. . . . Having turned upward one's own place, the circle of the horizon is midway of the sphere.

16. As covered with a casing (*vastra*) and as left uncovered, it is the sphere surrounded by Lokâloka. . . .

The simple direction to turn upward one's own situation upon the central wooden globe which represents the earth does not, it is evident, contemplate any very careful or exact adjustment of the instrument.

Verse 16 is very elliptical and obscure in its expressions, but their general meaning is plain, and is that which is attributed to them by the commentator. The proper elevation having been given to the pole of the sphere, a circle is by some means or other to be fixed about its midst, or equally distant from its zenith and nadir, to represent the horizon. Then the part below is to be encased in a cloth covering, the upper hemisphere alone being left open. As thus arranged, the sphere is, as it were, girt about by the Lokâloka mountains. Lokâloka is, as we have seen above (note to xii. 32–44), the name of the giant mountain-range which, in the Puranic geography, is made the boundary of the universe: it is apparently so called because it separates the world (*loka*) from the non-world (*aloka*); and as out of the Puranic Meru the new astronomical geography makes the axis and poles of the earth, so out of these mountains it makes the visible horizon.

The "wonder-working fabric of the terrestrial and stellar sphere" is now fully constructed, and only requires farther, in order to its completion as an edifying and instructive illustration of the relations of the heavens to the earth, to be set in motion about its fixed axis.

16. . . . By the application of water is made ascertainment of the revolution of time.

17. One may construct a sphere-instrument combined with quicksilver: this is a mystery; if plainly described, it would be generally intelligible in the world.

18. Therefore let the supreme sphere be constructed according to the instruction of the preceptor (*guru*). In each successive age (*yuga*), this construction, having become lost, is, by the Sun's

19. Favor, again revealed to some one or other, at his pleasure. . . .

Here we have another silly mystification of a simple and comparatively insignificant matter, like that already noticed at the end of the sixth chapter. The revolution of the machine of which the construction has now been explained, in imitation of the actual motion of the heavens about the earth, is something so calculated to strike the minds of the uninitiated with wonder, that the means by which it is to be accomplished must not be fully explained even in this treatise, lest they should become too generally known: they must be learned by each pupil directly from his teacher, as the latter has received them by successive tradition, from the original and superhuman source whence they came. It is perfectly evident that such a fabric could only be made to revolve in a rude and imperfect way; that it should have marked time, and continued for any period to correspond in position with the actual sphere, is impossible.

The word which, upon the authority of the commentator, we have rendered "water," in verse 16, is *amṛtasrâva*, literally "having an immortal flow": perhaps the phrase should be translated rather, "by managing a constant current of water."

19. . . . So also, one should construct instruments (*yantra*) in order to the ascertainment of time.

20. When quite alone, one should apply quicksilver to the wonder-causing instrument. By the gnomon (*çanku*), staff (*yashṭi*), arc (*dhanus*), wheel (*cakra*), instruments for taking the shadow, of various kinds,

21. According to the instruction of the preceptor (*guru*), is to be gained a knowledge of time by the diligent. . . .

The commentator interprets the first part of verse 20 in correspondence with the sense of the preceding passage: the application of mercury to a revolving machine, in order to give it the appearance of automatic motion, must be made privately, lest people, understanding the method too well, should cease to wonder at it. The instruments mentioned in the latter half of the same verse are explained in the commentary simply by citations from the *yantrâdhyâya*, "chapter of instruments," of the Siddhânta-Çiromaṇi (Golâdhy., pp. 111–136, published edition). We will state, as briefly as may be, their character:

The gnomon (*çanku*) needs no explanation: its construction and the method of using it have been fully exhibited in the third chapter of our treatise. The "staff-instrument" (*yashṭiyantra*) is described as follows. A circle is described upon a level surface with a radius proportioned to that of the sphere, or to tabular radius. Its cardinal points are ascertained, and its east and west and north and south diameters are drawn. From the former, at either extremity, is laid off the sine of amplitude (*agrâ*) ascertained by calculation for the given day: the points thus determined upon the circumference of the circle represent the points on the horizon at which the sun rises and sets. Another circle, with a radius proportioned to that of the calculated diurnal circle of the day (*dyujyâ*), is also described about the centre of the other, and is divided into sixty equal parts, representing the division of the sun's daily revolution into sixty nâḍîs. Into a depression at the centre, the foot of a staff (*yashṭi*), equal in length to the radius of the larger circle, is loosely inserted. When it is desired to ascertain the time of the day, this staff is pointed directly toward the sun, or in such manner that it casts no shadow; its extremity then represents the place of the sun at the moment upon the sphere. Measure, by a stick, the distance of that extremity from the point of sunrise or of sunset: this will be the chord of that part of the diurnal circle which is intercepted between the sun's actual position and the point at which he rose, or will set: the value of the corresponding arc in nâḍîs may be ascertained by applying the stick to the lesser graduated circle. The result is the time since sunrise, or till sunset.

The "wheel" (*cakra*) is a very simple instrument for obtaining, by observation, the sun's altitude and zenith-distance. It is simply a wheel, suspended by a string, graduated to degrees, having its lowest point and the extremities of its horizontal diameter distinctly marked, and with a projecting peg at the centre. When used, its edge is turned toward the sun, so that the shadow of the peg falls upon the graduated periphery, and the distances of the point where it meets the latter from the horizontal and lowest points of the wheel respectively are the required altitude and zenith-distance of the sun. From these, by the methods of the third chapter (iii. 37–39), the time may be derived.

The "arc" (*dhanus*) is the lower half of the instrument just described —or, we may also suppose, a quadrant of it; since only a quadrant is required for making the observations for which the instrument is employed.

21. By water-instruments, the vessel (*kapâla*) etc., by the peacock, man, monkey, and by stringed sand-receptacles, one may determine time accurately.

22. Quicksilver-holes, water, and cords, ropes (*çulba*), and oil and water, mercury, and sand are used in these: these applications, too, are difficult.

The instruments and methods hinted at in these verses are only partially and obscurely explained by the commentator. The *kapâla*, "cup" or "hemisphere," is doubtless the instrument which is particularly described below, in verse 23. The *nara*, "man," is also spoken of below, in verse 24, and is simply a gnomon; it is perhaps one of a particular construction and size, and so named from having about the height of a man. The peacock and monkey are obscure. The "sand-vessels" (*reṇugarbha*), which are "provided with cords" (*sasûtra*), are probably suspended instruments, of the general character of our hour-glasses. The commentator connects them also with the "peacock," as if the latter were a figure of the bird having such a vessel in his interior, and letting the sand pour out of his mouth. In illustration of the "quicksilver-holes" (*pâradârâ*) a passage is cited from the Siddhânta-Çiromaṇi (as above), giving the description of an instrument in which they are applied. It is a wheel, having on its outer edge a number of holes, of equal size, and at equal distances from one another, but upon a zig-zag line: these holes are filled half full of mercury, and stopped at the orifice: and it is claimed that the wheel will then, if supported upon an axis by a couple of props, revolve of itself. The application of this method may well enough be styled "difficult": if a machine so constructed would work, the Hindus would be entitled to the credit of having solved the problem of perpetual motion. The descriptions of one or two other somewhat similar machines are also cited in the commentary from the Siddhânta-Çiromani: the only new feature worthy of notice which they contain is the application of the siphon, or bent tube, in emptying a vessel of the water it contains.

It will have been noticed that, throughout the whole of this chapter, the different parts or passages end in the middle of a verse. In the twenty-first verse the coincidence between the end of a passage and the end of a verse is re-established, but it is at the cost of such an irregularity as is nowhere else committed in the treatise: the verse is made to consist of three half-çlokas, instead of two, the whole chapter being thus allowed to contain an uneven number of lines. There are two or three very superfluous half-verses at the beginning of the chapter, the omission of any one of which would seem an easier and preferable method of restoring the regular and connected construction of the text.

23. A copper vessel, with a hole in the bottom, set in a basin of pure water, sinks sixty times in a day and night, and is an accurate hemispherical instrument.

This instrument appears to have been the one most generally and frequently in use among the Hindus for the measurement of time: it is the only one described in the Âyîn-Akbari (ii. 302). One of the common names for the sixtieth part of the day, *ghaṭî* or *ghaṭikâ*, literally "vessel," is evidently derived from it: the other, *nâḍî* or *nâḍikâ*, "reed," probably designated in the first place, and more properly, a measure of length, and not of time. A verse cited in the commentary to this passage gives the form and dimensions of the vessel used: it is to be of ten *palas'* weight of copper, six digits (*angula*) high, and of twice that width at the mouth, and is to contain sixty *palas* of water: the hole in the bottom through which it is to fill itself is to be such as will just admit a gold pin four digits long, and weighing three and a third *mâshas*. The description of the Âyîn-Akbari does not precisely agree with this; and it is, indeed, sufficiently evident that an instrument intended for such a purpose could not be accurately constructed by Hindu workmen from measurements alone, but would have to be tested by comparison with some recognized standard, or by actual use.

24. So also, the man-instrument (*narayantra*) is good in the day-time, and when the sun is clear. The best determination of time by means of determinations of the shadow has been explained.

We have already noticed above, under verse 21, that the *nara* was a simple gnomon. The explanations here referred to are, of course, those which are presented in the third chapter.

The concluding verse of the chapter is an encouragement held out to the astronomical student.

25. He who thoroughly knows the system of the planets and asterisms, and the sphere, attains the world of the planets in the succession of births, his own possessor.

CHAPTER XIV.

OF THE DIFFERENT MODES OF RECKONING TIME.

CONTENTS:—1–2, enumeration of the modes of measuring time, and general explanation of their uses; 3, solar time; 4–6, of the periods of eighty-six days; 7–11, of points and divisions in the sun's revolution; 12–13, lunar time; 14, time of the Fathers; 15, sidereal time; 15–16, of the months and their asterisms; 17, of the twelve-year cycle of Jupiter; 18–19, civil, or mean solar, time; 20–21, time of the gods, Prajâpati, and Brahma; 22–26, conclusion of the work.

1. The modes of measuring time (*mâna*) are nine, namely those of Brahma, of the gods, of the Fathers, of Prajâpati, of Jupiter, and solar (*sâura*), civil (*sâvana*), lunar, and sidereal time.

2. Of four modes, namely solar, lunar, sidereal, and civil time, practical use is made among men; by that of Jupiter is to be determined the year of the cycle of sixty years; of the rest, no use is ever made.

This chapter contains the reply of the sun's incarnation to the last of the questions addressed to him by the original recipient of his revelation (see above, xii. 8). The word *mâna*, which gives it its title of *mânâdhyâya*, and which we have translated "mode of measuring or reckoning time," literally means simply "measure": it is the same term which we have already (iv. 2–3) seen applied to designate the measured disks of the sun and moon.

3. By solar (*sâura*) time are determined the measure of the day and night, the *shaḍaçîtimukhas*, the solstice (*ayana*), the equinox (*vishuvat*), and the propitious period of the sun's entrance into a sign (*sankrânti*).

The adjective *saura*, which we translate "solar," is a secondary derivative from *sûrya*, "sun." It is applied to those divisions of time which are dependent on and determined by the sun's actual motion along the ecliptic. The "day and night" measured by it are probably those of the gods and demons respectively; see above, xii. 48–50. The solar year, as already noticed (note to i. 12–13), is sidereal, not tropical; it commences whenever the sun enters the first sign of the immovable sidereal zodiac, or when he is 10 minutes east in longitude from the star ζ Piscium. The solar month is the time during which he continues in each successive sign, or arc of 30°, reckoning from that point. The length of the solar year and month is subject only to an infinitesimal variation, due to the slow motion, of 1′ in 517 years, assumed for the sun's line of apsides (see above, i. 41–44); but it is, as has been shown above (note to i. 29–34, near the end), somewhat differently estimated by different authorities. The precise length of the solar months, as reckoned according to the Sûrya-Siddhânta, is thus stated by Warren (Kâla Sankalita, p. 69):

Duration of the several Solar Months.

No.	Name.	Duration.					Sum of duration.				
		d	n	v	′′′	′′′′	d	n	v	′′′	′′′′
1	Vâiçâkha,	30	55	32	2	39	30	55	32	2	39
2	Jyâishṭha,	31	24	12	2	41	62	19	44	5	20
3	Âshâḍha,	31	36	38	2	44	93	56	22	8	4
4	Çrâvaṇa,	31	28	12	2	42	125	24	34	10	46
5	Bhâdrapada,	31	2	10	2	40	156	26	44	13	26
6	Âçvina,	30	27	22	2	38	186	54	6	16	4
7	Kârttika,	29	54	7	2	35	216	48	13	18	39
8	Mârgaçîrsha,	29	30	24	2	33	246	18	37	21	12
9	Pâusha,	29	20	53	2	31	275	39	30	23	43
10	Mâgha,	29	27	16	2	32	305	6	46	26	15
11	Phâlguna,	29	48	24	2	33	334	55	10	28	48
12	Câitra,	30	20	21	2	36	365	15	31	31	24

The former passage (i. 12–13) took no note of any solar day; in this chapter, however, such a division of time is distinctly contemplated: it is also recognized by the Siddhânta-Çiromaṇi (Gaṇitâdhy., ii. 8), and seems to be, for certain uses, generally accepted. The solar day is the time during which the sun traverses each successive degree of the ecliptic, with his true motion, and its length accordingly varies with the rapidity of his motion: three hundred and sixty such days compose the sidereal year. In order to determine the solar day corresponding to any given moment, it is, of course, only necessary to calculate, by the methods of the second chapter, the sun's true longitude for that moment. Hence it is a matter of very little practical account: all the periods regarded as determined by it may be as well derived directly from the sun's longitude, without going through the form of calling its degrees days. It is thus with the equinoxes, solstices, and entrances of the sun into a sign (*sankrânti*, "entrance upon connection with"): for the latter, and for the continuance of the propitious influences which are believed to attend upon it, see below, verse 11. The *shaḍaçîtimukhas* form the subject of the next following passage.

The manuscript without commentary inserts here the following verse: "the day and night of the gods and demons, which is determined by the sun's revolution through the circle of asterisms (*bhacakra*), and the number of the Golden (*kṛta*) and other Ages, as already stated, is to be known."

4. Beginning with Libra, the *shaḍaçîtimukha* is at the end of the periods of eighty-six (*shaḍaçîti*) days, in succession: there are four of them, occurring in the signs of double character (*dvisvabhâva*);

5. Namely, at the twenty-sixth degree of Sagittarius, at the twenty-second of Pisces, at the eighteenth degree of Gemini, and at the fourteenth of Virgo.

6. From the latter point, the sixteen days of Virgo which remain are suitable for sacrifices: anything given to the Fathers (*pitaras*) in them is inexhaustible.

We have not been able to find anywhere any explanation of this curious division of the sun's path into arcs of 86°, commencing from the autumnal equinox, and leaving an odd remnant of 16° at the end of Virgo. The commentary offers nothing whatever in elucidation of their character and significance. The epithet "of double character" (*dvisvabhâva*) belongs to the four signs mentioned in verse 5; judging from the connection in which it is applied to them by Varâha-Mihira (Laghujâtaka, i. 8, in Weber's Indische Studien, ii. 278), it designates them as either variable (*cara*) or fixed (*sthira*), in some astrological sense. The term *shaḍaçîtimukha* is composed of *shaḍaçîti*, "eighty-six," and *mukha*, "mouth, face, beginning." We do not understand the meaning of the compound well enough to venture to translate it.

7. In the midst of the zodiac (*bhacakra*) are the two equinoxes (*vishuvat*), situated upon the same diameter (*samasûtraga*), and likewise the two solstices (*ayana*); these four are well known.

8. Between these are, in each case, two entrances (*sankrânti*); from the immediateness of the entrance are to be known the two feet of Vishṇu.

9. From the sun's entrance (*sankrânti*) into Capricorn, six months are his northern progress (*uttarâyaṇa*); so likewise, from the beginning of Cancer, six months are his southern progress (*dakshiṇâyana*).

10. Thence also are reckoned the seasons (*ṛtu*), the cool season (*çiçira*) and the rest, each prevailing through two signs. These twelve, commencing with Aries, are the months; of them is made up the year.

The commentator explains *samasûtraga*, like *samasûtrastha* above (xii. 52), to mean situated at opposite extremities of the same diameter of the earth, or antipodal to one another.

The technical term for the sun's entrance into a sign of the zodiac is, as noticed already, *sankrânti* (the commentary also presents the equivalent word *sankramana*); of these there take place two between each equinox and the preceding or following solstice. The latter half of verse 8 is quite obscure. The commentator appears to understand it as signifying that, in each quadrant, the entrance (*sankrânti*) immediately following the solstice or equinox is styled "Vishṇu's feet." In the earliest Hindu mythology, Vishṇu is the sun, especially considered as occupying successively the three stations of the orient horizon, the meridian, and the occident horizon; and the three steps by which he strides through the sky are his only distinctive characteristic. These three steps, then, appear under various forms in the later Vâishṇava mythology, and there is plainly some reference to them in this designation of the sun's entrances into the signs. It would seem easiest and most natural to recognize in the three signs intervening between each equinox and solstice Vishṇu's three steps, and to regard the two intermediate entrances as the marks of his feet; this may possibly be the figure intended to be conveyed by the language of the text.

The word *ṛtu* means originally and literally any determined period of time, a "season" in the most general sense of the term; but it has also been employed from very early times to designate the various divisions of the year. They were anciently reckoned as three, five, six, or seven; but the prevailing division, and the only one in use in later times, is that into six seasons, named Çiçira, Vasanta, Grîshma, Varsha, Çarad, and Hemanta, which may be represented by cool season, spring, summer, rainy season, autumn, and winter. Çiçira begins with the month Mâgha, or about the middle of January (see note to i. 48–51, and the table given below, under vv. 15–16), and each season in succession includes two solar months.

11. Multiply the number of minutes in the sun's measure (*mâna*) by sixty, and divide by his daily motion: a time equal to half the result, in nâḍîs, is propitious before the sun's entrance into a sign (*sankrânti*), and likewise after it.

The propitious influences referred to above, in verse 3, as attending upon the sun's entrance into a sign, are regarded as enduring so long as any part of his disk is upon the point of separation between the two signs. This time is found by the following proportion: as the sun's actual daily motion, in minutes, is to a day, or sixty nâḍîs, so is the measure of his disk, in minutes, to the time which it will occupy in passing the point referred to.

12. As the moon, setting out from the sun, moves from day to day eastward, that is the lunar method of reckoning time (*mâna*): a lunar day (*tithi*) is to be regarded as corresponding to twelve degrees of motion.

13. The lunar day (*tithi*), the karaṇa, the general ceremonies, marriage, shaving, and the performance of vows, fastings, and pilgrimages, are determined by lunar time.

14. Of thirty lunar days is composed the lunar month, which is declared to be a day and a night of the Fathers: the end of the month and of the half-month (*paksha*) are at their mid-day and midnight respectively.

For the *tithi*, or lunar day, see above, ii. 66: for the karaṇa, see ii. 67–69. For the month considered as the day of the *pitaras*, or manes of the departed, see note to xii. 73–77. Manu (i. 66) pronounces the day of the Fathers to be the dark half-month, or the fortnight from full moon to new moon, and their night to be the light half-month, or the fortnight from new moon to full moon. With this mode of division might be made to accord that stated in the latter part of verse 14, by rendering *madhye* "between," instead of "at the middle point of": we have translated according to the directions of the commentator.

15. The constant revolution of the circle of asterisms (*bhacakra*) is called a sidereal day. The months are to be known by the names of the asterisms (*nakshatra*), according to the conjunction (*yoga*) at the end of a lunar period (*parvan*).

16. To the months Kârttika etc. belong, as concerns the conjunction (*samayoga*), the asterisms Kṛttikâ etc., two by two: but

three months, namely the last, the next to the last, and the fifth, have triple asterisms.

The subject of sidereal time, although one of prominent importance in the present treatise, since the subdivision of the day is regulated entirely by it, is here very summarily dismissed with half a verse, while we find appended to it in the same passage matters with which it has nothing properly to do.

We have already (note to i. 48–51) had occasion to notice that the months are regarded as having received their names from the asterisms (*nakshatra*) in which the moon became full during their continuance. According to Sir William Jones (As. Res., ii. 296), it is asserted by the Hindus "that, when their lunar year was arranged by former astronomers, the moon was at the full in each month on the very day when it entered the *nakshatra*, from which that month is denominated." Whether this assertion is strictly true admits of much doubt. Our text does not imply any such claim: it only declares that the month is to be called by the name of that asterism with which the moon is in conjunction (*yoga*) at the end of the *parvan:* this latter word might mean either half of a lunar month, but is evidently to be understood here, as explained by the commentary, of the light half (*çukla paksha*) alone, so that the end of the *parvan* (*parvânta*) is equivalent to the end of the day of full moon (*pûrṇimânta*), or to the moment of opposition in longitude. Now it is evident that, owing to the incommensurability of the times of revolution of the sun and moon, as also to the revolution of the moon's line of apsides, full moon is liable to occur in succession in all the asterisms, and at all points of the zodiac; so that although, at the time when the system of names for the months originated and established itself, they were doubtless strictly applicable, they would not long continue to be so. Instead, however, of being compelled to alter continually the nomenclature of the year, we are allowed, by verse 16, to call a month Kârttika in which the full of the moon takes place either in Kṛttikâ or in Rohiṇî, and so on; the twenty-seven asterisms being distributed among the twelve months as evenly as the nature of the case admits.

At what period these names were first introduced into use is unknown. It must have been, of course, posterior to the establishment of the system of asterisms, but it was probably not much later, as the names are found in some of the earlier texts which contain those of the *nakshatras* themselves. We can hardly suppose that they were not originally applied independently to the lunar months; and certainly, no more suitable derivation could be found for the name of a lunar period than from the asterism in which the moon attained during its continuance her full beauty and perfection. In later times, as we have already seen (note to i. 48–51), the true lunar months are entirely dependent for their nomenclature upon the solar months, according to the determination of the latter, as regards their commencement and duration, by the data and methods of the modern astronomical science. There has been handed down another system of names for the months (see Colebrooke in As. Res., vii. 284; Essays, i. 201), which have nothing to do with the asterisms: whether they are to be regarded as more ancient than the others we do not know. They are—commencing with the first month of the season Vasanta, or with that one which in the other system is called Câitra—as follows: Madhu, Mâdhava, Çukra, Çuci, Nabhas, Nabhasya, Isha, Ûrja, Sahas, Sahasya, Tapas, Tapasya.

For the sake of a clearer understanding of the relations of the asterisms, months, and seasons, we present their correspondences below in a tabular form:

Season.	Month.	Asterisms in which full moon may occur.
Çarad.	Kârttika. (Oct.-Nov.)	Kṛttikâ. Rohiṇî.
Hemanta.	Mârgaçîrsha. (Nov.-Dec.)	Mṛgaçîrsha. Ârdrâ.
	Pâusha. (Dec.-Jan.)	Punarvasu. Pushya.
Çiçira.	Mâgha. (Jan.-Feb.)	Âçleshâ. Maghâ.
	Phâlguna. (Feb.-Mar.)	P.-Phalgunî. U.-Phalgunî. Hasta.
Vasanta.	Câitra. (Mar.-Apr.)	Citrâ. Svâtî.
	Vâiçâkha. (Apr.-May.)	Viçâkhâ. Anurâdhâ.
Grîshma.	Jyâishṭha. (May-June.)	Jyeshṭhâ. Mûla.
	Âshâḍha. (June-July.)	P.-Ashâḍhâ. U.-Ashâḍhâ.
Varsha.	Çrâvaṇa. (July-Aug.)	Çravaṇa. Çravishṭhâ.
	Bhâdrapada. (Aug.-Sept.)	Çatabhishaj. P.-Bhâdrapadâ. U.-Bhâdrapadâ.
Çarad.	Âçvina. (Sept.-Oct.)	Revatî. Açvinî. Bharaṇî.

Davis (As. Res., iii. 218) notices that some of the ancient astronomers have divided the asterisms somewhat differently, giving to Çrâvaṇa the three beginning with Çravaṇa, to Bhâdrapada the three beginning with Pûrva-Bhâdrapadâ, and to Âçvina only Açvinî and Bharaṇî. It seems, indeed, that the selection of the three months to which three asterisms, instead of two, were assigned, must have been made somewhat arbitrarily.

It will be noticed that in this passage Kârttika is treated as the first of the series of months, while above (v. 10) Çiçira was mentioned as the first season, and while in practice (see note to i. 48–51) Vâiçâkha is treated as the first of the solar months, and Câitra of the lunar. Another name for Mârgaçîrsha, also, is Agrahâyaṇa, which appears to mean "commencement of the year." How much significance these variations of usage may have, and what is their reason, is not known to us.

As regards Vâiçâkha and Câitra, indeed, the case is clear, and we may also regard the rank assigned to Kârttika as due to the ancient position of Kṛttikâ, as first among the lunar mansions.

17. In Vâiçâkha etc., a conjunction (*yoga*) in the dark half-month (*kṛshṇa*), on the fifteenth lunar day (*tithi*), determines in like manner the years Kârttika etc. of Jupiter, from his heliacal setting (*asta*) and rising (*udaya*).

We have already, in an early part of the treatise (i. 55), made acquaintance with a cycle of the planet Jupiter, composed of sixty years; in this verse we have introduced to our notice a second one, containing twelve years, or corresponding to a single sidereal revolution of the planet. The principle upon which its nomenclature is based is very evident. Jupiter's revolution is treated as if, like that of the sun, it determined a year, and the twelve parts, each quite nearly equalling a solar year (see note to i. 55), into which it is divided, are, by the same analogy, accounted as months, and accordingly receive the names of the solar months. The appellations thus applied to the years, in their order, we are directed to determine by the asterism (*nakshatra*) in which the planet is found to be at the time of its disappearance in the sun's rays, and its disengagement from them: for it would, of course, set and rise heliacally twelve times in each revolution, and each time about a month later than before. The name of the year, however, will not agree with that of the month in which the rising and setting occur, but will be the opposite of it, or six months farther forward or backward, since the month is named from the asterism with which the sun is in opposition, but the year of the cycle from that with which he is in conjunction. The terms in which the rule of the text is stated are not altogether unambiguous: there is no expressed grammatical connection between the two halves of the verse, and we are compelled to add in our translation the important word "determines," which links them together. The meaning, however, we take to be as follows: if, in any given year, the heliacal setting of Jupiter takes place in the month Vâiçâkha, then the asterism with which the moon is found to be in conjunction at the end of that month—which will be, of course, the asterism in which the sun is at the same time situated—will determine the name of the year, which will be Kârttika: and so on, from year to year. The expression "in like manner," in the second half of the verse, is interpreted as implying that to the years of this cycle is made the same distribution of the asterisms as to the months in the preceding passage: the second and third columns of the last table, then, will apply to the cycle, if we alter their headings respectively, from "month" to "year of the cycle," and from "asterisms in which full moon may occur" to "asterisms in which Jupiter's heliacal setting and rising may occur."

There is one untoward circumstance connected with this arrangement which is not taken into account by the text, and which appears to oppose a practical difficulty to the application of its rule. The amount of Jupiter's motion during a solar year is not precisely one sign, but perceptibly more than that, so that the mean interval between two successive heliacal settings is a little more than a solar month; and this difference accumulates so rapidly that the thirteenth setting would take place about four degrees farther eastward than the first, so that, without some system of periodical omissions of a month, the correspondence between the names of the years, if applied in regular succession, and the asterisms in which the planet disappeared would, after a few revolutions, be altogether dislocated and broken up. If the cycle were of more practical consequence, or if it were contemplated as one of the proper subjects of this treatise, we might expect to find some method of obviating this difficulty prescribed. Warren, however, in his brief account of the cycle of twelve years (Kâla Sankalita, p. 212 etc.), states that he knows of no nation or tribe making any use of it, but only finds it mentioned in the books. According to both him and Davis (As. Res., iii. 217 etc.), the cycle of twelve years is subordinate to that of sixty, the latter being divided into five such cycles, to which special names are applied, and of each of which the successive years receive in order the titles of the solar months. The appellations of the cycles themselves are those which properly belong to the years of the lustrum (*yuga*), or cycle of five years, by which, as already noticed (note to i. 56–58), the Hindus appear first to have regulated time, and effected by intercalation the coincidence of the solar and lunar years: they are Samvatsara, Parivatsara, Idâvatsara, Idvatsara, (or Anuvatsara), and Vatsara (or Idvatsara, or Udravatsara). It would appear, then, either that the cycle of sixty years was derived from and founded upon the ancient lustrum, being an imitation of its construction in time of the planet Jupiter, of which a month equals a solar year, or else that the already existing cycle had been later fancifully compared with the lustrum, and subdivided after its model into sub-cycles for years, and years for months: of these two suppositions we are inclined to regard the latter as decidedly the more probable.

18. From rising to rising of the sun, that is called civil (*sâvana*) reckoning. By that are determined the civil days (*sâvana*), and by these is the regulation of the time of sacrifice;

19. Likewise the removal of uncleanness from child-bearing etc., and the regents of days, months, and years: the mean motion of the planets, too, is computed by civil time.

The term *sâvana* we have translated "civil," as being a convenient way of distinguishing this from the other kinds of time, and as being very properly applicable to the day as reckoned in practical use from sunrise to sunrise: in the more general sense, as denoting the mode of reckoning the mean motions of the planets, and the regency of successive periods, *sâvana* corresponds to what we call "mean solar" time. The word itself seems to be a derivative from *savana*, "libation," the three daily *savanas*, or the sunrise, noon, and sunset libations, being determined by this reckoning.

20. The mutually opposed day and night of the gods (*sura*) and demons (*asura*), which has been already explained, is time of the gods, being measured by the completion of the sun's revolution.

21. The space of a Patriarchate (*manvantara*) is styled time of Prajâpati: in it is no distinction of day from night. An Æon (*kalpa*) is called time of Brahma.

It may well be said that the mode of reckoning by time of the gods has been already explained: the length of a day of the gods, with the method of its determination, has been stated and dwelt upon, in almost identical language, over and over again (see i. 13–14; xii. 45–50, 67, 74; and the interpolated verse after xiv. 3), almost as if it were so new and striking an idea as to demand and bear repeated inculcation. For the Patriarchate (*manvantara*), or period of 308,448,000 years, see above, i. 18: this is the only allusion to it as a unit of time which the treatise contains. For the Æon (*kalpa*), of 4,320,000,000 years, as constituting a day of Brahma, see above, i. 20.

The remaining verses are simply the conclusion of the treatise.

22. Thus hath been told thee that supreme mystery, lofty and wonderful, that sacred knowledge (*brahman*), most exalted, pure, all guilt destroying;

23. And the highest knowledge of the heaven, the stars, and the planets hath been exhibited: he who knoweth it thoroughly obtaineth in the worlds of the sun etc. an everlasting place.

24. With these words, taking leave of Maya, and being suitably worshipped by him, the part of the sun ascended to heaven, and entered his own disk.

25. So then Maya, having personally learned from the sun that divine knowledge, regarded himself as having attained his desire, and as purified from sin.

26. Then, too, the sages (*rshi*), learning that Maya had received from the sun this gift, drew near and surrounded him, and reverently asked the knowledge.

27. And he graciously bestowed upon them the grand system of the planets, of mysteries in the world the most wonderful, and equal to the Scripture (*brahman*).

The Sûrya-Siddhânta, in the form in which it is here presented, as accepted by Ranganâtha and fixed by his commentary, contains exactly five hundred verses. This number, of course, cannot plausibly be looked upon as altogether accidental: no one will question that the treatise has been intentionally wrought into its present compass. We have often found occasion above to point out indications, more or less distinct and unequivocal, of alterations and interpolations; and although in some cases our suspicions may not prove well-founded, there can be no reasonable doubt that the text of the treatise has undergone since its origin not unimportant extension and modification. Any farther consideration of this point we reserve for the general historical summary to be presented at the end of the Appendix.

APPENDIX:

CONTAINING ADDITIONAL NOTES AND TABLES, CALCULATIONS OF ECLIPSES, A STELLAR MAP, ETC.

---

1. p. ii. The name *siddhânta*, by which the astronomical text-books are generally called, has, by derivation and original meaning, nothing to do with astronomy, but signifies simply "established conclusion;" and it is variously applied to other uses in the Sanskrit literature. It may not be uninteresting to present here a summary view of the existing astronomical literature of the Hindus, as derived from such sources of information upon the subject as are accessible to us, even though such a view must necessarily be imperfect and incomplete. We commence by giving a list of works furnished to the translator, at his request, by the native Professor of Mathematics in the Sanskrit College at Pûna, and which may be taken as representing the knowledge possessed, and the opinions held, by the learned of Western India at the present time. Along with it is offered the list of nine treatises given in the modern Sanskrit Encyclopedia, the Çabdakalpadruma, as entitled to the name of Siddhânta. The longer list was intended to be arranged chronologically; the remarks appended to the names of treatises are those of its compiler.

1. Brahma-Siddhânta.
2. Sûrya-Siddhânta.
3. Soma-Siddhânta.
4. Vâsishṭha-Siddhânta.
5. Romaka-Siddhânta.
6. Pâulastya-Siddhânta.
7. Bṛhaspati-Siddhânta.
8. Garga-Siddhânta.
9. Vyâsa-Siddhânta.
10. Pârâçara-Siddhânta.
11. Bhoja-Siddhânta; earlier than the Çiromaṇi.
12. Varâha-Siddhânta; earlier than the Çiromaṇi.
13. Brahmagupta-Siddhânta; earlier than the Çiromaṇi.
14. Siddhânta-Çiromaṇi; *çake* 1072 [A.D. 1150].
15. Sundara-Siddhânta; about 400 years ago.
16. Tattva-Viveka-Siddhânta; in the time of the reign of Jaya Sinha, about 250 years ago.
17. Sârvabhâuma-Siddhânta; in the time of the reign of Jaya Sinha.
18. Laghu-Ârya-Siddhânta } earlier than the Çiromaṇi.
19. Bṛhad-Ârya-Siddhânta } earlier than the Çiromaṇi.

1. Brahma-Siddhânta.
2. Sûrya-Siddhânta.
3. Soma-Siddhânta.
4. Bṛhaspati-Siddhânta.
5. Garga-Siddhânta.
6. Nârada-Siddhânta.
7. Pârâçara-Siddhanta.
8. Pâulastya-Siddhânta.
9. Vasishṭha-Siddhânta.

It is obvious that these lists are uncritically constructed, and that neither of them is of a nature to yield valuable information without additional explanations. The one is most unreasonably curt, and seems founded on the principle of allowing the title of Siddhânta to no work which is the acknowledged composition of a merely human author, while the other contains treatises of very heterogeneous character and value: and neither list distinguishes works now actually in existence from those which have become lost, and those of which the existence at any period is questionable. A more satisfactory account of the Siddhânta literature may be drawn up from the notices contained in the writings of Western scholars, and especially from the various essays of Colebrooke. For what we shall here offer, he is our main authority.

In the present imperfect state of our knowledge of the subject, there is perhaps no better method of classifying the Hindu astronomical treatises than by dividing them into four classes, as follows: first, those which profess to be a revelation on the part of some superhuman being; second, those which are attributed to ancient and renowned sages, or to other supposititious or impersonal authors; third, those regarded as the works of actual authors, astronomers of an early and uncertain period; fourth, later texts, of known date and authorship, and mostly of a less independent and original character.

I. The first class comprises the Brahma, Sûrya, Soma, Bṛhaspati, and Nârada Siddhântas.

1. *Brahma-Siddhânta.* The earliest treatise bearing this name is said to have formed a part of the Vishṇudharmottara Purâṇa, a work which seems to be long since lost, and scarcely remembered except in connection with the Siddhânta. The latter, too, is only known by a few citations in astronomical writings, and by the treatise of Brahmagupta (see below, third class) founded upon it. Another work laying claim to the same title is that which we have many times cited above as the Çâkalya-Sanhitâ. Sanhitâ, "text, comprehensive work," is a term employed to denote a complete course of astronomy, astrology, horoscopy, etc.: this treatise, according to the manuscript in our possession, forms the second division (*praçna*) of such a course. It professes to be revealed by Brahma to the semi-divine personage Nârada. Of its relation to the Sûrya-Siddhânta we have spoken above (note to viii. 10–12). It does not appear to be referred to as an independent work in either of the native lists we have given.

2. *Sûrya-Siddhânta.* This is the treatise of which the translation has been given above, and of which, accordingly, we do not need to speak here more particularly.

3. *Soma-Siddhânta.* Judging from its title, this work must profess to derive its origin from the moon (*soma*), as the preceding from the sun (*sûrya*). Bentley speaks of it as following in the main the system of the Sûrya-Siddhânta. There is a manuscript of it in the Berlin Library (Weber's Catalogue, No. 840), and Colebrooke seems also to have had it in his hands.

4. *Bṛhaspati-Siddhânta.* Bṛhaspati is the name of a divine personage, priest and teacher of the gods, as also of the planet Jupiter. No work bearing this name is mentioned, so far as we can ascertain, by any European scholar, although Bṛhaspati is not infrequently referred to in native writings as an authority in astronomical matters.

5. *Nârada-Siddhânta.* A Nâradî-Sanhitâ, or course of astrology, in the Berlin Library (Weber, No. 862), and an occasional reference to

Nârada, among other divine or mythical personages, as an astronomical authority, are all the indications we find justifying the introduction of this name into the list of the Çabdakalpadruma.

II. In the second class we include the Gârga, Vyâsa, Pârâçara, Pâuliça, Pâulastya, and Vâsishṭha Siddhântas. Garga, Parâçara, Vyâsa, Pulastya, and Vasishṭha are prominent among the sages of the ancient period of Hindu history: the two latter are of the number of those who give name to the stars in Ursa Major (they are β and ζ respectively). They cannot possibly have been the veritable authors of Siddhântas, or works presenting the modern astronomical system of the Hindus: but—and this seems to be especially the case with regard to Garga and Parâçara—one and another of them may have distinguished themselves in connection with the older science, and so have furnished some ground for the part attributed to them by the later tradition, and for the fathering of astronomical works upon them.

1. *Garga-Siddhânta.* Astronomical treatises and commentaries upon them occasionally offer citations from Garga (see, for instance, Colebrooke's Essays, ii. 356; Sir William Jones in As. Res., ii. 397), but of a Siddhânta, or text-book of astronomy, bearing his name, we find nowhere any mention excepting in these lists.

2. *Vyâsa-Siddhânta.* This name, too, is known to us only from the list above given.

3. *Pârâçara-Siddhânta.* According to Bentley, the second chapter of the Ârya-Siddhânta contains an extract from this work, in which are stated the elements of the mean motions of the planets adopted by it. The work itself appears to be lost; unless, indeed, it may have been contained in a manuscript of the Mackenzie Collection, which in Wilson's Catalogue (i. 120) is called Vriddha-Parâsara, and said to be "a system of astrology, attributed to Parâsara, the father of Vyâsa."

4. *Pâuliça-Siddhânta.* The planetary elements of this treatise also are preserved in later commentaries, and are stated by Bentley and Colebrooke. We have noticed above (note to i. 4–6) that al-Bîrûnî attributes it to Paulus the Greek; whence Weber (Ind. Lit., p. 226) conjectures that it was founded upon the *Εἰσαγωγή* of Paulus Alexandrinus. If this account of its origin be correct, the Puliça to whom the later Hindus attribute it is a fictitious personage, whose name is manufactured out of Pâuliça. The work, it will be seen, is not mentioned in either of the lists we have given, its place appearing to be taken by the Pulastya-Siddhânta. According to the Hindu tradition, the school represented by the Pâuliça-Siddhânta was the rival of that of Âryabhaṭṭa.

5. *Pulastya-Siddhânta.* Of this Siddhânta we find mention only in such native lists as omit the preceding. Hence we are led to conjecture that the two names may indicate the same work; an attempt, founded upon the similarity of the names, having been made by some to attribute the Pâuliça-Siddhânta to a known and acknowledged Hindu sage.

6. *Vasishṭha-Siddhânta.* This work is spoken of as actually in existence by both Colebrooke and Bentley, and the latter states its system to correspond with that of the Sûrya-Siddhânta. More than one treatise bearing the name is referred to, the older one being of unknown authorship, and the other a later compilation founded upon this, by

Vishṇu-candra, who is said also to have derived his material in part from Âryabhaṭṭa. A copy of a Vṛddha-Vasishṭha-Siddhânta formed a part of the Mackenzie Collection (Wilson's Catalogue, i. 121).

III. To the third class may be assigned the Siddhântas of Âryabhaṭṭa, Varâha-mihira, and Brahmagupta, and the Romaka-Siddhânta, as well as the later version of the Vasishṭha-Siddhânta, last spoken of. The first three names are those of greatest prominence and highest importance in the history of Hindu astronomical science, and there is every reason to believe that the sages who bore them lived about the time when the modern system may be supposed to have received its final and fully developed form, or during the fifth and sixth centuries of our era.

1. *Ârya-Siddhânta.* The two principal works of Âryabhaṭṭa appear to have been originally entitled the Âryâshṭaçata, "work of eight hundred verses," and Daçagîtikâ, "work of ten cantos." Colebrooke knew neither of them excepting by citations in other astronomical text-books and commentaries. Bentley had in his hands two treatises which he calls the Ârya-Siddhânta and the Laghu-Ârya-Siddhânta, which are possibly identical with those above named.* The Berlin Library also contains (Weber, No. 834) a work which professes to be a commentary on the Daçagîtikâ.

2. *Varâha-Siddhânta.* The only distinctively astronomical work of Varâha-mihira appears to have been his Pañca-siddhântikâ, or Compendium of Five Astronomies, of which we have already spoken (note to i. 2–3), and which was founded upon the Brahma, Sûrya, Pâuliça, Vasishṭha, and Romaka Siddhântas. It is supposed to be no longer in existence, although the astrological works of the same author have been carefully preserved, and are without difficulty accessible.

3. *Brahma-Siddhânta.* The proper title of the work composed by Brahmagupta, upon the foundation of an earlier treatise bearing this name, is Brahma-sphuṭa-Siddhânta, "corrected Brahma-Siddhânta," but the word *sphuṭa*, "corrected," is frequently omitted in citing it, as has been our own usage in the notes to the Sûrya-Siddhânta. Colebrooke possessed an imperfect copy of it, and it was also in Bentley's possession. Upon it was professedly founded, in the main, the Siddhânta-Çiromaṇi of Bhâskara.

4. *Romaka-Siddhânta.* Of the name of this treatise, the only one we have thus far met with which is not derived from a real or supposed author, we have spoken in the note to i. 4–6. It is said by Colebrooke to be by Çrîsheṇa, and to have been founded in part upon the original Vasishṭha-Siddhânta; its early date is proved by its being one of those treated as authorities by Varâha-mihira. No copy of it seems to have have been discovered in later times.

Our list also mentions a Bhoja-Siddhânta, probably referring to some astronomical work published during the reign, and under the patronage, of Râja Bhoja Deva, of Dhârâ, in the tenth or eleventh century of our era.

---

* See an article by Fitz-Edward Hall, Esq., On the Ârya-Siddhânta, in the Journal of the American Oriental Society, vol. vi, 1860.

IV. Our fourth class is headed by the Siddhânta-Çiromaṇi, written in the twelfth century by Bhâskara Âcârya, and founded upon the Brahma-Siddhânta of Brahmagupta. Our numerous references to it and citations from it indicate the prominent and important position which it occupies in the modern astronomical literature of India. For a description of the numerous commentaries upon it, see Colebrooke's Hindu Algebra, note A (Essays, ii. 450 etc.).

The longer of the lists given above mentions two or three other works of yet later date. Among them the Siddhânta-Sundara is the most ancient, having been composed by Jñâna-râja at the beginning of the sixteenth century. The Graha-Lâghava is a treatise of the same class, and is highly considered and much used throughout India, although omitted from the Pûna list. It is of nearly the same date with the work last spoken of, being the composition of Ganeça, and dated *çake* 1442 (A. D. 1520). The Siddhânta Tattva-Viveka, more usually styled the Tattva-Viveka simply, is a century later: it was written by Kamalâkara, about A. D. 1620. The Siddhânta-Sârvabhâuma dates from very nearly the same period, and is the work of Muniçvara, who is also the author of a commentary on the Çiromaṇi, and the son of Ranganâtha, the commentator on the Sûrya-Siddhânta.

This class of astronomical writings might be almost indefinitely extended, but the works which have been mentioned appear to be the most authoritative and important.

Of all the treatises whose names we have cited, we know of but three which have as yet been published—the Sûrya-Siddhânta, the Siddhânta-Çiromaṇi, and the Graha-Lâghava; the two latter under the auspices of the School-Book Society of Calcutta. Prof. Hall's edition of the Sûrya-Siddhânta, to which reference is made in our Introductory Note, has been completed by the addition of a fourth Fasciculus since our own publication was commenced, so that we have been able to avail ourselves of its valuable assistance throughout.

2. p. ii. Ranganâtha, in the verses with which he closes his commentary, states it to have been completed on the same day with the birth of his son Muniçvara, in the *çâka* year 1525, or A. D. 1603. For his relationship to other well-known authors or commentators of astronomical treatises, see Colebrooke's Essays, ii. 452 etc. Other commentators on the Sûrya-Siddhânta mentioned by Colebrooke are Nṛsinha, who wrote but a few years later than Ranganâtha, and Bhûdhara and Dâdâ Bhâï, whose age is not stated. The Mackenzie collection (see Wilson's Catalogue, p. 118 etc.) contained commentaries on the whole or parts of the same text by Mallikârjuna, Yellaya, an Âryabhaṭṭa, Mammabhaṭṭa, and Tammaya.

3. p. iii. As no especially suitable opportunity has hitherto offered itself for giving in our notes the synonymy of the names of the planets, we present here all the appellations by which they are known in the text of the Sûrya-Siddhânta.

The sun is called by the following names derived from roots signifying "to shine"; *arka, bhânu, ravi, vivasvant, sûrya;* also *savitar*, literally "enlivener, generator"; *bhâskara*, "light-maker"; *dinakara* and *divâkara*, "day-maker"; and *tigmânçu* and *tîkshṇânçu*, "having hot or piercing rays."

The moon, besides her ordinary names *indu*, *candra*, *vidhu*, is styled *niçâkara*, "night-maker"; *niçâpati*, "lord of night"; *anushṇagu*, *çîtagu*, *çîtânçu*, *çîtadîdhiti*, *himaraçmi*, *himânçu*, *himadîdhiti*, "having cool rays"; and *çaçin* and *çaçânka*, "marked with a hare": the Hindu fancy sees the figure of this animal in the spots on the moon's disk. The name *soma* nowhere directly occurs, but it is implied in the title *sâumya* given to Mercury.

Mercury is styled *jña* and *budha*, "wise, knowing"; also *çaçija* and *sâumya*, "son of the moon." The reason of neither appellation is obvious. It will be seen below that the moon, the sun, and the earth have each of them one of the lesser planets assigned to it as its son: why Mercury, Saturn, and Mars were selected, and on what grounds their respective parentage was given them, is hitherto entirely unknown.

Venus has one name, *çukra*, "brilliant," which is derived from her actual character: she is also known as *bhṛgu*, which is the name of one of the most noted of the ancient sages, or as *bhṛguja* or *bhârgava*, "son of Bhṛgu."

Mars has likewise a single appellation, *angâraka*, "coal," which is given him on account of his fiery burning light: all his other titles, namely *kuja*, *bhûputra*, *bhûmiputra*, *bhûsuta*, *bhâuma*, mark him as "son of the earth."

Jupiter is known as *bṛhaspati*, which is, as already more than once noticed, the name of a divine personage, priest and teacher among the gods; the word means originally "lord of worship." The planet also receives some of his titles, namely *guru*, "preceptor," and *amarejya*, "teacher of the immortals." The only other name given to it, *jîva*, "living," is of doubtful origin.

Saturn has two appellations, each represented by several forms; namely "son of the sun," or *arkaja*, *ârki*, *sûryatanaya;* and "the slow-moving," or *manda*, *çani*, *çanâiçcara*.

All these names, it will be noticed, are of native Hindu origin, and have nothing to do with the appellations given by other nations to the planets. In the Hindu astrological writings, however, even those of a very early period (see Weber's Ind. Stud., ii. 261), appear, along with these, other titles which are evidently derived from those of the Greeks.

4. p. 2. We have everywhere cited Bentley's work on Hindu astronomy according to the London edition of it (8vo., 1825), the only one to which we have had access.

In a few instances, where we have not specified the part of Bhâskara's Siddhânta-Çiromaṇi to which we refer, the Gaṇitâdhyâya, or properly astronomical portion of it, is intended.

5. p. 17. For the convenience of any who may desire to make a more detailed examination of the elements of the mean motions of the planets adopted in this treatise, and to work out the results deducible from them, we present them in the following table in a more exact form. We give the mean time of sidereal revolution, in mean solar days, and the amount of mean motion, in seconds, during a day, and also during a Julian year, of 365¼ mean solar days.

Mean Motions of the Planets.

Planet.	Time of sidereal revolution.	Mean daily motion.	Mean yearly motion.
	d	"	"
Sun,	365.25875648	3,548.16956	1,295,968.931
Mercury,	87.96970228	14,732.34496	5,380,988.996
Venus,	224.69856755	5,767.72702	2,106,662.295
Mars,	686.99749394	1,886.46976	689,033.081
Jupiter,	4,332.32065235	299.14683	109,263.381
Saturn,	10,765.77307461	120.38151	43,969.346
Moon,			
sider. rev.,	27.32167416	47,434.86773	17,325,585.437
synod. rev.,	29.53058795	43,886.69817	16,029,616.507
apsis,	3,232.09367415	400.97848	146,457.389
node,	6,794.39983121	190.74532	69,669.730

6. p. 17. The system of the Sûrya-Siddhânta, so far as concerns the mean motions of the planets, the date of the last general conjunction, and the frequency of its recurrence, is also that of the Çâkalya-Sanhitâ. It is likewise presented, according to Bentley (Hind. Astr., p. 116), by the Soma and Vasishtha Siddhântas. So far as can be gathered from the elements of the Pâuliça and Laghu-Ârya Siddhântas, as reported by Colebrooke and Bentley, these treatises, too, followed a similar system; the revolutions of the planets in an Age, as stated by them, where they differ from those of the Sûrya-Siddhânta, always differ by a number which is a multiple of four. Some of the astronomical text-books, however, have constructed their systems in a somewhat different manner. Thus the Siddhânta-Çiromaṇi, following the authority of Brahmagupta and of the earlier Brahma-Siddhânta, makes the planets commence their motions together at the star ζ Piscium at the very commencement of the Æon, and return to a general conjunction at the same point only after the lapse of the whole period of 4,320,000,000 years. The same is the case with the Ârya and Pârâçara Siddhântas: they too, as reported by Bentley (Hind. Astr., pp. 148, 150), state the revolutions of the planets for the whole Æon only, and in numbers which have no common divisor, so that they assume no briefer cycle of conjunction. But they all, at the same time, take special notice of the commencement of the Iron Age, which they make to begin at the moment of mean sunrise at Lankâ, and manage to effect very nearly a general conjunction at the time of its occurrence, as is shown by the table at the end of this note, in which are presented the positions of all the planets, and of the moon's apsis and node, as stated by them for that moment.

We insert these data here, because they seem to us to furnish ground for important conclusions respecting the comparative antiquity of the two systems. The commencement of the Iron Age, which to the one is of cardinal importance as an astronomical epoch, is to the other simply a chronological era, having no astronomical significance. Now if, as has been shown in our notes to be altogether probable, that epoch is in fact of astronomical origin, being arrived at by retrospective calculation of the planetary motions, we can hardly avoid the conclusion that the system which presents it in its true character is the more ancient and original. This conclusion is strengthened by the notice taken of the epoch by the Siddhânta-Çiromaṇi and its kindred treatises. We do not see how their treatment of it is to be explained, excepting upon the supposition that a general conjunction at that time was already so firmly established as a fundamental dogma of the Hindu astronomy, that they were compelled, even while rejecting the theory of brief cycles and recurring conjunctions, to pay it homage by so constructing their elements that these should exhibit at least a very near approach to a conjunction at the moment. We are clearly of opinion, therefore, that, apart from all consideration of the relative age of the separate treatises, the system represented by the Sûrya-Siddhânta is the more ancient.

Mean Places of the Planets, 6 *o'c A.M. at Ujjayinî, Feb.* 18*th, B.C.* 3102.

Planet.	Siddhânta-Çiromaṇi.				Ârya-Siddhânta.				Pârâçara-Siddhânta.			
	s	°	'	''	s	°	'	''	s	°	'	''
Sun,	0	0	0	0	0	0	0	0	0	0	0	0
Mercury,	11	27	24	29	11	21	21	36	11	21	17	17
Venus,	11	28	42	14	11	27	7	12	11	26	58	34
Mars,	11	29	3	50	0	0	0	0	11	29	14	38
Jupiter,	11	29	27	36	11	27	7	12	11	27	2	53
Saturn,	11	28	46	34	0	0	0	0	11	28	57	22
Moon,	0	0	0	0	0	0	0	0	0	0	10	48
" apsis,	4	5	29	46	4	3	50	24	4	5	12	29
" node,	5	3	12	58	5	2	38	24	5	2	49	12

7. p. 20. We present in the annexed table, in the same form as above (note 5), the elements of the mean motions of the planets as corrected by the *bîja.*

Mean Motions of the Planets, as corrected by the bîja.

Planet.	Time of sidereal revolution.	Mean daily motion.	Mean yearly motion.
	d	''	''
Mercury,	87.96978075	14,732.33182	5,380,984.196
Venus,	224.69895152	5,767.71717	2,106,658.695
Jupiter,	4,332.41581277	299.14026	109,260.981
Saturn,	10,764.89171783	120.39136	43,972.946
Moon's apsis,	3,232.12015592	400.97519	146,456.189
" node,	6,794.28280845	190.74861	69,670.930

8. p. 22. At the time when we wrote our note, we had not observed that Bentley himself explains, in a foot-note to page 117 of his work, this apparent error. In the case of Mercury, since the number of revolutions as stated by the text of our treatise did not yield him the result which he desired, he has quietly taken the liberty of altering it from 17,937,060 to 17,937,024, assuming, as his justification, an error of the copyists which has not the slightest plausibility, and ignoring the fact that the correctness of the former number is avouched by its occurrence in other treatises. It is highly characteristic of Bentley, that he has thus arbitrarily amended one of the data upon which he rests the most important of his general conclusions, a conclusion which, but for such emendation, would be not a little weakened or modified. Any one can see for himself, upon referring to our table given on page 44, with how much plausibility Bentley is able to deduce, from the dates of its fourth column, the year A.D. 1091 as that of the composition of the Sûrya-Siddhânta. We have been solicitous to allow Bentley all the credit we possibly could for his labors upon the Hindu astronomy, but we cannot avoid expressing here our settled conviction that, as an authority upon the subject, he is hardly more to be trusted than Bailly himself, that his work must be used with the extremest caution, and that his determination of the successive epochs in the history of astronomical science in India is from beginning to end utterly worthless.

9. p. 23. We have not fulfilled our promise to recur in the eighth chapter to the subject of the sun's error of position, because we felt ourselves incompetent to cast at present any valuable light upon it. Nothing but a careful and thorough sifting and comparison of all the earliest treatises, together with the traditions preserved by the commentators, and the practical methods of construction of the calendar, is likely to settle the question as to the manner in which the elements of the planetary orbits were originally made up.

10. p. 24. In making out our comparative table of sidereal revolutions, we have calculated the column for Ptolemy as we conceive that he would himself have calculated it, had he been called upon to do so. M. Biot, having in view an object different from ours, has carefully revised Ptolemy's processes (see his Traité Élémentaire d'Astronomie Physique, 3me éd., v. 37–71), and has deduced from the latter's original data what he regards as the true times of sidereal revolution of the primary planets furnished by them; his periods are accordingly slightly different from those presented in our table.

Colebrooke (As. Res., xii. 246; Essays, ii. 412) has also given a comparative table of the daily motions of the planets, but has committed in it the gross error of setting side by side the sidereal rates of motion of the Hindu text-books and the tropical rates of Ptolemy and Lalande. Of course, his data being incommensurable, the conclusions he draws from their comparison are erroneous.

11. p. 27. We add, in the following table, a comparison of the positions of the apsides and nodes of the planets as stated in our treatise—being those which are adopted, with unimportant variations, by all the schools of Hindu astronomy—with those laid down by Ptolemy in his Syntaxis. The latter we give as stated by Ptolemy for his own period, without reducing them to their value in distances from the initial point of the Hindu sphere. The actual distance of that point, or of the vernal equinox of A.D. 560, from the vernal equinox of Ptolemy's time, is about 5½°. We should remark also that Ptolemy does not state expressly and distinctly the positions of the nodes: we derive them from the rules given by him, in the sixth chapter of his thirteenth Book, for calculating the latitude of the planets: not being, however, altogether confident of our correct understanding and interpretation of those rules.

Positions of the Apsides and Nodes of the Planets.

Planet.	Sûrya-Siddhânta.		Ptolemy.		Difference.	
Apsides:	°	′	°	′	°	′
Sun,	77	15	65	30	+ 11	45
Mercury,	220	26	190	0	+ 30	26
Venus,	79	49	55	0	+ 24	49
Mars,	130	1	115	30	+ 14	31
Jupiter,	171	16	161	0	+ 10	16
Saturn,	236	38	233	0	+ 3	38
Nodes:						
Mercury,	20	44	10	0	+ 10	44
Venus,	59	45	55	0	+ 4	45
Mars,	40	4	25	30	+ 14	34
Jupiter,	79	41	51	0	+ 28	41
Saturn,	100	25	183	0	− 82	35

It will be perceived that the differences here are not so great as to exclude the supposition of a connected origin. We do not ourselves believe that the Hindus were ever sufficiently skilled in observation, or in the discussion of the results of observation, to be able to derive such data for themselves, or even intelligently to modify and improve them, when obtained from other sources. In order, however, fully to understand the relation of the Hindu to the Greek science in this part, we require to know, first, what were the positions assigned to the apsides and nodes by Greek astronomers prior to Ptolemy, and secondly, what were their actual positions at the periods in question. Upon the first point no information appears to have been handed down to our times; and as regards the other, we have not found any modern determination of the desired data, and are not ourselves at present in a situation to undertake so intricate and laborious a calculation.

12. p. 29. The era of the *kali yuga*, or Iron Age, is not in practical use among the Hindus of the present day: two others, of a less remote date, are ordinarily employed by them in the giving of dates. These are styled the eras of Çâlivâhana and of Vikramâditya respectively, from two sovereigns so named: their origin and historical significance are matters of much doubt and controversy. The years of the era of Çâlivâhana are, according to Warren (Kâla Sankalita, p. 381 and elsewhere), solar years: their reckoning commences after the lapse of 3179 complete years of the Iron Age, or early in April, A. D. 78: the 1782nd year, accordingly, coinciding with the 4961st of the Iron Age, commenced, as is shown by the table on p. 30, April 12th, 1859, and ended April 11th, 1860. The years of this era are generally cited as *çaka* or *çâka* years. In the other era, the luni-solar reckoning is followed (Warren, as above, p. 391 and elsewhere); and its first year began with the 3045th of the Iron Age, or early in 58 B. C.: its 1962nd year, coinciding with the 4961st of the remoter era, commenced (see table on p.

30) April 4th, 1859, and ended March 22nd, 1860. The years of this era are called and quoted as *samvatsara* years, or, by abbreviation, simply *samvat.*

13. p. 39. M. Vivien de St. Martin (in Julien's Mémoires de Hiouen-Thsang, ii. 258) supposes the value of the *li* in use in China during the seventh century to have been about 329 metres, or 1080 English feet. This would make the values of the three kinds of yojana mentioned by the Buddhist traveller to be $8\frac{1}{6}$, $6\frac{1}{8}$, and $3\frac{1}{8}$ English miles respectively.

14. p. 44. In the first table upon this page, we have, by an oversight, given the earth's heliocentric longitude, instead of the sun's geocentric longitude. To the sun's place as stated, accordingly, should be added 180°.

15. p. 52. M. Biot (Journal des Savants, 1859, p. 409) suggests that the Hindus, like Albategnius, obtained their sines directly from the chords of Hipparchus or Ptolemy. This may not be an altogether impossible supposition, but it is at least an unnecessary one, for they certainly had geometry enough, at the time of the elaboration of their astronomical system, to construct their table independently. Our notes have presented Delambre's view of the method of its construction and the reason of its limitation to arcs which are multiples of 3° 45′. We cannot but feel, however, upon maturer consideration, that the correctness of that view is very questionable; that the Hindus could probably have made out a more complete table if they had chosen to do so; and that a sufficient reason is found for their selection of the arc of 3° 45′ in the fact that it is a natural subdivision of a recognized unit, the arc of 30°, while the series of twenty-four sines was sufficiently full and accurate for their uses. We have been at the pains to calculate the complete series of Hindu sines from Ptolemy's table of chords, assuming the value of radius to be 3438′, in order to test the question whether there were any correspondence of errors between them which should prove the one to be derived from the other: our results are as follows. In five of the instances (the 14th, 15th, 19th, 22nd, and 23rd sines of the table) in which the value of the Hindu sine exceeds the truth, Ptolemy supports the error; in the other three cases (the 16th, 17th, and 18th sines), Ptolemy affords the correct value; to the 6th sine, also, which by the Hindus is made too small, Ptolemy's table gives its true value, but the next following sine he makes too great (namely 1520.59, which would give 1521, instead of 1520); this is his only independent error. The evidence yielded by the comparison may be regarded as not altogether unequivocal.

For the benefit of any who may desire to make practical use of the Hindu sines, in calculations conducted according to the processes of the Sûrya-Siddhânta, we give, upon the opposite page, a more detailed table of them than has been presented hitherto, with such sets of differences annexed as will enable the calculator readily to find the sine of any given arc, or the reverse, without resorting to the laborious proportions by which the text contemplates that they should in each case be determined. Such a table we have ourselves found highly useful, and even almost indispensable, in connection with our own calculations.

Table of Hindu Sines, with Differences.

Arc.	Sine.	Diff.	Arc.	Sine.	Diff.	Arc.	Sine.	Diff.
° ′	′	′ ′	° ′	′	′ ′	° ′	′	′ ′
0	0		30	1719	1 0.849 2 1.698 3 2.547 4 3.396 5 4.244	60	2978	1 0.471 2 0.942 3 1.413 4 1.884 5 2.356
1	60		31	1769.93		61	3006.27	
2	120	1 1.000	32	1820.87		62	3034.58	
3	180		33	1871.80		63	3062.80	
3 45	225		33 45	1910		63 45	3084	
4	239.93	1 0.996 2 1.991 3 2.987 4 3.982 5 4.978	34	1922.20	1 0.813 2 1.627 3 2.440 4 3.253 5 4.067	64	3090.20	1 0.413 2 0.827 3 1.240 4 1.653 5 2.067
5	299.67		35	1971		65	3115	
6	359.40		36	2019.80		66	3139.80	
7	419.13		37	2068.60		67	3164.60	
7 30	449		37 30	2093		67 30	3177	
8	478.60	1 0.987 2 1.973 3 2.960 4 3.947 5 4.933	38	2116.20	1 0.773 2 1.547 3 2.320 4 3.093 5 3.867	68	3187.53	1 0.351 2 0.702 3 1.053 4 1.404 5 1.756
9	537.80		39	2162.60		69	3208.60	
10	597		40	2209		70	3229.67	
11	656.20		41	2255.40		71	3250.74	
11 15	671		41 15	2267		71 15	3256	
12	714.80	1 0.973 2 1.947 3 2.920 4 3.893 5 4.867	42	2299.80	1 0.729 2 1.458 3 2.187 4 2.916 5 3.644	72	3269	1 0.289 2 0.578 3 0.867 4 1.156 5 1.444
13	773.20		43	2343.52		73	3286.33	
14	831.60		44	2387.27		74	3303.67	
15	890		45	2431		75	3321	
16	947.33	1 0.956 2 1.911 3 2.867 4 3.822 5 4.778	46	2472.07	1 0.684 2 1.369 3 2.053 4 2.738 5 3.422	76	3334.60	1 0.227 2 0.453 3 0.680 4 0.907 5 1.133
17	1004.67		47	2513.14		77	3348.20	
18	1062		48	2554.21		78	3361.80	
18 45	1105		48 45	2585		78 45	3372	
19	1119	1 0.933 2 1.867 3 2.800 4 3.733 5 4.667	49	2594.53	1 0.636 2 1.271 3 1.907 4 2.542 5 3.178	79	3374.47	1 0.164 2 0.329 3 0.493 4 0.658 5 0.822
20	1175		50	2632.67		80	3384.33	
21	1231		51	2670.80		81	3394.20	
22	1287		52	2708.93		82	3404.07	
22 30	1315		52 30	2728		82 30	3409	
23	1342.33	1 0.911 2 1.822 3 2.733 4 3.644 5 4.556	53	2745.47	1 0.582 2 1.164 3 1.747 4 2.329 5 2.911	83	3411.93	1 0.098 2 0.196 3 0.293 4 0.391 5 0.489
24	1397		54	2780.40		84	3417.80	
25	1451 67		55	2815.33		85	3423.67	
26	1506.33		56	2850.27		86	3429 53	
26 15	1520		56 15	2859		86 15	3431	
27	1559.80	1 0.884 2 1.769 3 2.653 4 3.538 5 4.422	57	2882.80	1 0.529 2 1.058 3 1.587 4 2.116 5 2.644	87	3432.40	1 0.031 2 0.062 3 0.093 4 0.124 5 0.156
28	1612.87		58	2914.53		88	3434.27	
29	1665.94		59	2946.26		89	3436.13	
30	1719		60	2978		90	3438	

In explaining how the Hindus may have arrived at their empirical rule, as laid down in verses 15 and 16, for the development of the series of sines, we have also, as mentioned in our note, followed the guidance of Delambre. Prof. Newton, however, is of opinion that the rule in question was probably obtained by direct geometrical demonstration, in some such method as the following, which is much more in accordance with the mathematical processes exhibited or implied in other parts of the Sûrya-Siddhânta.

In the quadrant A B (Fig. 34), let B F, B D, and B E be three arcs, of which each exceeds its predecessor by the equal increment D F or D E; and let F *m*, D *l*, and E *k* be their sines, increasing by the unequal differences D *h* and E *g*. Now as E D and D F are small arcs (they are shown in the figure of three times the proportional length of the arcs of difference of the Hindu table), E D *g* and D F *h* may be regarded as plane triangles, and the angles made by C D at D as right angles: hence the angles E D *g* and C D *l* are equal, the triangles E D *g* and C D *l* are similar, and E D : E *g* : : C D : C *l*; or E *g* = E D.C *l* ÷ C D. In like manner, D *h* = E D.C *m* ÷ C D. Therefore D *h* – E *g* = E D.*l m* ÷ C D; and E *g*, which is the amount by which E *k* exceeds D *l*, equals D *h* – (E D.*l m* ÷ C D). But, by similarity of the triangles C D *l* and D F *h*, F *h*, or *l m*, equals E D.D *l* ÷ C D; and hence E D.*l m* ÷ C D = (E D² ÷ C D²) D *l*, or (E D ÷ C D)² D *l*. Now when E D equals 225′ and C D 3438′, E D ÷ C D = $\frac{1}{15}$ nearly (or exactly $\frac{1}{15.28}$), and (E D ÷ C D)² = $\frac{1}{225}$ nearly (more exactly, $\frac{1}{233.48}$). Hence E *k* = D *l* + D *h* – $\frac{1}{225}$ D *l*, which is equivalent to the Hindu rule.

Fig. 34.

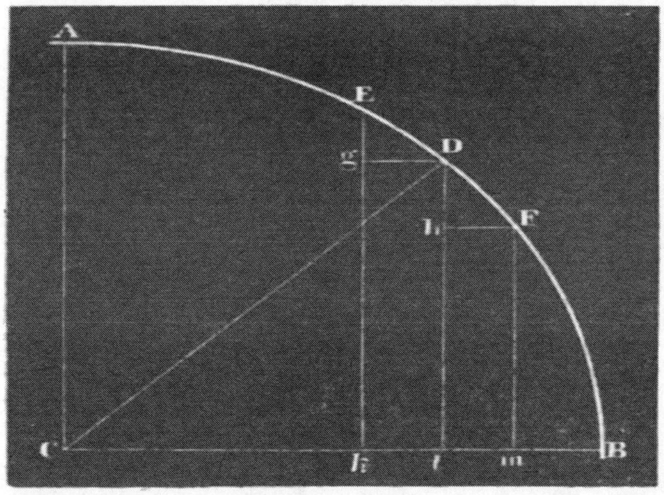

When we wrote the note to the passage of the text relating to the sines, we assumed that the rule as there stated would give the series of sines, having found upon trial that it held good for the first few terms of the series. But, it having been pointed out to us by Prof. Newton that the adoption of $\frac{1}{15}$ as the value of E D ÷ C D could not but lead to palpably erroneous results, we carried our calculations farther, and found that only five of the sines following the first one can be deduced from it by the processes prescribed; that with the seventh sine begins a discordance between the table and the result of calculation by the rule, which goes on increasing to the end, where it amounts to as much as 70′ in the value obtained for radius.

This untoward circumstance, which may be regarded as a trait highly characteristic of a Hindu astronomical treatise, seems to us rather to favor the opinion that the rule is the result of construction and demonstration, and not empirically deduced from a consideration of the actual second differences. In the latter case we should more naturally suppose that it would have been tested throughout by actual trial; while, if it had been arrived at in the manner above explained, an application of it to the first few members only of the series might more easily have been accepted as a sufficient test of its correctness.

16. p. 59. We are not sure that the name *bhuja* may not originally and properly belong rather to the arc than to its chord or sine. It comes from a root *bhuj*, "bend," and signifies primarily "a bend, curve," being applied also to designate the arm on account of the latter's suppleness or flexibility. The word *koṭi* also most frequently means "the end or horn of a bow." We might, then, look upon the relations of the arc (*dhanus, câpa, kârmuka*) and its parts and appurtenances as follows. The whole arc taken into account is (Fig. 2, p. 59) QRS: of this, BRC is the *bhuja*, curve or bow proper, while BQ and CS are its two *koṭis* or horns: BC is the chord or bow-string (*jyâ* etc.), or, more distinctively, the *bhujajyâ;* which name, by substitution for *jyârdha*, is also applied to either of its halves, BH or HC: BF or CL is in like manner the *koṭijyâ;* RH, finally, the versed sine, is the "arrow" (*çara, ishu*); by this name it is often known in other treatises, although not once so styled in this Siddhânta. If this view be correct, the terms *bhuja* and *koṭi* as applied to the base and perpendicular of a right-angled triangle, are given them on account of their relation to one another as sine and cosine, while the synonyms of *bhuja*, namely *bâhu* and *dos*, are employed on account only of their agreement with it in the signification "arm," and not in that which gives it its true application. For *koṭi* the treatise affords no synonyms.

17. p. 63. M. Delambre, in his History of Ancient Astronomy (i. 462 etc.), has subjected to a detailed examination the rules of the Sûrya-Siddhânta for the calculation of the equations of the centre for the sun and moon, has reduced them to a single formula, and has calculated for each degree of a quadrant the values of the equations, comparing them with those furnished by the Hindu tables, as reported by Davis (As. Res., ii. 255–256). M. Biot has more recently, in the Journal des Savants for 1859 (p. 384 etc.), taken up the same subject anew, especially pointing out, and illustrating by figures and calculations, the error of the Hindus in assuming the variation of the equation to be the same in all the four quadrants of mean revolution.

18. p. 76. Neither Delambre nor Biot (both as above cited), nor any other western savant who has treated of the Hindu astronomy, has found any means of accounting for the variation of dimensions of the planetary epicycles. In its present form and extent, indeed, it seems to defy explanation: we can only conjecture that it may be an unintelligent and reasonless extension to all the planets, and to both classes of epicycles, of a correction originally devised and applied only in one or two special cases. According to Colebrooke (As. Res., xii. 235 etc.; Essays, ii. 400 etc.), there is discordance among the different Hindu authorities upon this point. Âryabhaṭṭa agrees with the Sûrya-Siddhânta throughout; Brahmagupta and Bhâskara make the epicycles only of Venus and Mars variable; Muniçvara, in the Siddhânta-Sârvabhâuma, regards all the epicycles as invariable.

19. p. 92. Our suggestion of a possible derivation of the term *yoga* from the "sum" of the longitudes of the sun and moon is unquestionably erroneous. That term is to be understood here in the sense of "junction, conjunction," and the conception upon which is founded its application to the periods in question is that of a conjunction (*yoga*) of the moon with the twenty-seven asterisms (*nakshatra*) in their order, or her successive continuance in their respective portions. Only the system is divorced from any actual connection with the asterisms; for while the latter are stellar groups, having fixed positions in the heavens, they are here treated as if the twenty-seven-fold division of the ecliptic founded upon them had no natural limits, but was to be reckoned from the actual position of the sun at any given moment.

According to Warren (Kâla Sankalita, p. 74), the names of the twenty-seven yogas, as given by us on page 92, are also applied by the Hindus to the junction-stars (*yogatârâ*) of the asterisms (with the omission, of course, of Abhijit): for which see the notes to the eighth chapter. This fact we do not find noticed elsewhere; possibly the usage is a local one only.

Of the twenty-eight yogas of the other system, to which the Sûrya-Siddhânta makes no reference, the names are given by Colebrooke as follows:

1. Ânanda.	10. Mudgara.	19. Siddhi.
2. Kâladaṇḍa.	11. Chattra.	20. Çubha.
3. Dhûmra.	12. Mâitra.	21. Amṛta.
4. Prajâpati.	13. Mânasa.	22. Musala.
5. Sâumya.	14. Padma.	23. Gada.
6. Dhvânksha.	15. Lambaka.	24. Mâtanga.
7. Dhvaja.	16. Utpâta.	25. Râkshasa.
8. Çrîvatsa.	17. Mṛtyu.	26. Cara.
9. Vajra.	18. Kâṇa.	27. Sthira.
		28. Pravardha.

Colebrooke says farther: "The foregoing list is extracted from the Ratnamâlâ of Çrîpati. He adds the rule by which the yogas are regulated. On a Sunday, the nakshatras answer to the yogas in their natural order; viz. Açvinî to Ânanda, Bharaṇî to Kâladaṇḍa, etc. But, on a Monday, the first yoga (Ânanda) corresponds to Mṛgaçiras, the second to Ârdrâ, and so forth. On a Tuesday, the nakshatra which answers to the first yoga is Âçleshâ; on Wednesday, Hasta; on Thursday, Anurâdhâ; on Friday, Uttara-Ashâḍhâ; and on Saturday, Çatabhishaj."

This is by no means a clear and sufficient explanation of the character and use of the system, yet we seem to see distinctly from it that this, no less than the other system, is cut off from any actual connection with the twenty-eight asterisms, since the succession of the yogas is made to depend upon the day of the week, while the week stands in no constant and definable relation to the motion of the moon.

20. p. 102. In stating that the Sûrya-Siddhânta furnished no hint of the precession excepting in this passage, we failed to notice that in one other place, namely in connection with the rules for finding the time when the declinations of the sun and moon are equal (xi. 6), the precession is distinctly ordered to be calculated, and in terms which contain an evident reference to those in which the fact of the precession is here stated. The exception, however, is one which goes to prove, rather than overthrow, the general rule: the process in which we are for once favored with explicit directions upon the point in question is the one of all others in the work the most trivial, and the chapter which contains it furnishes, as pointed out by us in the notes, good reason to suspect late alterations and interpolations. We do not, then, regard the statement made in our note as requiring to be either retracted or seriously modified. Nor do we, although fully appreciating the difficulty of assuming that the original elaborators of the general Hindu system can have been ignorant of, or ignored, the precession, regret the force and distinctness with which we have stated the circumstances which appear to favor that assumption. Whether it be true or false, there is much in connection with the subject which is strange, and demands explanation: and that can only be satisfactorily given when there shall have been attained a more thorough comprehension of the early history and the varying forms of the science in India.

21. p. 114. The commentary frequently styles the sine of altitude *mahâçanku*, "great gnomon," to distinguish it from the *çanku*, "gnomon."

22. p. 131. Our statement that the Sûrya-Siddhânta employs only the term *graha* to designate the planets requires a slight modification. In one instance (ii. 69) they are called *khacârin*, and in one other (ix. 9) *khacara*, both words signifying "moving in the ether" (see xii. 23, 81).

23. p. 138. This use of the word *prâcî*, "east, east point," appears to be taken from the projections of eclipses, as directed to be drawn in the sixth chapter. Thus, in the figure there given (Fig. 27, p. 157), E M and *v* M represent the directions of the equator and ecliptic with reference to one another at the moment of first contact, and E and *v* are the east-points (*prâcî*) of those lines respectively: the arc E *v*, or the "interval of the two east-points," is the measure of the angle which the two lines make with one another at the given time.

24. p. 141. As promised above, we present here, by way of appendix to the fourth chapter of our translation and notes, a

Calculation, according to the Data and Methods of the Sûrya-Siddhânta, of the Lunar Eclipse of February 6th, 1860,

for the latitude and longitude of Washington.

Bailly, in his work on the Hindu astronomy (p. 355 etc.), presents several calculations of eclipses by Hindu methods, namely of the lunar eclipse of July 29th, 1730, of the lunar eclipse of June 17th, 1704, and of the solar eclipse of Nov. 29th, 1704. But, owing to his imperfect comprehension of the character and meaning of many of the processes, and owing to his incessant use of Hindu terms in the most barbarous transcriptions, without explanations, his intended illustrations are only with difficulty intelligible, and are exceedingly irksome to study. Davis, in his first valuable article in the Asiatic Researches (ii. 273 etc.), has also furnished a calculation of a lunar eclipse, as made by native astronomers, comparing their results, obtained by several different methods, with the actual elements of the eclipse, as given by the Nautical Almanac. As it seemed desirable to give a like practical illustration of the Hindu methods of calculation, in connection with this fuller exposition of their foundation and meaning, and by way of an additional test of the accuracy of the results which the system is in condition to furnish, we have selected for the purpose the partial eclipse of the moon which occurred on the evening of Feb. 6th, 1860. Our calculations are made according to the elements of our text alone, without adding, like Davis, the correction of the *bîja*, since our object is to illustrate the text itself, and not the modern system as altered from it. The course of the successive steps of our processes may not everywhere strictly accord with that which would be pursued by a native astronomer, as we take the rules of the text and apply them according to our own conception of their connection.

We omit the preliminary tentative processes, and conceive ourselves to have ascertained that, at the time of full moon in the month Mâgha, I. A. 4961 (see page 30), or *samvat* 1917 (see add. note 12), the moon will be eclipsed.

I. To find the sum of days (*ahargaṇa, dinarâçi*) for mean midnight next preceding full moon.

The sixth day of February, 1860, being the day of full moon (*pûrṇimâ*), is the fifteenth day of the first, or light, half of the lunar month Mâgha, the eleventh month of the year, as is shown by the table on page 30. The time, then, for which we are to find the sum of days, is $4960^{y}\ 10^{m}\ 14^{d}$, reckoning (i. 56) only from the commencement of the Iron Age. For this period the sum of days, as found by the processes already sufficiently illustrated in the notes to i. 48–51, is 1,811,981 days.

II. To find the mean longitude of the sun and moon, and of the moon's apsis.

The proportions (i. 53)

$$1{,}577{,}917{,}828 : 1{,}811{,}981 :: \begin{cases} 4{,}320{,}000 : 4960^{rev}\ 9^{s}\ 23^{\circ}\ 17'\ 1'' \\ 57{,}753{,}336 : 66{,}320^{rev}\ 3^{s}\ 9^{\circ}\ 44'\ 19'' \\ 488{,}203 : 561^{rev}\ 1^{s}\ 13^{\circ}\ 43'\ 1'' \end{cases}$$

give us—rejecting whole revolutions, and deducting 3^{s} from the motion of the moon's apsis, for its position at the epoch (see note to i. 56–58)—the mean longitudes required. These are for the time of mean midnight at Ujjayinî: to find them for mean midnight at Washington, which is distant from Ujjayinî $1671^{y}.28$, upon a parallel of latitude $3936^{y}.75$ in circumference (note to i. 63–65), we add to the position of each $\frac{1671.28}{3936.75}$ or .42453 of its mean motion during a sidereal day. This correction is styled the *deçântaraphala*. We have, then,

	Long. at Ujjay.		Correction.		Long. at Wash'n.
Sun,	$9^{s}\ 23^{\circ}\ 17'\ 1''$	+	$25'\ 2''$	=	$9^{s}\ 23^{\circ}\ 42'\ 3''$
Moon,	$3^{s}\ 9^{\circ}\ 44'\ 19'$	+	$5^{\circ}\ 34'\ 43''$	=	$3^{s}\ 15^{\circ}\ 19'\ 2''$
Moon's apsis,	$10^{s}\ 13^{\circ}\ 43'\ 1''$	+	$2'\ 50''$	=	$10^{s}\ 13^{\circ}\ 45'\ 51''$

The place of the sun's apsis remains as already found for Jan. 1st (note to ii. 39):

Longitude of sun's apsis,	2ˢ 17° 17′ 24″

In applying here the correction for difference of meridian, as well as in all other processes of the whole calculation into which the amounts of motion of the planets etc. during fractions of a day enter as elements, we have derived those amounts from the motions during a sidereal day, and not, as in the illustrative processes of our notes, during a mean solar day. The divisions of the day given in the text (i. 11–12) are distinctly stated to be those of sidereal time, and all the rules of the treatise are constructed accordingly (see, for instance, ii. 59). It is evident, then, that in making any proportion in which is involved the amount of motion during 60 nâḍîs, that amount is to be regarded as the motion during a sidereal day only. In overlooking in our notes the difference between the two, we have followed the example of all the illustrations of Hindu methods of calculation known to us. The difference is, indeed, in a Hindu process, of very small account; but we have preferred, in making this calculation, to follow what we conceive to be the exacter method. The mean motions during a sidereal day of the bodies concerned in a lunar eclipse are as follows:

Sun,	58′ 58″ 28‴ 55⁗
Moon,	13° 8′ 25″ 21‴ 21⁗
Moon's apsis,	6′ 39″ 53‴ 1⁗
Moon's node,	3′ 10″ 13‴ 28⁗

III. To find the true longitudes and motions of the sun and moon:

1. To find the sun's true longitude (note to ii. 39):

Longitude of sun's apsis,	2ˢ 17° 17′ 24″
deduct sun's mean longitude (ii. 29),	9ˢ 23° 42′ 3″
Sun's mean anomaly (*kendra*),	4ˢ 23° 35′ 21″
Arc determining the sine (*bhuja*—ii. 30),	36° 25′
Sine of sun's mean anomaly (*bhujajyâ*),	2040′
Corrected epicycle (ii. 38),	13° 48′
Equation (*bhujajyâphala*—ii. 39),	+ 1° 18′
add to sun's mean longitude,	9ˢ 23° 42′
Sun's true longitude,	9ˢ 25° 0′

2. To find the moon's true longitude (note to ii. 39):

Longitude of moon's apsis,	10ˢ 13° 45′ 51″
deduct moon's mean longitude,	3ˢ 15° 19′ 2″
Moon's mean anomaly,	6ˢ 28° 26′ 49″
Arc determining the sine,	28° 27′
Sine of moon's mean anomaly,	1637′
Corrected epicycle,	31° 50′
Equation,	− 2° 25′
deduct from moon's mean longitude,	3ˢ 15° 19′
Moon's true longitude,	3ˢ 12° 54′

3. To find the sun's true rate of motion (ii. 48–49):

Sun's mean motion in 60 nâḍîs,		58′ 58″
Sine of sun's mean anomaly,	2040′	
Difference of sines,	183′	
Daily increase of sine of anomaly,		47′ 58″
Equation of motion,	+	1′ 50″
add to sun's mean motion,		58′ 58″
Sun's true motion,		60′ 48″

4. To find the moon's true rate of motion (ii. 47–49):

Moon's mean motion in 60 nâḍîs,		788′ 25″
deduct motion of apsis (ii. 47),		6′ 40″
Daily increase of moon's mean anomaly,		781′ 45″
Sine of moon's mean anomaly,	1637′	
Difference of sines,	199′	
Daily increase of sine of anomaly,		691′ 25″
Equation of motion,	+	61′ 8″
add to moon's mean motion,		788′ 25″
Moon's true motion,		849′ 33″

IV. To find the interval between the given instant of midnight and the end of the half-month, or the moment of opposition in longitude of the sun and moon, which is the middle of the eclipse.

At the instant of mean midnight preceding full moon, we have found the true longitudes of the sun and moon, and their distance in longitude, to be as follows:

Sun's true longitude,	$9^s\ 25^\circ\ 0'$
Moon's do.,	$3^s\ 12^\circ\ 54'$
Distance in longitude,	$6^s\ 12^\circ\ 6'$

Hence we see that the moon has still 12° 6′ to gain upon the sun. We have also found their true rates of motion, and the difference of those rates, to be as follows:

Moon's true motion,	849′ 33″
Sun's do.,	60′ 48″
Moon's daily gain,	788′ 45″

Now we make the proportion: if the moon in 60 nâḍîs gains upon the sun 788′ 45″, in how many nâḍîs will she gain her present distance in longitude from the sun? or

$$788'\ 45'' : 60^n :: 726' : 55^n\ 13^v\ 3^p$$

It thus appears that the time of opposition is $55^n\ 13^v\ 3^p$ after mean midnight of Feb. 5–6. This result, however, requires correction, for the moon's motion has become sensibly accelerated during so long an interval, and we find, upon calculation, that she is then 2′ past the point of opposition. A repetition of the same process shows that it is necessary to deduct $10^v\ 3^p$ from the time stated. Then, at $55^n\ 3^v$ after mean midnight, we have as follows:

Sun's mean longitude,	$9^{s}\ 24^{\circ}\ 36'$
Equation of place,	$+\ \ 1^{\circ}\ 20'$
Sun's true longitude,	$9^{s}\ 25^{\circ}\ 56'$
Moon's mean longitude,	$3^{s}\ 27^{\circ}\ 22'$
Longitude of apsis,	$10^{s}\ 13^{\circ}\ 52'$
Equation of moon's place,	$-\ \ 1^{\circ}\ 26'$
Moon's true longitude,	$3^{s}\ 25^{\circ}\ 56'$

By the same process as before, the true motions of the two planets at the moment of opposition are found to be :

Sun's true motion,	60′ 48″
Moon's do.	854′ 36″

It would have been better to adopt, as the starting-point of our calculations, the mean midnight following, instead of that preceding, the opposition of the sun and moon, because in that case, the interval to the moment of opposition being so much less, it might have been found by a single process, not requiring farther correction. The same change would have enabled us to follow strictly the rule given in ii. 66 for finding the end of the lunar day ; which rule we were obliged above to apply in a somewhat modified form, because a little more than one whole lunar day was found to intervene between the given midnight and the moment of opposition.

V. To determine the instant of local time corresponding to the middle of the eclipse.

What we have thus far found is the interval between mean midnight and the moment of opposition. But since Hindu time is practically reckoned from true sunrise to true sunrise, we have now, in order to determine at what time the eclipse will take place, to ascertain the interval between mean midnight and true sunrise.

In order to this, we require first to know the equation of time, or the difference between mean midnight and true or apparent midnight, which is the moment when the sun actually crosses the inferior meridian. As concerns this correction, we have deviated somewhat from the method contemplated by the text. It is there prescribed (ii. 46) that, so soon as the sun's equation of the centre has been determined, there should at once be calculated from it, and applied to the longitude of the two planets, a correction representing, in terms of their motion, the equation of time ; so that the distance of the moment of opposition from mean midnight does not directly enter into account at all. We have preferred to follow the course we have taken, in order to bring out and illustrate more fully the utter inadequacy of the prescribed method of making allowance for the equation of time, to which we have already briefly referred in the note to ii. 46. The method in question is virtually as follows : the sun being found at the given midnight to be 1° 18′, or 78′, in advance of his mean place, the equation of time may be ascertained by this proportion : as a whole circle is to a sidereal day, so is the sun's equation of place to the time by which his true transit will precede or follow his mean transit ; or, in the present case,

$$21,600' : 60^{n} :: 78' : 0^{n}\ 13^{v}$$

which gives us 13 vinâḍîs, or $5\frac{1}{2}$ minutes, as the value of the equation. But this is assuming that the sun's motion takes place along the equator, instead of along the ecliptic, which is so grossly and palpably erroneous that we wonder how the Hindus could have tolerated a process which implied it. Their own methods furnish the means of making a vastly more correct determination of the equation in question. The mean longitude of the sun at the given midnight is—after adding to it the amount of the precession, as determined farther on—$10^s\ 14°\ 7'$: hence, if the sun were $10^s\ 14°\ 7'$ distant upon the equator from the vernal equinox, or if he had that amount of right ascension, mean and true midnight would coincide. But he is actually at $10^s\ 15°\ 25'$ of longitude. If, then, we ascertain what point on the equator will pass the meridian at the same time with that point of the ecliptic, its distance from the sun's mean place in right ascension will be the equation of time required. This may be accomplished as follows. The sun is in the eleventh sign, of which the equivalent in right ascension (iii. 42–45) is 1795^p: his distance from its commencement is $15°\ 25'$, or $925'$. Hence the proportion (ii. 46)

$$1800' : 1795^p :: 925' : 922^p$$

gives us 922^p as the ascensional equivalent of the part of the eleventh sign traversed by the sun (*bhuktâsavas*). Now add together the

Ascensional equivalents	of three quadrants,	$16,200^p$
do.	of the tenth sign,	$1,935^p$
do.	of the part of the eleventh sign traversed,	922^p
their sum is		$19,057^p$

which is equal to $10^s\ 17°\ 37'$; this, then, is the sun's true right ascension. The difference between it and his mean right ascension, $10^s\ 14°\ 7'$, is $3°\ 30'$, of which the equivalent in sidereal time is 210^p or 35^v, or 14 minutes. This, which is more than two and a half times as much as the value formerly found for the equation, is quite nearly correct; its actual amount for Feb. 6th being given by the Nautical Almanac as $14^m\ 20^s$.

There is not, among all the processes taught in the Sûrya-Siddhânta, another one of so inexcusably bungling a character as this, while the means lay so ready at hand for making it tolerably exact.

In going on to calculate the local time of the eclipse, we shall adopt the valuation of the equation of time given by the Hindu method, or 13^v, but we shall reserve the distance of the phases of the eclipse from midnight, free from this constant error of about 10^m, for final comparison with the like data given by our modern tables.

To find the local time, we must first ascertain (ii. 59) the length of the sun's day, from midnight to midnight, and in order to this we need to know in what sign the sun is. Hence we require

1. To determine the amount of precession for the given date.

By iii. 9–12, the proportion

$$1,577,917,828^d : 600^{rev} :: 1,811,981^d : 0^{rev}\ 8^s\ 8°\ 2'\ 14''.6$$

gives us $248°\ 2'\ 14''.6$ as the part of a revolution accomplished by the movable point. Of this, the part determining the sine is 68° 2′ 14″.6. Then the farther proportion

10 : 3 : : 68° 2′ 14″.6 : 20° 24′ 44″

gives us 20° 24′ 44″ as the amount of the precession. Now, then, to the

Sun's true longitude,	9ˢ 25° 56′
add the precession,	20° 25′
Sun's distance from vernal equinox,	10ˢ 16° 21′

This quantity is often called *sâyana sûrya;* that is to say, "the sun's longitude with the precession (*ayana*) added."

The sun is accordingly in the eleventh sign, of which the ascensional equivalent is 1795p. His daily motion has been found to be 60′ 48″. Hence the proportion (ii. 59)

1800′ : 1795p : : 60′ 48″ : 60p.64

gives us 61p, or 10v 1p, as the excess of the sun's day over a true sidereal day of 60 nâḍîs: its length is accordingly 60n 10v 1p, or 21,661p.

Next we desire to know how much of this day passed between midnight and sunrise, and for this purpose we have

2. To find the sun's ascensional difference (*cara*).

a. To ascertain the sun's declination, and its sine and versed sine.

The sun's longitude, with precession added (*sâyana sûrya*),	10ˢ 16° 21′
Arc determining the sine (*bhuja*),	43° 39′
Sine,	2372′

Now, then, the proportion (ii. 28)

3438′ : 1397′ : : 2372′ : 964′

gives us 964′ as the sine of declination (*krântijyâ*); the corresponding arc (ii. 33) is 16° 17′ S; its versed sine (ii. 31–32) is 139′.

b. To find the radius of the sun's diurnal circle (ii. 60).

From radius,	3438′
deduct versed sine of declination,	139′
Radius of diurnal circle (*dinavyâsadala, dyujyâ*),	3299′

c. To find the earth-sine (ii. 61).

The measure of the equinoctial shadow at Washington is (see note to ii. 61–63) 9d.68. The proportion, then,

12d : 9d.68 : : 964′ : : 778′

shows the value of the earth-sine (*kshitijyâ, kujyâ*) to be 778′.

d. To find the sun's ascensional difference (ii. 61–62).

The proportion

3299′ : 3438′ : : 778′ : 811′

gives the sine of ascensional difference (*carajyâ*), which is 811′. The corresponding arc, or the sun's ascensional difference (*cara, caradala*), is 13° 39′, or 819p.

3. To find the time from midnight to sunrise.

The sun's declination being south, the ascensional difference is to be added (ii. 62–63) to the quarter of the sun's complete day, to give the length of the half-night. That is to say,

Quarter of sun's complete day (21,661p ÷ 4),	5,415p
Sun's ascensional difference,	819p
Sun's half-night,	6,234p

The interval between true midnight and true sunrise is therefore 6,234p, or 17^n 19^v. That from sunrise till noon (a quantity required in later processes) is found in like manner by subtracting the ascensional difference from the quarter-day: it is 4596p.

Now then, finally,

Time of opposition, reckoned from mean midnight,		55^n 3^v
	deduct equation of time,	13^v
do.	reckoned from true midnight,	54^n 50^v
	deduct interval till sunrise,	17^n 19^v
do.	reckoned from sunrise,	37^n 31^v

The time at which the opposition of the sun and moon in longitude takes place, or the middle of the eclipse, is accordingly, by civil reckoning at Washington, 37^n 31^v.

VI. To determine the diameters of the sun, moon, and shadow.

1. To find the sun's apparent diameter.

The sun's mean motion in a sidereal day being 58′ 58″, his true motion at the time of the eclipse being 60′ 48″, and his mean diameter 6500 yojanas, we find, by the proportion (iv. 2)

$$58' \ 58'' : 60' \ 48'' :: 6500y : 6702y.81$$

that the sun covers of his mean orbit, at the time of the eclipse, 6702.81 yojanas. This is reduced to its value upon the moon's mean orbit by the proportion (iv. 2)

$$57{,}753{,}336 : 4{,}320{,}000 :: 6702y.81 : 501y.37$$

And upon dividing the result, 501.37 yojanas, by 15 (iv. 3), we find the sun's apparent diameter to be 33′ 25″.

2. To find the moon's apparent diameter.

In like manner as before, the proportion (iv. 2)

$$788' \ 25'' : 854' \ 36 :: 480y : 520y.3$$

shows us that the moon's corrected diameter is 520.3 yojanas. This also, divided by 15 (iv. 3), gives the value of the moon's apparent diameter in arc: it is 34′ 41″.

3. To find the diameter of the earth's shadow.

The following proportion (iv. 4),

$$788' \ 25'' : 854' \ 36'' :: 1600y : 1734y.3$$

determines the value of the earth's corrected diameter (*sûci*) to be 1734.3 yojanas.

Again, from the

Sun's corrected diameter,	6702y.81
deduct the earth's diameter (iv. 4),	1600
remains	5102y.81

and this remainder, when reduced by the following proportion (iv. 5),

6500y : 480y : : 5102y.81 : 376y.8

gives us the excess of the earth's corrected diameter (*sûcî*) over the diameter of the shadow on the moon's mean orbit. Hence, from the

Earth's corrected diameter,	1734y.3
deduct last result,	376y.8
Diameter of shadow,	1357y.5
divide by	15
Diameter of shadow in arc,	90′ 30″

VII. To determine the moon's latitude at the middle of the eclipse, and the amount of greatest obscuration.

The proportion (i. 53)

1,577,917,828 : 232,238 : : 1,811,981 : 266rev 8s 7° 28′ 25″

gives us the amount of retrograde motion of the moon's node since the commencement of the Iron Age. Deducting from this 6s, for the position of the node at that time (note to i. 56–58), and taking the complement to a whole circle, we have

Longitude of moon's node, mean midnight, at Ujj.,	9s 22° 31′ 35″
deduct for difference of meridian,	1′ 21″
Longitude of moon's node, mean midnight, at Wash'n,	9s 22° 30′ 14″
deduct motion during 55n 8v,	2′ 55″
Longitude of moon's node at moment of opposition,	9s 22° 27′ 19″
subtract from moon's longitude (ii. 57),	3s 25° 56′
Moon's distance from node,	6s 3° 29′
Arc determining the sine (*bhuja*),	3° 29′
Sine,	209′

Hence the proportion

3438′ : 270′ : : 209′ : 16′ 25″

gives us, as the moon's latitude at the moment of opposition, 16′ 25″ S.

Now, then, by iv. 10–11,

Semi-diameter of eclipsed body (34′ 41″÷2),	17′ 22″
do. of eclipsing body (90′ 30″÷2),	45′ 15″
their sum,	62′ 37″
deduct moon's latitude,	16′ 25″
Amount of greatest obscuration (*grâsa*),	46′ 12″

and since this amount is greater than the diameter of the eclipsed body, it is evident that the eclipse is a total one.

This is a most unfortunate result for the Hindu calculation to yield; for, in point of fact, the eclipse in question is only a partial one, obscuring about four-fifths of the diameter of the moon's disk. The source of the error lies mainly in the misplacement, relatively to the sun and moon, of the moon's node, and the consequent false value found for the moon's latitude. The latter quantity actually amounts, at the time of opposition, to 35′ 42″, or more than twice the value given it by the Hindu processes. And it will be seen, on referring to the table on p. 44, that the relative error in the place of the moon's node, having been accumulating for seven centuries, is now about 3½°, and so reduces, by more than half, the true distance of the moon from her node. We have tried whether the admission of the correction of the *bîja* would better the result, but that is not the case: the error of position is still (see the table) nearly 2°, and the moon's latitude is increased only to 24′ 11″, so that the eclipse still appears to be total. It is evidently high time that a new correction of *bîja* be applied by the Hindu astronomers to their elements, at least to such as enter into the calculation of eclipses.

VIII. To find the duration of the eclipse, and of total obscuration, and the times of contact, immersion, emergence, and separation.

Diameter of the eclipsing body, the shadow,	90′ 30″	90′ 30″
do. eclipsed body, the moon,	34′ 41″	34′ 41″
Sum and difference,	125′ 11″	55′ 49″
Half-sum and half-difference (C M and C N, Fig. 21, p. 133),	62′ 35″	27′ 55″
Squares of do.,	3919′	724′
deduct square of latitude,	269′	269′
remain,	3650′	455′
Square roots of remainders (C A and C B),	60′ 25″	21′ 19″

In order to reduce these quantities to time, we need first to ascertain the difference of the true daily motions of the sun and moon at the given moment:

Moon's true daily motion,	854′ 36″
Sun's do.,	60′ 48″
Moon's gain in a day,	793′ 48″

Hence the proportions (iv. 13)

$$793' \, 48'' : 60^{n} :: \begin{cases} 60' \, 25'' : 4^{n} \, 34^{v} \\ 21' \, 19'' : 1^{n} \, 36^{v} \, 4^{p} \end{cases}$$

give us the half-duration of the eclipse as $4^{n}\ 34^{v}$, and the half-time of total obscuration as $1^{n}\ 36^{v}\ 4^{p}$, supposing the moon's latitude to remain constant through the whole continuance of the eclipse. We now proceed to correct these results for the moon's motion in latitude. And first, as regards the half-duration. We calculate the amount of motion of the moon and of her node during the mean half-duration by the following proportions (iv. 14):

$$60^{n} : 854' \, 36'' :: 4^{n} \, 34^{v} : 1^{\circ} \, 5' \, 2''$$
$$60^{n} : 3' \, 10'' :: 4^{n} \, 34^{v} : 14''$$

Farther,

To and from moon's long. at opposition,	$3^s\ 25^\circ\ 56'$	$3^s\ 25^\circ\ 56'$
add and subtract motion during half-duration,	$1^\circ\ 5'$	$1^\circ\ 5'$
Moon's long. at end and beginning of eclipse,	$3^s\ 27^\circ\ 1'$	$3^s\ 24^\circ\ 51'$
From and to long. of node at opposition,	$9^s\ 22^\circ\ 27'\ 21''$	$9^s\ 22^\circ\ 27'\ 21''$
subtract and add motion during half-duration,	$14''$	$14''$
Long. of node at end and beginning of eclipse,	$9^s\ 22^\circ\ 27'$	$9^s\ 22^\circ\ 28'$
Moon's distance from node,	$6^s\ 4^\circ\ 34'$	$6^s\ 2^\circ\ 23'$
Arc determining sine,	$4^\circ\ 34'$	$2^\circ\ 23'$
Sine,	$274'$	$143'$
Moon's latitude at end and beginning of eclipse,	$21'\ 31''$ S.	$11'\ 14''$ S.

From these valuations of the latitude we now proceed to calculate anew, in the same manner as before, the half-durations, as follows:

Square of half-sum of diameters,	$3919'$	$3919'$
deduct squares of latitude,	$463'$	$126'$
remain,	$3456'$	$3793'$
Square roots of remainders,	$58'\ 47''$	$61'\ 35''$

And the proportions

$$793'\ 48'' : 60^n :: \begin{cases} 58'\ 47'' : 4^n\ 26^v\ 3^p \\ 61'\ 35'' : 4^n\ 39^v\ 2^p \end{cases}$$

give us the corrected values of the intervals between opposition and contact and separation respectively, or the former and latter half-durations, as $4^n\ 39^v\ 2^p$ and $4^n\ 26^v\ 3^p$.

The text contemplates the repetition of this corrective process, if still greater accuracy be required in the results attained: we have not thought it worth while to carry the calculation any farther, as a second correction would be of altogether insignificant amount.

By a like process, the former and latter half-times of total obscuration, and the moon's latitude at immersion and emergence, are found to be as follows:

Moon's latitude at immersion and emergence,	$14'\ 36''$	$18'\ 13''$
Half-times of total obscuration,	$1^n\ 42^v\ 3^p$	$1^n\ 29^v\ 4^p$

By adding the two halves we obtain

Duration of the eclipse (*sthiti*),	$9^n\ 5^v\ 5^p$
do. of total obscuration (*vimarda*),	$3^n\ 12^v\ 1^p$

And by subtracting and adding the half-times of duration and of total obscuration from and to the time of opposition (iv. 16–17), we obtain the following scheme for the successive phases of the eclipse:

Phase.	Time of occurrence: after mean midnight.	Time of occurrence: after sunrise.
First contact,	$50^n\ 23^v\ 4^p$	$32^n\ 51^v\ 4^p$
Immersion,	$53^n\ 20^v\ 3^p$	$35^n\ 48^v\ 3^p$
Middle of eclipse,	$55^n\ 3^v\ 0^p$	$37^n\ 31^v\ 0^p$
Emergence,	$56^n\ 32^v\ 4^p$	$39^n\ 0^v\ 4^p$
Last contact,	$59^n\ 29^v\ 3^p$	$41^n\ 57^v\ 3^p$

The proper calculation of the eclipse is now completed. If, however, we desire to project it, we have still to determine the *valana*, or deflection of the ecliptic from an east and west line, for its different phases, as also the scale of projection. We will therefore proceed to calculate them, deferring to the end of the whole process any comparison of the results we have obtained with those given by modern astronomical science.

IX. To calculate the deflection of the ecliptic from an east and west line (*valana*) for the middle, beginning, and end of the eclipse.

1. For the middle of the eclipse.

a. To find the length of the moon's day and night respectively at the given time.

Moon's longitude at opposition,	3ˢ 25° 56′
Precession,	20° 25′
Moon's distance from vernal equinox,	4ˢ 16° 21′
Arc determining sine,	43° 39′
Sine,	2372′

The moon's declination is then found by the following proportion (ii. 28):

$$3438' : 1397' :: 2372' : 964' = \sin 16° 17'$$

Now, from

Moon's declination,	16° 17′ N.
deduct her latitude (ii. 58),	16′ S.
Moon's true declination,	16° 1′ N.
Sine of do.,	948′
Versed sine of do.,	135′
deduct from radius (ii. 60),	3438′
Moon's day-radius,	3303′

Again, to find the earth-sine, we say (ii. 61),

$$12^{d} : 9^{d}.68 :: 948' : 765' = \text{earth-sine}.$$

and to find the ascensional difference (ii. 61–62),

$$3303' : 3438' :: 765' : 796' = \sin 13° 24' \text{ or } 804'.$$

The excess of the moon's complete revolution over a sidereal day is found by the proportion (ii. 59)

$$1800' : 1795^{p} :: 849' 33'' : 848^{p}$$

Adding this to a sidereal day, or 21,600ᵖ, we find that the moon's day is of 22,448ᵖ, of which one quarter is 5612ᵖ. Increase and diminish this by the moon's ascensional difference (ii. 62), and the half-day and half-night are found to be 6416ᵖ and 4808ᵖ respectively.

All this laborious process of ascertaining the length of the moon's half-day, or the time which, with the given declination, she would occupy in rising from the horizon to the meridian, is rendered necessary by the correction which the commentary applies to the rule of the text in which the moon's hour-angle is involved, as pointed out in the note to iv. 24–25 (p. 140, above). We now proceed

b. To find the hour-angle, and the corrected hour-angle.

At the moment of opposition, the moon's hour-angle is evidently the same with that of the sun. Hence it may be found as follows:

Time of opposition reckoned from sunrise, $37^n\ 31^v$, or	13,506p
deduct the whole day,	9,192p
remains	4,314p
deduct from the half-night,	6,235p
Sun's distance in time from inferior meridian,	1,921p

The moon's distance eastward from the upper meridian is accordingly 1921p. This is corrected, or reduced to its proportional value as a part of the moon's arc of revolution from the horizon to the meridian, by the following proportion:

$$6416^p : 90^\circ :: 1921^p : 26^\circ\ 57'$$

The moon's corrected hour-angle, then, is 26° 57′ : its sine is 1557′.

c. To determine the amount of deflection for latitude (*valanânçâs*, or *âksha valana*—iv. 24).

The sine of the latitude of Washington, 38° 54′, is 2158′. Hence the proportion

$$3438' : 1557' :: 2158' : 977' = \sin 16^\circ\ 31'$$

gives us 16° 31′ as the value of the quantity sought. The moon being in the eastern hemisphere, it is to be reckoned as north in direction.

d. To determine the amount of deflection for ecliptic-deviation (*âyana valana*—iv. 25).

Moon's distance from vernal equinox,	$4^s\ 16^\circ\ 21'$
add a quadrant,	3^s
their sum,	$7^s\ 16^\circ\ 21'$
arc determining sine,	$46^\circ\ 21'$
sine,	2486′

Hence, by ii. 28, the proportion

$$3438' : 1397' :: 2486' : 1010' = \sin 17^\circ\ 6'$$

gives us 17° 6′ as the amount of declination of the point of the ecliptic which is a quadrant in advance of the moon, and this is the deflection required. Its direction is south. We are now ready for the final process.

e. To ascertain the net amount of deflection (*valana*), in digits.

From the ecliptic-deflection,	17° 6′ S.
deduct the deflection for latitude,	16° 31′ N.
remains the net deflection, in arc,	35′ S.
divide (iv. 25) by	70
Deflection in digits,	$0^d.50$ S.

It thus appears that, at the moment of opposition, the part of the ecliptic in which the moon is situated very nearly coincides in direction with an east and west circle. The amount of deflection is so small that in our projection, given in connection with the sixth chapter, we were obliged to exaggerate it somewhat, in order to make it perceptible.

2. For the beginning of the eclipse.

As, owing to the moon's motion in latitude and longitude, her declination, and so also her ascensional difference, are not precisely the same at the beginning and end of the eclipse as at the moment of opposition, we ought in strictness to repeat the first part of the preceding calculation, determining anew the length of the moon's half-day, as it would be if she made her whole revolution about the earth with those declinations respectively. This we take the liberty of omitting to do, as the modification thus introduced into the process would be of very small importance.

a. To find the moon's corrected hour-angle.

And first, for the sun's hour-angle:

Time of first contact, reckoned from sunrise, 32n 51v 4p, or	11,830p
deduct the whole day,	9,192p
remain	2,638p
deduct from the half-night,	6,235p
Sun's distance in time from inferior meridian,	3,597p

This, then, is the hour-angle of the centre of the shadow at the time of contact. The distance of the centre of the moon in longitude from that of the shadow was found above (under VIII) to be 61′ 35″. This is reduced to its value in right ascension by the proportion

1800′ : 1795p : : 61′ 35″ : 61p.4

Now, then,

from the hour-angle of the shadow,	3,597p
deduct the difference of the moon's right ascension,	61p
Moon's hour-angle at beginning of eclipse,	3,536p

This is virtually an application of the process taught in iii. 50.

The moon's hour-angle is now corrected, as before, by the proportion

6416p : 90° : : 3536p : 49° 36′

The sine of 49° 36′ is 2617′.

b. To find the deflection for latitude.

The proportion

3438′ : 2158′ : : 2617′ : 1643′ = sin 28° 34′

gives us the deflection for latitude as 28° 34′, which is north, as before.

c. To find the ecliptic-deflection.

Moon's distance from vernal equinox at opposition,	4s 16° 21′
deduct motion during 4n 39v 2p,	1° 6′
do. at time of contact,	4s 15° 15′
add a quadrant,	3s
sum,	7s 15° 15′
arc determining sine,	45° 15′
sine,	2441′

Next, the proportion

$$3438' : 1397' :: 2441' : 992' = \sin 16° \, 47'$$

shows us that the ecliptic-deflection is 16° 47′; it is, as in the former case, south.

d. To find the deflection, in digits.

From the deflection for latitude,	28° 34′ N.
deduct the ecliptic-deflection,	16° 47′ S.
remains the net deflection, in arc,	11° 47′ N.
its sine is	702′
divide by	70
Deflection, in digits,	10d.03 N.

3. For the end of the eclipse.

Of this process, which is throughout closely analogous to the last, we shall present only a brief statement of the results.

Hour-angle of the centre of the shadow,	322p E.
Distance of the centre of the moon in right ascension,	59p E.
Moon's hour angle,	381p E.
do. corrected,	5° 20′
Sine,	320′
Deflection for latitude,	3° 21′ N.
Moon's distance from vernal equinox + 3s,	7s 17° 24′
Arc determining sine,	47° 24′
Sine,	2530′
Ecliptic-deflection,	17° 24′ S.
Net deflection, in arc,	14° 3′ S.
do. in digits,	11d.93 S.

The mode of application of these quantities in making a projection of an eclipse is sufficiently explained in the notes to the sixth chapter, and illustrated by the figure there given, which is adapted to the conditions of the eclipse here calculated. All the quantities entering into the projection, however, of which the value has been stated in minutes, require also to be reduced to digits, according to a scale determined by the following process.

X. To determine the scale of projection of the disks and latitudes (iv. 26).

This process we will perform only for the moment of opposition, or for the middle of the eclipse. At this time, as has been seen above, we have

Moon's half-day,	6416p
do. hour-angle (*nata*),	1921p
do. altitude in time (*unnata*),	4495p
add 6416p × 3	19,248p
the sum is	23,743p
divide by	6,416p
the quotient is	3.7

At the elevation, then, which the moon has when in opposition, 3′.7 make a digit, and by this amount the values of the disk of the moon, the shadow, and the latitudes, are to be divided, in order to reduce them to a scale upon which they may be plotted. It is evident that, in strictness, the same calculation requires to be made also for the time of contact and the time of separation, or the time of any other phase of which the projection is to serve as an illustration: but it is evident also that this is wellnigh impracticable, since one projection could then be used to illustrate only a single phase, unless several different scales should be employed in the same figure.

It now only remains for us to present a comparison of the elements of the eclipse, as thus calculated, with their true values as determined by modern astronomical science. This is done in the annexed table. The true elements we take from the American Nautical Almanac for 1860. In comparing the time of the middle of the eclipse, we take, as already mentioned, the value of it given by the Hindu process as calculated from mean midnight.

	Sûrya-Siddhânta.	Am. Naut. Almanac.	Hindu error.	
Time of opposition in long.,	9h 57m 35s P. M.	9h 27m 10s.8 P. M.	+	30m 24s
Moon's long. at opposition,	136° 21′	137° 35′ 53″.7	−	1° 15′
" lat. at "	16′ 25″ S.	35′ 42″.1 S.	−	19′ 17″
" hourly motion in long.,	35′ 37″	38′ 0″.6	−	2′ 24″
Semi-diameter of sun,	16′ 42″	16′ 15″.2	+	27″
do. of moon,	17′ 20″	16′ 42″.6	+	37″
do. of shadow,	45′ 15″	45′ 16″	−	1″
Amount of obscuration,	1.33	0.812	+	0.518
Whole duration of eclipse,	3h 37m 44s	2h 52m 24s	+	45m 20s

25. p. 155. Our next note is a

Calculation, according to Hindu Data and Methods, of the Solar Eclipse of May 26th, 1854,

for the latitude and longitude of Williams' College, Williamstown, Mass.

As has been already mentioned in the closing note to the fifth chapter, the following calculation of a solar eclipse was mainly made for the translator, while in India, by his native assistant. Some additional calculations have been appended here by us, in order to render the whole process a more complete illustration of the rules as given in the text of our treatise; and we have also had to reject and replace certain parts of the work actually done, on account of their inaccuracy. For the most part, we present the work as it was made, although involving some repetitions which might be regarded as superfluous, after the explanations and illustrations already given in the notes and in the preceding calculation of a lunar eclipse. The eclipse selected is the one calculated and delineated in Prof. James H. Coffin's useful work, entitled "Solar and Lunar Eclipses familiarly illustrated and explained, with the method of calculating them, according to the theory of Astronomy as taught in New-England Colleges" (New York, 1845).

I. To find the sum of days (*ahargaṇa*) from the commencement of the planetary motions to the time of calculation.

The eclipse in question occurs at the close of the month Vâiçâkha, the second month of the luni-solar year, in the 1777th year of the era of Çâlivâhana (see add. note 12). To compute, then, the number of whole years, and to reduce them, with the remaining part of a year, to mean solar days, we proceed as follows:

Sandhi at the beginning of the *kalpa*,	1,728,000
Six *manvantaras*,	1,850.688,000
Twenty-seven *mahâyugas* of the seventh Manu,	116,640,000
	1,969,056,000
deduct the time spent in creation,	17,064,000
From creation to beginning of 28th *mahâyuga*,	1,951 992,000
Kṛta yuga of 28th or current *mahâyuga*,	1,728,000
Tretâ yuga of "	1,296,000
Dvâpara yuga of "	864,000
Kali yuga, to era of Çâlivâhana,	3,179
Complete years elapsed of the era,	1,776
From the creation to end of March, 1854, complete years,	1,955,884,955
to reduce to solar months, multiply by	12
Solar months,	23,470,619,460
add month of current year elapsed,	1
Whole number of solar months,	23,470,619,461

Now, to find the intercalary months, we make the proportion

51,840,000 : 1,593,336 :: 23,470,619,461 : 721,384,701

Then, to

Solar months elapsed,	23,470,619,461
add intercalary months,	721,384,701
Lunar months elapsed,	24,192,004,162
to reduce to lunar days, multiply by	30
Lunar days,	725,760,124,860
add for current month,	29
Whole number of lunar days,	725,760,124,889

Farther, to find the number of *tithikshayas*, or omitted lunar days, in this period, we say

1,603,000,080 : 25,082,252 :: 725,760,124,889 : 11,356,018,362

Next, from

Lunar days elapsed,	725,760,124,889
deduct omitted lunar days,	11,356,018,362
Mean solar days elapsed,	714,404,106,527

This, then, is the required *ahargaṇa*, or sum of days from the commencement of the planetary motions to about the time of new moon, May, 1854. The processes by which it is found are in all respects the same with those illustrated by us in the notes to i. 21–23, 24, 48, 48–51, above. It will be noticed that the Hindu astronomer, at least when working out an illustrative process, like the one in hand, scorns to make use of any of the means for reducing the labor of computation which the text directly or impliedly permits, and of which, in our own calculations, we have been glad to avail ourselves.

II. To ascertain the mean longitudes of the sun, the moon, the sun's apsis, the moon's apsis, and the moon's node, for mean midnight on the Hindu meridian, at the given interval from the creation.

The amount of motion, since the creation, of the bodies named, in their order, is found by the following series of proportions :

1,577,917,828 :	714,404,106,527 ::	4,320,000 :	$1{,}955{,}884{,}955^{rev}$	$1^s\ 12^\circ\ 14'\ 14''$
1,577,917,828 :	714,404,106,527 ::	57,753,336 :	$26{,}147{,}889{,}118^{rev}$	$1^s\ 9^\circ\ 44'\ 29''$
1,577,917,828,000 :	714,404,106,527 ::	387 :	175^{rev}	$2^s\ 17^\circ\ 17'\ 23''$
1,577,917,828 :	714,404,106,527 ::	488,203 :	$22{,}134{,}467^{rev}$	$2^s\ 21^\circ\ 56'\ 9''$
1,577,917,828 :	714,404,106,527 ::	232,238 :	$105{,}146{,}020^{rev}$	$10^s\ 17^\circ\ 11'\ 50''$

Rejecting whole revolutions, and, in the case of the moon's node, subtracting the fraction from a whole revolution, we have, as the mean longitudes required :

Sun,	$1^s\ 12^\circ\ 14'\ 14''$
Moon,	$1^s\ 9^\circ\ 44'\ 29''$
Sun's apogee,	$2^s\ 17^\circ\ 17'\ 23''$
Moon's apogee,	$2^s\ 21^\circ\ 56'\ 9''$
Moon's node,	$1^s\ 12^\circ\ 48'\ 10''$

The Hindu calculator has taken, in the case of the moon's apsis and node, the numbers of revolutions given by the text, omitting the correction of the *bîja.* We have not, in order to test the accuracy of his arithmetical operations, worked over again the proportions, excepting in two instances, the first and last : our results differ but slightly from those above given (we find the seconds of the sun's place to be 40″, and the minutes and seconds of the node's motion to be 12′ 43″)—not enough to render any modification necessary.

III. To ascertain the values of the same quantities at mean sunrise on the equator, or 6 o'clock.

In order to this, we must add to each planet's longitude one fourth the amount of its mean motion in a day. We require, then, the mean daily motions. They are found as follows, taking the sun as an example :

$$1{,}577{,}917{,}828^d : 4{,}320{,}000^{rev} :: 1^d : 59'\ 8''\ 10'''\ 10''''.4$$

We omit the other proportions and their results, as the latter have been fully stated in the table of mean motions of the planets (note to i. 29–34). Adding a quarter of the daily motion, we have as follows :

	Long. at midnight.		Correction.		Long. at sunrise.
Sun,	$1^s\ 12^\circ\ 14'\ 14''$	+	$14'\ 47''$	=	$1^s\ 12^\circ\ 29'\ 1''$
Moon,	$1^s\ 9^\circ\ 44'\ 29''$	+	$3^\circ\ 17'\ 39''$	=	$1^s\ 13^\circ\ 2'\ 8''$
Sun's apogee,	$2^s\ 17^\circ\ 17'\ 23''$	+	0	=	$2^s\ 17^\circ\ 17'\ 23''$
Moon's apogee,	$2^s\ 21^\circ\ 56'\ 9''$	+	$1'\ 40''$	=	$2^s\ 21^\circ\ 57'\ 49''$
Moon's node,	$1^s\ 12^\circ\ 48'\ 10''$	—	$48''$	=	$1^s\ 12^\circ\ 47'\ 22''$

IV. To ascertain the values of the same quantities at mean sunrise upon the equator, on the meridian of the given place.

Adopting 75° 50′ as the longitude of the Hindu meridian east from Greenwich, we have, as the interval in longitude of Williams' College from it, 149° 2′ 30″, which is equal to 24^{n} 50^{v} 2^{p}. The latitude is 42° 42′ 51″. We have, then, first, to determine the distance of the place in question, upon its own parallel of latitude, from the Hindu meridian.

The equatorial circumference of the earth has been found above (note to i. 59–60) to be 5059.64 yojanas. Its circumference upon the parallel of latitude of Williams' College is found (i. 60) by the following proportion :

3438′ (=R) : 2525′ (= cos 42° 42′ 51″) : : 5059y.64 : 3715y.97

The *deçântara*, or difference of longitude in yojanas, is then determined thus :

60n : 24^{n} 50v 2p : : 3715y.97 : 1538y.41

And the *deçântaraphala*, or correction for difference of longitude, is calculated from the daily motion of each body, by such a proportion as the one subjoined, which gives the sun's correction :

3715y.97 : 1538y.41 : : 59′ 8″ : 24′ 27″

We omit the other proportions, and merely present their results in the following table :

	Sunrise at Lankâ.		Correction.		Sunrise on giv. merid.
Sun,	1^{s} 12° 29′ 1″	+	24′ 27″	=	1^{s} 12° 53′ 28″
Moon,	1^{s} 13° 2′ 8″	+	5 27 12	=	1^{s} 18° 29′ 20″
Sun's apogee,	2^{s} 17° 17′ 23″	+	0	=	2^{s} 17° 17′ 23″
Moon's apogee,	2^{s} 21° 57′ 49″	+	2′ 45″	=	2^{s} 22° 0′ 34″
Moon's node,	1^{s} 12° 47′ 22″	−	1′ 19″	=	1^{s} 12° 46′ 3″

We have already (note to i. 63–65) called attention to the excessively awkward and cumbrous character of this process for making the correction for difference of meridian.

V. To find the sun's true longitude.

From the longitude of the sun's apsis,	2^{s} 17° 17′ 23″
deduct sun's mean longitude (ii. 29),	1^{s} 12° 53′ 28″
Sun's mean anomaly,	1^{s} 4° 23′ 55″
Sine,	1927′

The diminution of the sun's epicycle is now found by the following proportion (ii. 38) :

3438′ : 20′ : : 1927′ : 11′ 12″

The dimensions of the epicycle are, then (ii. 34), 14° − 11′ 12″, or 13° 48′ 48″. Next, the proportion (ii. 39)

360° : 13° 48′ 48″ : : 1927′ : 74′ 11″

gives us the sun's equation of the centre, which, by ii. 45, is additive. Hence to the

Sun's mean longitude,	1^{s} 12° 53′ 28″
add the equation,	1° 14′ 11″
Sun's true longitude,	1^{s} 14° 7′ 39″

This calculation exhibits a rather serious error: the sine of 34° 24′, the anomaly, is 1942′, not 1927′. The final result, however, is not perceptibly modified by it: the equation ought to be 1° 14′ 30″, and the true longitude 1ˢ 14° 7′ 58″.

VI. To find the moon's true longitude.

From the longitude of the moon's apsis,	2ˢ 22° 0′ 34″
deduct moon's mean longitude,	1ˢ 18° 29′ 20″
Moon's mean anomaly,	1ˢ 3° 31′ 14″
Sine,	1898′
Diminution of epicycle,	11′ 2″
Dimensions of epicycle,	31° 48′ 58″
Equation of the centre,	+ 2° 47′

Hence, to the

Moon's mean longitude,	1ˢ 18° 29′ 20″
add the equation,	2° 47′
Moon's true longitude,	1ˢ 21° 16′ 20″

VII. To calculate the true daily motions of the sun and moon.

The equations of motion for the sun and moon have been found by the calculator of the eclipse by the following proportion: as the whole orbit of either planet is to its epicycle, so is its mean daily motion to the required equation. That is to say, for the sun,

360° : 13° 48′ 48″ : : 59′ 8″ : 2′ 16″

which, by ii. 49, is subtractive. Hence the sun's true motion is 59′ 8″ — 2′ 16″, or 56′ 52″.

Again, for the moon,

360° : 31° 48′ 58″ : : 790′ 35″ : 69′ 36″

And the moon's true motion is 790′ 35″ — 69′ 36″, or 720′ 59″.

These calculations are exceedingly incomplete and erroneous, as may readily be seen by referring to the corresponding process in the other eclipse, or to that given as an illustration in the note to ii. 47–49. The actual value of the sun's equation of motion, as fully calculated by the method of our treatise, is only 1′ 51″; that of the moon is only 58′ 49″: whence the true motions are 57′ 17″ and 731′ 46″ respectively. These are elements of so much importance, and they enter so variously into the after operations, that we have hesitated as to whether it would not be better to cancel the whole work of the Hindu calculator from this point onward, and to perform it anew in a more exact manner; but we have finally concluded to present the whole as it is, as a specimen—although, we hope, not a favorable one—of native work; pointing out, at the same time, its deficiences, and cautioning against its results being accepted as the best that the system is capable of affording.

We have thus far found the true longitudes of the sun and moon for the moment of mean sunrise at the equator, upon the meridian of the given place. We desire now farther to find the same data for the moment of sunrise upon the same meridian in latitude 42° 42′ 51″ N.

VIII. To find the longitudes of the sun and moon at sunrise in long. 149° 2′ 30″, lat. 42° 42′ 51″ N.

1. To calculate the precession of the equinoxes (iii. 9–12).
The proportion

$$1,577,917,828^{d} : 600^{rev} :: 714,404,106,527 : 271,650^{rev}\ 8^{s}\ 7^{\circ}\ 45'\ 22''$$

gives us the amount of the motion of the equinox in its own circle of libratory revolution, since the beginning of things. Rejecting complete revolutions, and deducting 6^{s} from the fraction of a revolution, we have the distance of the equinox from the origin of the sidereal sphere, in terms of its own revolution, as 67° 45′ 22″: three tenths of this, or 20° 19′ 36″, is the amount of the precession.

2. To calculate the sun's declination.

Sun's longitude,	$1^{s}\ 14^{\circ}\ 7'\ 39''$
Precession,	$20^{\circ}\ 19'\ 36''$
Sun's distance from vernal equinox,	$2^{s}\ 4^{\circ}\ 27'\ 15''$
Sine,	$3101'$

Then, by ii. 28,

$$3438' : 1397' :: 3101' : 1260' = \sin 21^{\circ}\ 31'\ 3''$$

the sun's declination is therefore 21° 31′ 3″.

3. To calculate the sun's ascensional difference.

The radius of the sun's diurnal circle (*dyujyâ*—ii. 60) is 3199′.

The equinoctial shadow in the given latitude is $11^{d}.07$, being found by the proportion (iii. 17)

$$\text{cos lat.} : \text{sin lat.} :: \text{gnom.} : \text{eq. shad.}$$

or

$$2525' : 2330' :: 12^{d} : 11^{d}.07$$

Again, to find the earth-sine (*kujyâ*—ii. 61),

$$12^{d} : 11^{d}.07 :: 1260' : 1162'$$

And, to find the sine of ascensional difference,

$$3199' : 3438' :: 1162' : 1249'$$

The corresponding arc is 21° 19′, or 1279′; and since a minute of arc is equivalent to a respiration of time, the sun's ascensional difference in time is 1279^{p}, or 213^{v}, or $3^{n}\ 33^{v}$, rejecting the odd respiration.

4. To calculate the length of the sun's day.

The sun being in the third sign, of which the equivalent in right ascension (iii. 42–45) is 1935^{p}, the excess of his day over 60 nâḍîs is found by the proportion

$$1800' : 1935^{p} :: 59'\ 8'' : 63^{p}$$

whence the length of his day is $21,663^{p}$.

In this calculation of the length of the sun's day, the operator has taken the mean, instead of the true, motion of the sun, which is obviously less accurate, and which is contrary to the meaning of the rule of the text (ii. 59), as explained by the commentator.

Now, in order to find the difference between the sun's longitude at sunrise on the equator and sunrise on the given parallel of north latitude, we make a proportion, as follows: if in his whole day the sun moves an amount equal to his daily motion, how much will he move during an interval corresponding to his ascensional difference? or

$$21,663^{p} : 59'\ 8'' :: 1279^{p} : 3'\ 29''$$

The sun's declination being north, sunrise on the given parallel precedes sunrise on the equator, and hence this result—which is called the *carakalâs*, "minutes (*kalâ*) of longitude corresponding to the ascensional difference (*cara*)"—is to be subtracted from the sun's longitude as formerly found. That is to say,

Sun's longitude at equatorial sunrise,	$1^s\ 14^\circ\ 7'\ 39''$
deduct the correction (*carakalâs*),	$3'\ 29''$
Sun's longitude at sunrise, lat. $42^\circ\ 42'\ 51''$ N., long. $149^\circ\ 2'\ 30''$ W. from Lankâ,	$1^s\ 14^\circ\ 4'\ 10''$

In finding the corresponding value of the moon's longitude we apply first a correction for the sun's equation of place; it is, in fact, the equation of time, calculated after the entirely insufficient method which we have already fully exposed, in connection with part V of the preceding process. The proportion is (ii. 46) as follows:

$$21{,}600' : 790'\ 35'' :: 1^\circ\ 14'\ 11'' : 2'\ 43''$$

Here, again, bad is made worse by taking as the second term of the proportion the moon's mean, instead of her true, rate of motion. It is to be noticed that a like correction should have been applied also to the sun's longitude, but was omitted by the calculator. We have, then,

Moon's longitude, mean equatorial sunrise,	$1^s\ 21^\circ\ 16'\ 20''$
add the correction for the equation of time,	$2'\ 43''$
Moon's longitude, true equatorial sunrise,	$1^s\ 21^\circ\ 19'\ 3''$

Now we apply farther the correction for the sun's ascensional difference (*carasanskâra*); it is calculated in the same manner with that of the sun, and its amount is found to be $47'\ 51''$.

Moon's longitude, true equatorial sunrise,	$1^s\ 21'\ 19'\ 3''$
deduct the correction for the sun's asc. diff.,	$47'\ 51''$
Moon's longitude at sunrise, lat. $42^\circ\ 42'\ 51''$ N., long. $149^\circ\ 2'\ 30''$ W. from Lankâ,	$1^s\ 20'\ 31'\ 12''$

On comparing the longitudes of the sun and moon, as thus determined, it is seen that the time of conjunction is already past. Hence the calculation is carried a day backward, by subtracting from the longitude of each body its motion during a day. That is to say,

	Longitude, sunrise following eclipse.		day's motion.		Longitude, sunrise preceding eclipse.
Sun,	$1^s\ 14^\circ\ 4'\ 10''$	−	$56'\ 52''$	=	$1^s\ 13^\circ\ 7'\ 18''$
Moon,	$1^s\ 20^\circ\ 31'\ 12''$	−	$12^\circ\ 0'\ 59''$	=	$1^s\ 8^\circ\ 30'\ 13''$
Moon's node,	$1^s\ 12^\circ\ 46'\ 3''$	+	$3'\ 11''$	=	$1^s\ 12^\circ\ 49'\ 14''$

This is an entirely uncalled-for, and a highly inaccurate proceeding. By the rule given in our text (ii. 66), it is just as easy and regular a process to find from any given time the interval to the beginning of the current lunar day by reckoning backward, as that to the end of the day by reckoning forward. And to assume that the whole calculation may be transferred from one sunrise back to the preceding by simply deducting the amount of motion in a day as determined for the former time is to take a most unwarrantable liberty, and to ignore the change during the interval of many of the elements of the calculation, as the sun's and moon's rates of motion, the sun's declination and ascensional difference, etc. In making the transfer, moreover, the longitude of the moon's node has been taken as found for mean equatorial sunrise, without any correction for the equation of time, or for the sun's ascensional difference.

IX. To find the time of true conjunction, and the longitudes of the sun, moon, and moon's node at that time. By ii. 66, from the

Moon's true longitude,	1ˢ 8° 30′ 13″
deduct the sun's do.,	1ˢ 13° 7′ 18″
remains	11ˢ 25° 22′ 55″
divide by the portion of a lunar day,	720′
the quotient is	29ᵈ and 442′ 55″
deduct the remainder from a whole portion,	720′
remains	277′ 5″

This process shows us that the moon has still 277′ 5″ to gain upon the sun, in order to arrive at the end of the thirtieth or last day of the lunar month, or at conjunction with the sun.

Next, from the

Moon's true daily motion,	720′ 59″
deduct the sun's do.,	56′ 52″
Moon's daily gain in longitude,	664′ 7″

Hence the proportion

$$664' \; 7'' : 60^{n} :: 277' \; 5'' : 25^{n} \; 2^{v}$$

gives us the time of conjunction, reckoned from sunrise, as $25^{n}\ 2^{v}$.

Now, by iv. 8, we proceed to find the longitudes for that time. The amounts of motion during $25^{n}\ 2^{v}$ are found by the following proportions:

$$60^{n} : 25^{n}\ 2^{v} :: \begin{cases} 56'\ 52'' : 23'\ 43'' \\ 720'\ 59'' : 300'\ 48'' \\ 3'\ 11'' : 1'\ 19'' \end{cases}$$

Then, to the

Sun's longitude at sunrise,	1ˢ 13° 7′ 18″
add the correction,	23′ 43″
Sun's longitude at conjunction,	1ˢ 13° 31′ 1″
Moon's longitude at sunrise,	1ˢ 8° 30′ 13″
add the correction,	5° 0′ 48″
Moon's longitude at conjunction,	1ˢ 13° 31′ 1″
Node's longitude at sunrise,	1ˢ 12° 49′ 14″
deduct the correction,	1′ 19″
Node's longitude at conjunction,	1ˢ 12° 47′ 55″

The mode of proceeding adopted by us above, in the lunar eclipse, for finding the time of the middle of the eclipse, and the longitudes of the sun and moon at that time, is, as will not fail to be observed, quite different from that of the native calculator of this eclipse. That followed by Davis, or his native assistants (As. Res., ii. 273 etc.), varies considerably from both. Our own method, though varying in some respects from that contemplated by the text, is a not less legitimate application of its general methods than either of the others, and it possesses this important advantage over both, that we were able to verify it, and to show, by calculating the mean and true places for the given instant, that the latter was actually the one at which the system made the opposition of the sun and moon to take place: while, on the contrary, in the process now in hand, so many errors have been involved, that, were the same test to be applied, we should find the centres of the sun and moon many minutes apart at the moment fixed upon as that of conjunction, and the place of conjunction as far removed from the point of longitude above determined for it.

X. To find the apparent diameters of the sun and moon.

These quantities are determined by means of the following proportion: as the mean daily motion in yojanas is to the mean diameter in yojanas, so is the true motion in minutes to the true diameter in minutes. That is to say, for the sun and moon respectively,

$$11,858\frac{3}{4}y : 6500y :: 56' 52'' : 31' 10''$$
$$11,858\frac{3}{4}y : 480y :: 720' 59'' : 29' 2''$$

This method is in appearance quite different from that which is prescribed by our text (iv. 2–3), but it is in fact only a simplification, or reduction, of the rules there given. Thus, for the moon, the text gives

m. mot. in minutes : true mot. in min. : : m. diam. in yoj. : true diam. in min. × 15

Transposing, now, the middle terms, transferring the factor 15 from the fourth term to the first, and noting that the mean motion in minutes, when multiplied by 15, gives the value of the same in yojanas, we have the former proportion, namely,

m. mot. in yoj. : m. diam. in yoj. : : true mot. in min. : true diam. in min.

Again, in the case of the sun, the rules of the text give

m. mot. in min. : true mot. in min : : m. diam. in yoj. : true diam. in yoj.
and true diam. in yoj. = true diam. in min. × 15 × (sun's orbit ÷ moon's orbit)

Now transposing the second and third terms of the proportion, substituting for the fourth its equivalent as here stated, and transferring to the first term the last two factors of that equivalent, we have

$$\text{m. mot. in min.} \times 15 \times \frac{\text{sun's orbit}}{\text{moon's orbit}} : \text{m. d. in y.} :: \text{true mot. in min.} : \text{true diam. in min.}$$

But the first term, as thus constructed, is, by the method of determination of the planetary orbits (see xii. 81–83), equal to the sun's mean daily motion upon his orbit reckoned in yojanas: hence the proportion becomes for the sun, as for the moon,

m. mot. in yoj. : m. diam. in yoj. : : true mot. in min. : true diam. in min.

XI. To calculate the parallax in longitude (*lambana*), and the time of apparent conjunction (v. 3–9).

1. To find the orient ecliptic-point (*lagna*) at the moment of true conjunction (iii. 46–48).

In order to this, we require to have first the equivalents in oblique ascension (*udayâsavas*) of the several signs of the zodiac for the latitude of Williams' College, 42° 42′ 51″ N. We present annexed their values as employed by the calculator of the eclipse, and also as calculated by ourselves according to the method taught in our text (iii. 42–45). It will be noticed that the differences are not inconsiderable, and evince much carelessness on the part of the native astronomer; who, moreover, employs vinâḍîs only in his processes, rejecting the odd respirations, which is an inaccuracy not countenanced by the Sûrya-Siddhânta.

	Equivalent in oblique ascension: acc. to calculator.	acc. to us.	
1st sign	....	1008p	12th sign
2nd "	....	1238p	11th "
3rd "	287v or 1722p	1699p	10th "
4th "	359v or 2154p	2171p	9th "
5th "	387v or 2322p	2352p	8th "
6th "	388v or 2328p	2332p	7th "

The equivalents assigned by the Hindú calculator to the 3rd and 4th signs are moreover, it may be remarked, inconsistent with one another, since the one ought to fall short of 1935p by as much as the other exceeds that quantity.

Now, then, to the

Sun's longitude at conjunction,	1s 13° 31′ 1″
add the precession,	20° 19′ 36″
Sun's distance from the equinox,	2s 3° 50′ 37″

It appears, accordingly, that the sun is in the 3rd sign, and 26° 9′ 23″ from the beginning of the fourth. Hence the proportion (iii. 46)

30° : 287v : : 26° 9′ 23″ : 250v

give us 250v as the ascensional equivalent of the part of a sign to be traversed (*bhogyâsavas*). The time of the day, or the sun's distance in time from the eastern horizon, is 25n 2v, or 1502v. Then, from the

Time of conjunction,	1502v
deduct asc. equiv. of part of 3rd sign,	250v
remains	1252v
deduct asc. equiv. of 4th, 5th, and 6th signs,	1134v
remains	118v

This remainder of time, or of ascension, is reduced to its value in arc of the ecliptic by the proportion (iii. 49)

388v : 30° : : 118v : 9° 7′ 25″

Add this result to the whole signs preceding, and the longitude of the orient ecliptic-point (*lagna*) is found to be 6s 9° 7′ 25″: its sine is 544′ (more correctly, 545′).

2. To find the orient-sine (*udayajyâ*—v. 3).

This is found by the proportion

2525′ : 1397′ : : 544′ : 301′

2525′ being the cosine of the latitude, and 1397′ the sine of the inclination of the ecliptic (ii. 28).

3. To find the meridian ecliptic-point (*madhyalagna*—iii. 49).

In order to this, we must first know the sun's hour-angle (*nata*), or distance in time from the meridian; it is determined as follows:

A quarter of the complete day,	15n 0v
add the sun's ascensional difference,	3n 33v
The sun's half-day,	18n 33v
deduct from time of conjunction,	25n 2v
Sun's hour-angle, west,	6n 29v

The sun's distance from the beginning of the fourth sign was found above to be 26° 9′ 23″. Its equivalent in right ascension (*lankodayâsavas*) is found by the following proportion (iii. 49):

30° : 323v : : 26° 9′ 23″ : 285v

Now, from the

Sun's hour-angle, 6n 29v, or	389v
deduct the result of the last proportion,	285v
remains	104v

and this remainder, being less than the equivalent of a sign, is reduced to its value as longitude by the proportion (iii. 48)

323v : 30° : : 104v : 9° 3′ 57″

The longitude of the meridian ecliptic-point is accordingly 3s 9° 3′ 57″: its sine is 3393′.

In criticism of the process as thus conducted, we would only remark that the quarter of the sun's day should have been called 15n 2v 4p (see above, VIII. 4), and that to take 323v as the equivalent in right ascension of the third and fourth signs is inaccurate, the value given it by our treatise being 1935p, or 322½v.

4. To find the meridian-sine (*madhyajyâ*—v. 4–5).

First, the declination of the meridian ecliptic-point is determined by the proportion (ii. 28)

3438′ : 1397′ : : 3393′ : 1378′ = sin 23° 39′ 37″

Its value being north, it is deducted from the latitude of the place for which the calculation is made, since this, though by us reckoned as north, is to the Hindu apprehension (iii. 14) always south, being measured south from the zenith to the equator. That is to say,

From the given latitude,	42° 42′ 51″
deduct decl. of merid. ecliptic-point,	23° 39′ 37″
Meridian zenith-distance (*natânçâs*),	19° 3′ 14″

The sine of this arc, which is 1117′, is the meridian-sine.

Here is another blunder of the calculator: the sine of 19° 3′ 14″ is actually 1122′.

5. To find the sine of ecliptic zenith-distance (*dṛkkshepa*), and the sine of ecliptic-altitude (*dṛggati*).

First, by v. 5,

3438′ : 301′ : : 1117′ : 97′ 48″

Now, then, by v. 6,

Square of last result,	9,564′
deduct from square of mer.-sine,	1,247,689′
remains	1,238,125′
Square-root,	1113′

This, then, is the sine of ecliptic zenith-distance. The sine of ecliptic-altitude is found by subtracting its square from that of radius, and taking the square-root of the remainder; it is found to be 3253′.

6. To find the divisor (*cheda*), and the sun's parallax in longitude (*lambana*).

The sine of one sign, or 30°, is 1719′.

Square of sin 30°,	2,954,961
divide by	3,253
Divisor (*cheda*),	908

Next, to find the interval on the ecliptic between the sun's place and the meridian:

Longitude of meridian ecliptic-point,	3ˢ 9° 3′ 57″
do. of sun,	2ˢ 3° 50′ 37″
Interval in longitude,	1ˢ 5° 13′ 20″

Of this the sine is 1950′, and, upon dividing it by 908, the divisor (*cheda*) above found, the value of the parallax in longitude (*lambana*) is ascertained to be 2ⁿ 21ᵛ.

Here is some of the worst blundering which we have yet met with. The sine of 35° 13′ is actually 1982′, not 1950′; and upon dividing it by 908, we find the quotient to be only 2ⁿ 11ᵛ.

The calculator assumes the time of apparent conjunction to be determined by this single correction. As the text, however (v. 9), directs that the process be repeated, to insure a higher degree of accuracy, we shall finally quit at this point the guidance of his computations, and go on to apply in full the rules of the Sûrya-Siddhânta.

The sun being west of the meridian, or his longitude being less than that of the meridian ecliptic-point (v. 9), the correction for parallax is additive to the time of true conjunction. Hence, to the

Time of true conjunction,	25ⁿ 2ᵛ
add the correction,	2ⁿ 11ᵛ
Time of conjunction once equated,	27ⁿ 13ᵛ

For the time thus found, we now proceed to calculate again the value of the parallax. The results of the calculation are briefly presented below:

Sun's longitude at corrected time of conjunction,	2ˢ 3° 52′ 41″
Orient ecliptic-point (*lagna*),	6ˢ 18° 50′
Its sine,	1110′
Orient-sine (*udayajyâ*),	614′
Sun's hour-angle,	3103ᵖ

Meridian ecliptic-point (*madhyalagna*),	$3^s\ 21°\ 59'$
Its sine,	3188′
Its declination,	22° 9′ N.
Its zenith-distance,	20° 34′ S.
Meridian-sine (*madhyajyá*),	1207′
Sine of ecliptic zenith-distance (*dṛkkshepa*),	1188′
Sine of ecliptic-altitude (*dṛggati*),	3226′
Divisor (*cheda*),	916′
Sine of sun's dist. in long. from meridian,	2558′
Parallax in longitude (*lambana*),	$2^n\ 48^v$
add to time of true conjunction,	$25^n\ 2^v$
Time of conjunction twice equated,	$27^n\ 50^v$

Once more, we repeat the same calculation; its principal results are as follows:

Orient ecliptic-point,	$6^s\ 21°\ 41'$
Orient-sine,	702′
Meridian ecliptic-point,	$3^s\ 25°\ 26'$
Meridian-sine,	1241′
Sine of ecliptic zenith-distance,	1215′
Sine of ecliptic-altitude,	3216′
Divisor,	919′
Parallax in longitude,	$2^n\ 55^v$
add to time of true conjunction,	$25^n\ 2^v$
Time of apparent conjunction,	$27^n\ 57^v$

A farther repetition of the process would still yield an appreciable correction, but as so many errors have been involved in the preceding parts of the calculation as to render any exactness of result unattainable, and as enough has been done to illustrate the method of correction by successive approximation and the comparative value of the results it yields, we stop here, and rest content with the last time obtained, as that of the apparent conjunction of the sun and moon, or of the middle of the eclipse, at Williams' College.

XII. To calculate the parallax in latitude (*nati*) for the middle of the eclipse.

This is given us by the proportion (v. 10)

$$3438' : 731'\ 27'' \div 15 :: 1215' : 17'\ 14''\ \text{S.}$$

in which 1215′ is the sine of ecliptic zenith-distance, as found in the last process.

XIII. To calculate the moon's latitude, and her apparent latitude, for the middle of the eclipse.

We require first to find the longitude of the moon, and that of her node, for the moment of apparent conjunction, by adding to their longitudes, as already found (above, IX) for the time of true conjunction, their motion during $2^n\ 55^v$. The amount of motion is found by the proportions

$$60^n : 2^n\ 55^v :: \begin{cases} 720'\ 59'' : 35'\ 3'' \\ 3'\ 11'' : 0'\ 9'' \end{cases}$$

Now, then, to the

Moon's longitude at true conjunction,	$1^s\ 13^\circ\ 31'\ 1''$
add the correction,	$35'\ 3''$
Moon's longitude at apparent conjunction,	$1^s\ 14^\circ\ 6'\ 4''$

Farther, from the

Node's longitude at true conjunction,	$1^s\ 12^\circ\ 47'\ 55''$
deduct the correction,	$9''$
Node's longitude at apparent conjunction,	$1^s\ 12^\circ\ 47'\ 46''$
deduct from moon's longitude,	$1^s\ 14^\circ\ 6'\ 4''$
Moon's distance from node,	$1^\circ\ 18'\ 18''$
Sine,	$78'$

Hence the proportion (ii. 57)

$$3438' : 270' :: 78' : 6'\ 8''$$

gives us the

Moon's true latitude,	$6'\ 8''$ N.
deduct from parallax in latitude (v. 12),	$17'\ 14''$ S.
Moon's apparent latitude,	$11'\ 6''$ S.

XIV. To find the amount of obscuration (*grâsa*) at the moment of apparent conjunction.

By iv. 10, we add to the

Diameter of the eclipsing body, the moon,	$29'\ 2''$
Diameter of the eclipsed body, the sun,	$31'\ 10''$
Sum of diameters,	$60'\ 12''$
Half-sum of diameters,	$30'\ 6''$
deduct moon's apparent latitude,	$11'\ 6''$
Amount of greatest obscuration,	$19'\ 0''$

This remainder being less than the sun's diameter, the eclipse (iv. 11) is partial only.

XV. To determine the times of the beginning and end of the eclipse respectively.

As the eclipse is a partial one only, we have not to calculate the times of the beginning and end of total obscuration ; and indeed, we may well suppose that the Hindus would never venture to calculate those times in a solar eclipse : it is even questionable whether the accuracy of their methods would justify them in ever predicting with confidence that an eclipse would be total.

In the first place, we assume that the moon's apparent latitude, as calculated for the moment of conjunction, remains unchanged during the whole duration of the eclipse, and calculate, by iv. 12–13, what would be, upon that assumption, the interval between the middle of the eclipse and either contact or separation of the disks. That is to say (iv. 12), from the

Square of sum of semi-diameters (30′ 6″),	906′ 1″
deduct square of moon's latitude (11′ 6″),	123′ 13″
remains	782′ 48″
Square root of remainder,	27′ 59″

This result represents the distance, as rudely determined, of the two centres at the moments of contact and separation. To ascertain the corresponding interval of time, we say (iv. 13)

$$664' \ 7'' : 60^n :: 27' \ 59'' : 2^n \ 32^v$$

Now, then, from and to the

Time of apparent conjunction,	$27^n \ 57^v$
subtract and add the half duration,	$2^n \ 32^v$
Beginning of eclipse,	$25^n \ 25^v$
End of eclipse,	$30^n \ 29^v$

This is as far as the operation was carried by the native calculator, and with data and results somewhat different from those here given, owing to his neglect to repeat the process of determination of the parallax in longitude in finding the time of apparent conjunction. Unfortunately, however, the text (iv. 14–15; v. 13–17) prescribes a long and tedious series of modifications and corrections of the results so far obtained, of which we shall proceed to perform at least enough to illustrate the method of the process, and the comparative importance of the corrections which it furnishes.

We have first to find the longitude of the sun, moon, and node, at the moments thus determined as those of contact and separation; they are as follows:

Sun's long. at true conj. ($25^n \ 2^v$),	$1^s \ 13° \ 31' \ 1''$	$1^s \ 13° \ 31' \ 1''$
add for his motion	$22''$	$5' \ 10''$
Sun's long. at beg. and end of eclipse,	$1^s \ 13° \ 31' \ 23''$	$1^s \ 13° \ 36' \ 11''$
add the precession,	$20° \ 19' \ 36''$	$20° \ 19' \ 36''$
Sun's distance from the vernal equinox,	$2^s \ 3° \ 50' \ 59''$	$2^s \ 3° \ 55' \ 47''$
Moon's long. at app. conj.	$1^s \ 14° \ 6' \ 4''$	$1^s \ 14° \ 6' \ 4''$
subtract and add motion in $2^n \ 32^v$,	$30' \ 26''$	$30' \ 26''$
Moon's long. at beg. and end of eclipse,	$1^s \ 13° \ 35' \ 38''$	$1^s \ 14° \ 36' \ 30''$
Node's long. at app. conj.,	$1^s \ 12° \ 47' \ 46''$	$1^s \ 12° \ 47' \ 46''$
add and subtract	$8''$	$8''$
Node's long. at beg. and end of eclipse,	$1^s \ 12° \ 47' \ 54''$	$1^s \ 12° \ 47' \ 38''$

To find, then, the moon's true latitude at contact and separation, we have

Moon's distance from node,	47′ 44″	1° 48′ 52″
Sine,	48′	109′
Moon's latitude,	3′ 46″ N.	8′ 34″ N.

Next are calculated the moon's parallax in latitude, and her apparent latitude, at the beginning and end of the eclipse, by a process of which the main results are the following:

Orient ecliptic-point,	$6^s\ 10^\circ\ 28'$	$7^s\ 3^\circ\ 59'$
Sine,	$625'$	$1921'$
Orient-sine,	$345'$	$1063'$
Sun's hour-angle,	2455^p	4279^p
Meridian ecliptic-point,	$3^s\ 11^\circ\ 54'$	$4^s\ 11^\circ\ 7'$
Sine of do.,	$3363'$	$2590'$
Zenith-distance of do.,	$19^\circ\ 16'$	$24^\circ\ 53'$
Meridian-sine,	$1134'$	$1445'$
Sine of ecliptic zenith-distance,	$1128'$	$1374'$
Parallax in latitude,	$16'\ 0''$ S.	$19'\ 29''$ S.
deduct true latitude,	$3'\ 46''$ N.	$8'\ 34''$ N.
Moon's apparent lat. at beg. and end of eclipse,	$12'\ 14''$ S.	$10'\ 55''$ S.

Finally, from the

Square of sum of semi-diameters,	$906'\ 1''$	$906'\ 1''$
deduct squares of app. latitude,	$150'\ 39''$	$119'\ 11''$
remain	$755'\ 22''$	$786'\ 50''$
Distance of centres in longitude,	$27'\ 29''$	$28'\ 3''$
Corresponding interval,	$2^n\ 29^v$	$2^n\ 32^v$
Corrected times of beginning and end of eclipse,	$25^n\ 28^v$	$30^n\ 29^v$

It is evidently unnecessary to carry any farther this part of the process; at the time of the eclipse, the increase of the moon's latitude northward, and the increase of her parallax southward, so nearly balance one another, that the additional correction yielded by a new computation would be quite inappreciable—as, indeed, has been, in one of the two cases, that already obtained. In making this corrective calculation we have not followed with exactness the directions given in the commentary under v. 14–17. It is there taught that, after making the first rough determination of the half-duration, based upon the moon's apparent latitude at apparent conjunction, we must turn back to the true conjunction, find the positions of the planets and node at intervals of the half-duration from that point, and make these positions the data of our farther approximative processes. The text itself, as already remarked by us in the notes, shows an utter and provoking want of explicitness with regard to the whole matter, and may be regarded as favoring equally the method of the commentary, our own, or any other that might be devised. We have taken our own course, then, because we were unable to see any sufficient reason for reverting from apparent to true conjunction, as directed by the commentator.

With regard to the next steps, the language of the text is less ambiguous: it distinctly orders us to deduct from and add to the time of true conjunction (*tithyanta*) the intervals found as the former and latter half-duration, and from the moments thus determined to compute anew, by a repeated process, the parallax in longitude. This is a very laborious operation, and not altogether accurate, although perhaps as much so as any which the Hindu methods admit. As we are supposed to have already ascertained how far apart the two centres must be at the moments of contact and separation, the problem is, evidently, to determine at what moment of time they will, allowing for the parallax in longitude, be at that distance from one another. Now as formerly, to find the time of apparent conjunction, we started from that of true conjunction, and arrived at the desired result by a series of approximative calculations of the parallax in longitude, so now, starting from points removed from true conjunction by the given intervals, we shall ascertain, by a similar series of approximations, the times when the distances represented by those intervals will be apparent, or the moments to which contact and separation of the disks will be deferred by parallax in longitude. The results of the calculations, as made by us, are as follows:

Time of true conjunction,	$25^n\ 2^v$	$25^n\ 2^v$
subtract and add,	$2^n\ 29^v$	$2^n\ 32^v$
Times of true contact and separation,	$22^n\ 33^v$	$27^n\ 34^v$
Sun's longitude, with precession,	$2^s\ 3^\circ\ 48'\ 16''$	$2^s\ 3^\circ\ 53'\ 1''$
Orient ecliptic-point,	$5^s\ 27^\circ\ 9'$	$6^s\ 20^\circ\ 27'$
Orient-sine,	$95'$	$664'$
Meridian ecliptic-point,	$2^s\ 25^\circ\ 52'$	$3^s\ 23^\circ\ 56'$
Meridian-sine,	$1107'$	$1226'$
Sine of ecliptic zenith-distance,	$1106'$	$1203'$
Sine of ecliptic-altitude,	$3255'$	$3219'$
Divisor,	908	918
Moon's longitude,	$2^s\ 3^\circ\ 21'$	$2^s\ 4^\circ\ 21'$
Distance from meridian ecliptic-point,	$22^\circ\ 31'$	$1^s\ 19^\circ\ 35'$
Sine,	$1316'$	$2617'$
Parallax in longitude,	$1^n\ 27^v$	$2^n\ 51^v$

Again, we go on to correct these results by repeated calculations of the parallax, in the mode which has already been sufficiently illustrated. Annexed are the results only:

Times of contact and separation,	$22^n\ 33^v$	$27^n\ 34^v$
add correction for parallax,	$1^n\ 27^v$	$2^n\ 51^v$
Times of contact and separation, once equated,	$24^n\ 0^v$	$30^n\ 25^v$
Corresponding parallax,	$1^n\ 54^v$	$3^n\ 20^v$
add to times first obtained,	$22^n\ 33^v$	$27^n\ 34^v$
Times of contact and separation, twice equated,	$24^n\ 27^v$	$30^n\ 54^v$
Corresponding parallax,	$2^n\ 2^v$	$3^n\ 24^v$

Without taking the trouble to carry the calculations any farther, we may accept these as the finally determined values of the parallax in longitude at the times of apparent contact and separation. Then, by v. 16,

Parallax in longitude at contact and separation,	$2^n\ 2^v$	$3^n\ 24^v$
do. at apparent conjunction,	$2^n\ 55^v$	$2^n\ 55^v$
Difference of parallaxes,	53^v	29^v
add to former and latter mean half-duration,	$2^n\ 29^v$	$2^n\ 32^v$
True former and latter half-duration,	$3^n\ 22^v$	$3^n\ 1^v$
subtract and add from and to time of app. conj.,	$27^n\ 57^v$	$27^n\ 57^v$
Times of apparent contact and separation,	$24^n\ 35^v$	$30^n\ 58^v$

The calculation of the elements of the eclipse is thus completed. For the purpose, however, of illustrating the rules of the text (iv. 18–21) for determining, in the case of a solar eclipse, the amount of obscuration at any given moment during the continuance of the eclipse, we add also the following process:

XVI. To find the amount of obscuration of the sun, $2^n\ 38^v$ after first contact.

We make choice of this time, which is equivalent to $27^n\ 13^v$ after sunrise, because the data for finding the parallax in latitude at the moment have already been calculated (see above, XI). By iv. 18, from the

True former half-duration (*sphuṭa sparçasthityardha*),	$3^n\ 22^v$
deduct the given interval,	$2^n\ 38^v$
Interval to apparent conjunction (*madhyagrahaṇa*),	44^v

To reduce this interval in time to distance in longitude of the centres, we say (iv. 18)

$$60^n : 664'\ 7'' :: 44^v : 8'\ 7''$$

This, then, would be the interval in longitude between the two centres at the given moment, if there were no change of the moon's parallax in longitude during the eclipse, or if the moon actually gained in $2^n\ 29^v$, instead of in $3^n\ 22^v$, the distance intervening between her centre and the sun's at the moment of first contact. That, however, being not the case, we must reduce the result thus found in the ratio of $3^n\ 22^v$ to $2^n\ 29^v$, or of the true to the mean half-duration. That is to say (iv. 19),

$$3^n\ 22^v : 2^n\ 29^v :: 8'\ 7'' : 5'\ 59''$$

and this result, $5'\ 59''$, is the true distance of the two centres in longitude, $27^n\ 13^v$ after sunrise.

A briefer and more obvious method of obtaining the quantity in question would have been to make a proportion as follows: if, at the time of the eclipse, the moon gains upon the sun $27'\ 29''$ in $3^n\ 22^v$, what will she gain during 44^v? or

$$3^n\ 22^v : 27'\ 29'' :: 44^v : 5'\ 59''$$

Upon computation, we find the

Moon's parallax in latitude, $27^n\ 13^v$ after sunrise,	16′ 51″ S.
Moon's true latitude,	5′ 25″ N.
Moon's apparent latitude,	11′ 26″
Its square,	130′ 43″
Square of distance in longitude (5′ 59″),	35′ 59″
Their sum (iv. 20),	166′ 34″
Actual distance of centres,	12′ 54″
deduct from sum of semi-diameters,	30′ 6″
Amount of obscuration at given time,	17′ 12″

If it were desired to project the eclipse, we should now have to calculate (by iv. 24–25) the deflection (*valana*) for the moments of contact, conjunction, and separation, and likewise (by iv. 26) the scale of projection. As we do not, however, intend to present here a projection, and as the subject of the deflection has been sufficiently illustrated already, in the notes upon the text and in the calculation of the lunar eclipse, we regard it as unnecessary to go through with the labor required for making the computations in question. Finally, we annex, as in the case of the lunar eclipse formerly calculated, a summary comparison of the principal results of the Hindu processes with the elements of the eclipse in question as determined by Prof. Coffin, in his work referred to above. It must be borne in mind, however, that, owing to the faulty manner in which many of the computations of the native astronomer have been made, the comparison is not entirely trustworthy; a more careful adherence to the methods of the Siddhânta would have given somewhat different results: in the case of the daily motions of the sun and moon, the true calculations, as performed by us (see p. 308), give more correct values; in other instances, the contrary might perhaps have been the case.

	Sûrya-Siddhânta.	Prof. Coffin.	Hindu error.
Time of true conjunction in longitude,	$2^h\ 30^m$	$3^h\ 56^m$	$-\ 1^h\ 26^m$
Sun's and moon's longitude,	63° 50′ 37″	65° 12′ 37″	− 1° 22′
Moon's distance from node,	43′ 6″	4° 12′ 22″	− 3° 29′ 16″
Sun's daily motion in longitude,	56′ 52″	57′ 45″	− 53″
Moon's do. do.	12° 0′ 59″	12° 7′ 12″	− 6′ 13″
Sun's apparent diameter,	31′ 10″	31′ 37″	− 27″
Moon's do. do.	29′ 2″	29′ 45″	− 43″
Time of apparent conjunction,	$3^h\ 40^m$	$5^h\ 32^m$	$-\ 1^h\ 52^m$
Parallax in longitude, in time,	$1^h\ 10^m$	$1^h\ 36^m$	$-\ 26^m$
Amount of greatest obscuration,	19′	30′ 59″	− 11′ 59″
Time of first contact,	$2^h\ 20^m$	$4^h\ 15^m$	$-\ 1^h\ 55^m$
Time of separation,	$4^h\ 50^m$	$6^h\ 38^m$	$-\ 1^h\ 48^m$
Duration of eclipse,	$2^h\ 30^m$	$2^h\ 23^m$	$+\ 7^m$

26. pp. 183–200. Prof. Weber, of Berlin, has favored us in a private communication with a number of additional synonyms of the names of the asterisms, derived from the literature of the Brâhmaṇa period.

Mṛgaçiras, the fifth of the series, is also styled *andhakâ*, "the blind," apparently from its dimness; *âryikâ*, "honorable, worthy;" *invakâ*, of doubtful meaning: this latter epithet is also found in some manuscripts of the Amarakoça, as various reading for *ilvalâ*, which is there expressly declared (I. i. 2. 25) to designate the stars in the head of the antelope.

Ârdrâ, the sixth asterism, is called *bâhu*, "arm." Taking this name in connection with that of the preceding group, it seems probable that the Hindus figured to themselves the conspicuous constellation Orion as a running antelope, of which α, γ, β, and κ mark the feet: α, then, is the left fore-foot, or arm. Perhaps the name Mṛgavyâdha, "antelope-hunter," given to the neighboring Sirius (viii. 10), is connected with the same fancy.

The Maghâs are called in a hymn of the last book of the Rig-Veda (x. 85. 13) *aghâs*: the word means literally "evil, base, sinful," and its application to one of the asterisms is so strange that, if not found elsewhere, we should be inclined to conjecture a corrupted reading.

Phalgunî, or the Phalgunîs, forming the eleventh and twelfth groups, are styled also *arjunî*, "bright, shining."

Çravaṇa, the twenty-third asterism, receives the name *açvattha*, which is properly that of a tree, the Ficus religiosa; the reason of the appellation is altogether obscure.

Bhâdrapadâ, the last double asterism, is called *pratishṭhâna*, "stand, support," in evident allusion to the disposition of the four bright stars which compose it, like the four feet of a stand, table, bedstead, or the like.

27. p. 200. We offer herewith the stellar chart to which reference was made in the note on p. 205, and which is intended to illustrate the positions and mutual relations of the Hindu *nakshatras*, the Arab *manâzil al-ḳamar*, and the Chinese *sieu*. We add a brief explanation of the manner in which it has been constructed, and the form in which it is presented.

The form of the map is that of a plane projection, having the ecliptic as its central line. It would have better illustrated the Hindu method of defining the positions of the junction-stars, and the errors of the positions as defined by them, if the equator of A. D. 560, instead of the ecliptic, had been made the central line of the projection. This, however, would have involved the necessity of calculating the right ascension and declination of every star laid down, a labor which we were not willing to undertake. Moreover, the ecliptic is, in fact, the proper central line along which the groups of the Hindu and Arab systems, at least, are arranged, and the form given to the chart also facilitates the laying down of the equator of B. C. 2350, which we desired to add, for the purpose of enabling our readers to judge in a more enlightened manner of the plausibility of M. Biot's views respecting the origin of the Chinese system: it is drawn with a broken line, while the equator of A. D. 560 is also represented, by an entire line. As the zone of the heavens represented is, in the main, that bordering the ecliptic, the distances and the configuration of the stars are altered and distorted by the plane projection to only a very slight degree, not enough to be of any account in a merely illustrative chart, such as this is. As a general rule, we have laid down all the stars of the first four magnitudes which are situated near the ecliptic, or in that part of the heavens through which the line of the asterisms passes; stars of the fourth to fifth magnitude are also in many cases added; smaller ones are noted only when they enter into the groups of the several systems, or when there were other special reasons for introducing them. The positions are in all cases taken from Flamsteed's Catalogue, and the magnitudes are also for the most part from the same authority: in many individual cases, however, we have followed other authorities. We have endeavored so to mark the members of the three different series that these may readily be traced across the map; but, to assure and facilitate the comparison, we also place upon the page opposite it a conspectus of the nomenclature, constitution, and correspondence of the three systems, referring to pages 183–200 for a fuller discussion of these matters, and an exposition of what is certain, and what more or less hypothetical, or exposed to doubt, with regard to them.

Hindu asterism.	Arab *manzil.*	Chinese *sieu.*
1. Açvinî. β and γ Arietis.	1. ash-Sharaṭân. β and γ Arietis.	27. Leu. β Arietis.
2. Bharaṇî. 35, 39, and 41 Arietis.	2. al-Butain. 35, 39 and 41 Arietis.	28. Oei. 35 Arietis.
3. Kṛttikâ. η Tauri, etc. (Pleiades).	3. ath-Thuraiyâ. η Tauri etc. (Pleiades).	1. Mao. η Tauri.
4. Rohiṇî. α, ϑ, γ, δ, ε Tauri.	4. ad-Dabarân. α, ϑ, γ, δ, ε Tauri.	2. Pi. ε Tauri.
5. Mṛgaçiras. λ, φ^1, φ^2 Orionis.	5. al-Haḳ'ah. λ, φ^1, φ^2 Orionis.	3. Tse. λ Orionis.
6. Ârdrâ. α Orionis.	6. al-Han'ah. η, μ, ν, γ, ξ Geminorum.	4. Tsan. δ Orionis.
7. Punarvasu. β, α Geminorum.	7. adh-Dhirâ'. β, α Geminorum.	5. Tsing. μ Geminorum.
8. Pushya. ϑ, δ, γ Cancri.	8. an-Nathrah. γ, δ Cancri, and Præsepe.	6. Kuei. ϑ Cancri.
9. Âçleshâ. ε, δ, σ, η, ϱ Hydræ.	9. aṭ-Ṭarf. ξ Cancri, λ Leonis.	7. Lieu. δ Hydræ.
10. Maghâ. α, η, γ, ζ, μ, ε Leonis.	10. aj-Jabhah. α, η, γ, ζ Leonis.	8. Sing. α Hydræ.
11. Pûrva-Phalgunî. δ, ϑ Leonis.	11. az-Zubrah. δ, ϑ Leonis.	9. Chang. υ^1 Hydræ.
12. Uttara-Phalgunî. β, 93 Leonis.	12. aṣ-Ṣarfah. β Leonis.	10. Y. α Crateris.
13. Hasta. δ, γ, ε, α, β Corvi.	13. al-Auwâ'. β, η, γ, δ, ε Virginis.	11. Chin. γ Corvi.
14. Citrâ. α Virginis.	14. as-Simâk. α Virginis.	12. Kio. α Virginis.
15. Svâtî. α Bootis.	15. al-Ghafr. ι, κ, λ Virginis.	13. Kang. κ Virginis.
16. Viçâkhâ. ι, γ, β, α Libræ.	16. az-Zubânân. α, β Libræ.	14. Ti. α^2 Libræ.
17. Anurâdhâ. δ, β, π Scorpionis.	17. al-Iklîl. β, δ, π Scorpionis.	15. Fang. π Scorpionis.
18. Jyeshthâ. α, σ, τ Scorpionis.	18. al-Ḳalb. α Scorpionis.	16. Sin. σ Scorpionis.
19. Mûla. λ, υ, κ, ι, ϑ, η, ζ, μ, ε Scorp.	19. ash-Shaulah. λ, υ Scorpionis.	17. Uei. μ^2 Scorpionis.
20. Pûrva-Ashâḍhâ. δ, ε Sagittarii.	20. an-Na'âim. γ^2, δ, ε, η, φ, σ, τ, ζ Sagittarii.	18. Ki. γ^2 Sagittarii.
21. Uttara-Ashâḍhâ. σ, ζ Sagittarii.	21. al-Baldah. N. of π Sagittarii.	19. Teu. φ Sagittarii.
22. Abhijit. α, ε, ζ Lyræ.	22. Sa'd adh-Dhâbih. α, β Capricorni.	20. Nieu. β Capricorni.
23. Çravaṇa. α, β, γ Aquilæ.	23. Sa'd Bula'. ε, μ, ν Aquarii.	21. Nü. ε Aquarii.
24. Çravishthâ. β, α, γ, δ Delphini.	24. Sa'd as-Su'ûd. β, ξ Aquarii.	22. Hiü. β Aquarii.
25. Çatabhishaj. λ Aquarii etc.	25. Sa'd al-Akhbiyah. α, γ, ζ, η Aquarii.	23. Goei. α Aquarii.
26. Pûrva-Bhâdrapadâ. α, β Pegasi.	26. al-Fargh al-Muḳdim. α, β Pegasi.	24. Che. α Pegasi.
27. Uttara-Bhâdrapadâ. γ Pegasi, α Andromedæ.	27. al-Fargh al-Mukhir. γ Pegasi, α Andromedæ.	25. Pi. γ Pegasi.
28. Revatî. ζ Piscium, etc.	28. Baṭn al-Ḥût. β Andromedæ, etc.	26. Koei. ζ Andromedæ.

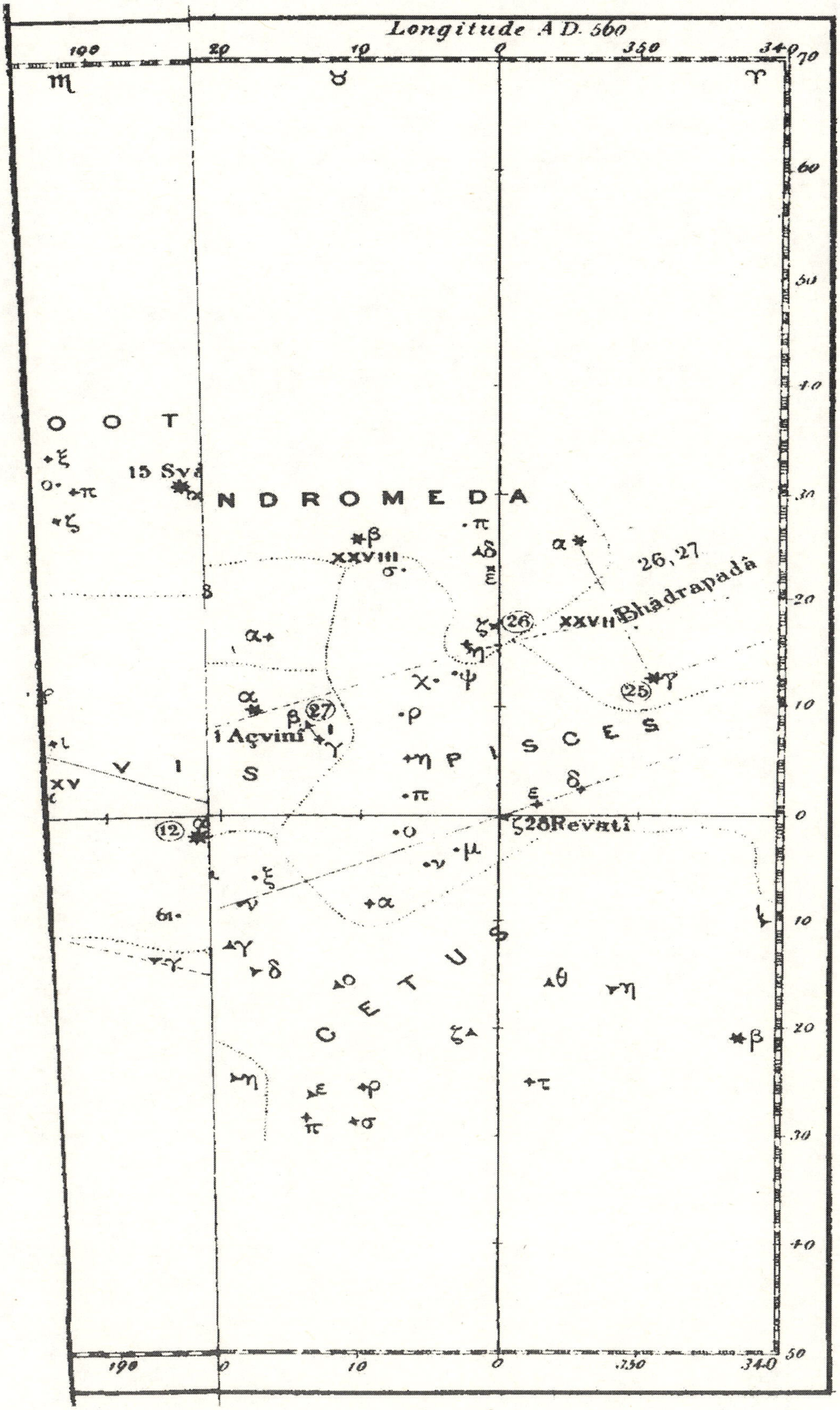
Longitude A.D. 560
190
20
10
0
350
340
70
60
50
40
30
20
10
0
10
20
30
40
50
BOOTES
15 Svâti
ANDROMEDA
XXVIII
26, 27
XXVII Bhâdrapadâ
PISCES
1 Açvinî
XV
VIRGO
28 Revatî
CETUS

28. p. 207. We have perhaps expressed ourselves in a manner liable to misconstruction as to the want of reason or authority for giving to the asterisms the name of "lunar mansions," "houses of the moon," and the like. We would by no means be understood as denying that in the Hindu science, especially its older forms, and in the Hindu mythology, they are brought into particular and conspicuous relations with the moon. Indeed, whether they were originally selected and established with reference to the moon's daily progress along the ecliptic, as has been, until lately, the universal opinion, or whether we are to believe with M. Biot that they had in the first instance nothing to do with the moon, and only came by chance to coincide in number with the days of her sidereal revolution—it is at any rate altogether probable that to the Hindu apprehension this coincidence formed the basis of the system. We may even conclude, from the fact that the asterisms are so frequently spoken of in the early literature of the Brâhmaṇa period, while nevertheless there is no distinct mention of the planets until later (Weber, Ind. Lit., p. 222), that for a long time the Hindus must have confined their attention and observations to the sun and moon, paying no heed to the lesser planets: and yet we cannot regard it as in any degree probable—hardly as possible, even—that any nation or people could establish a system of zodiacal asterisms without discovering and taking note of the planets; or that such a system could have been communicated to, and applied by, the Hindus, without a recognition on their part of those conspicuous and ever-moving stars. It may fairly be claimed, then, that the asterisms, as a Hindu institution, are an originally lunar division of the zodiac; but we object none the less to their being styled "lunar mansions," or called by any equivalent name; because, in the first place, the Hindus themselves have given them no name denoting a special relation to the moon, and no name signifying "house, mansion, station," or anything of the kind; and because, in the second place, as soon and as far as the Hindu astronomy extended itself beyond its limitation to observations of the moon, just so far and so soon did it employ the system of asterisms as a general method of division of the ecliptic; so that finally, as pointed out by us above, the asterisms have come to be divested, in the properly astronomical literature of India, of all special connection with the moon. With almost the same propriety might we call the Hindu signs "luni-solar mansions"—since they are, by origin, the parts of the ecliptic occupied by the sun during each successive synodical revolution of the moon—as denominate the *nakshatras* of the Siddhântas "lunar mansions."

29. p. 209. We should have mentioned farther, that an additional inducement—and one, probably, of no small weight—to the reduction of the number of asterisms from twenty-eight to twenty-seven, is to be recognized in the fact that the time of the moon's sidereal revolution in days, though intermediate between the two numbers, is yet decidedly nearer to twenty-seven, exceeding it by less than a third. M. Biot might even claim with some reason that the choice of the number twenty-eight tended to prove the whole system not a lunar one by origin: yet it might be replied that, the time of revolution being distinctly more than twenty-seven days, the larger number was fully admissible, and that it was also in some respects preferable, as being one that could be halved and quartered.

30. p. 273. In bringing this work to a close, we deem it advisable to present, in a summary manner, but more distinctly and connectedly than could properly be done in the notes upon the text, our conclusions as to certain points in the history of the Sûrya-Siddhânta, and of the astronomical science which it represents.

In the first place, Bentley's determination of the age of the treatise we conceive to be altogether set aside by the considerations which we have adduced against it (note to i. 29–34); there is no reasonable ground for questioning that the Sûrya-Siddhânta is, as the Hindus have long believed it to be, one of the most ancient and original of the works which present their modern astronomical science. How far the text of which the translation has been given above is identical in substance and extent with that of the original Sûrya-Siddhânta, is another question, and one not easy to solve. That it is not precisely the same is evident enough. Even the modern manuscripts differ from one another in single readings, in details of arrangement, in added or omitted verses. A comparison of the texts adopted and established by the different commentators would be highly interesting, as carrying the history of the treatise a step farther back: but to us only one commentary is accessible, nor do we find anywhere any notices respecting the versions given by the others: in the absence of such, we may conclude that all present substantially the same text, and so are alike posterior to the modelling of the work into its present form and with its present contents. But the indications of addition and interpolation, which we have had in so many cases to point out in our notes, are sometimes too telling to be misinterpreted. Farther than this we may not at present go: any detailed discussion of the subject must remain unsatisfactory, until a fuller acquaintance with other of the ancient treatises, and a more careful comparison of them with one another, shall throw upon it new light. A point of special interest connected with it is, whether the elements of mean motions of the planets do actually date from about the time pointed out by Bentley's calculations. With regard to this we are far from being confident; but we do not regard it as impossible, or even as very improbable, that those elements, as presented by our text, have been the same from the beginning, never having undergone correction until the application of the *bija*, about A. D. 1500 (p. 19 etc.). And the date of that correction is calculated at least to suggest the suspicion that Muslim science may have had something to do with it. That observation, and the improvement of their system by deductions from observation, were ever matters of such serious earnest with the Hindus that they should have been led to make such amendments independently, is yet to be proved. The most important alteration of which anything like direct proof is furnished is that which concerns the precession of the equinoxes (note to iii. 9–12); and even here we would not undertake to say confidently what is the conclusion to be drawn. All such inquiries must remain conjectural, mere gropings in the twilight, until the position of the Sûrya-Siddhânta in the Siddhânta literature shall be better understood. What has given it so much greater prominence and popularity than are enjoyed by the other works of its class, or from what period its preëminence dates, is unknown. There are treatises, like the Çâkalya-Sanhitâ (add. note 1), which agree with it in all essential features; there are yet others, like the Soma and Vasishṭha Siddhântas, which are said (add. note 6) to vary little from it: whether any one among them all is original—and if any, which—whether in each case the relation is one of co-ordination or of subordination—we must be content for the time to be ignorant.

One thing, however, is certain: underneath whatever variety may characterize the separate treatises, there exists a fundamental unity; their differences are of secondary importance as compared with their resemblances; they all represent essentially a single system. And this by no means in the same sense in which all modern astronomical works may be said to represent a single system. For the Hindu system is not one of nature; it is not even a peculiar method of viewing and interpreting nature, from which, after it had once been devised by some controlling intellect, others had not the force and originality to deviate: it is a thoroughly artificial structure, full of arbitrary assumptions, of absurdities even, which have no foundation in nature, and could be invented by one as well as another. We need only to refer, as instances, to the frame-work of monstrous chronological periods (i. 14–23)—to the common epoch of the commencement of the Iron Age (note to i. 29–34), with its exact or nearly exact (add. note 6) conjunction of all the planets—to the form of statement of the mean motions, yielding recurring conjunctions, at longer or shorter intervals—to the assumption of a starting-point for the planets from at or near ζ Piscium (note to i. 27)—to the revolutions of the apsides and nodes of the planets (i. 41–44)—to the double system of epicycles (ii. 34–38)—to the determination of the planetary orbits (xii. 80–90), etc., etc. These are plain indications that the Hindu science emanated from one centre; that it was the elaboration of a period and of a school, if not of a single master, who had power enough to impose his idiosyncracy upon the science of a whole nation. The question, then, of the comparative antiquity of single treatises is lost in the higher interest of the inquiry—when, where, and under what influence originated the system which they all agree in representing?

What our opinions are upon these points will not be a matter of doubt with any one who may have carefully looked through the preceding pages, although they have nowhere been explicitly stated. We regard the Hindu science as an offshoot from the Greek, planted not far from the commencement of the Christian era, and attaining its fully developed form in the course of the fifth and sixth centuries. The grounds of this opinion we will proceed briefly to state.

In considering such a question, it is fair to take first into account the general probabilities of the case. And there can be no question that, from what we know in other respects of the character and tendencies of the Hindu mind, we should not at all look to find the Hindus in possession of an astronomical science containing so much of truth. They have been from the beginning distinguished by a remarkable inaptitude and disinclination to observe, to collect facts, to record, to make inductive investigations. The old belief under the influence of which Bailly could form his strange theories—the belief in the immense antiquity of the Indian people, and its immemorial possession of a highly developed civilization—the belief that India was the cradle of language, mythology, arts, sciences, and religions—has long since been proved an error. It is now well known that Hindu culture cannot pretend to a remoter origin than 2000 B. C., and that, though marked by striking and eminent traits of intellect and character, the Hindus have ever been weak in positive science; metaphysics and grammar—with, perhaps, algebra and arithmetic, to them the mechanical part of mathematical science—being the only branches of knowledge in which they have independently won honorable distinction. That astronomy would come to constitute an exception to the general rule in this respect, there is no antecedent ground for supposing. The infrequency of references to the stars in the early Sanskrit literature, the late date of the earliest mention of the planets, prove that there was no special impulse leading the nation to devote itself to studying the movements of the heavenly bodies. All evidence goes to show that the Hindus, even after they had derived from abroad (p. 204) a systematic division of the ecliptic, limited their attention to the two chief luminaries, the sun and moon, and contented themselves with establishing a method of maintaining the concordance of the solar year with the order of the lunar months. If, then, at a later period, we find them in possession of a full astronomy of the solar system, our first impulse is to inquire, whence did they obtain it? A closer inspection does not tend to inspire us with confidence in it as of Hindu origin. We find it, to be sure, thoroughly Hindu in its external form, wearing many strange and fantastic features which are to be at once recognized as of native Indian growth; but we find it also to contain much true science, which could only be derived from a profound and long-continued study of nature. The whole system, in short, may be divided into two portions, whereof the one contains truth so successfully deduced that only the Greeks, among all other ancient nations, can show anything worthy to be compared with it; the other, the framework in which that truth is set, composed of arbitrary assumptions and absurd imaginings, which betray a close connection with the fictitious cosmogonies and geographies of the philosophical and Puranic literature of India. The question presses itself, then, strongly upon us, whether these two portions can possibly have the same origin: whether the scientific habit of mind which could lead to the discovery of the one is compatible with those traits which would permit its admixture with the other. But most especially, could a system founded—as this, if original, must have been—upon sagacious, accurate, and protracted observation of the heavenly bodies, so entirely ignore the ground-work upon which it rested, and refuse and deny all possibility of future improvement by like means, as does this Hindu system, in whose text-books appears no record of an observation, and no confessed deduction from observations; in which the astronomer is remanded to his text-book as the sole and sufficient source of knowledge, nor ever taught or counselled to study the heavens except for the purpose of determining his longitude, his latitude, and the local time? Surely, we have a right to

say that the system, in its form as laid before us, must come from another people or another generation than that which laid its scientific foundation; that it must be the work of a race which either had never known, or had had time to forget, the observing habits and the inductive methods of those who gave it origin. But the hypothesis that an earlier generation in India itself performed the labors of which the later system-makers reaped the fruit, is well-nigh excluded by the absence, already referred to, of all evidence in the more ancient literature of deep astronomical investigation: the other alternative, of derivation from a foreign source, remains, if not the only possible, at least the only probable one. We come, then, next to consider the direct evidences of a Greek origin.

First in importance among these is the system of epicycles for representing the movement, and calculating the positions, of the planets. This, the cardinal feature in both systems, is (ii. 34–45) essentially alike and the same in both. Now, notwithstanding the fact that such secondary circles do in fact represent, to a certain degree, true quantities in nature, there is yet too much that is strange and arbitrary in them to leave any probability to the supposition that two nations could have devised them independently. But there are sufficient grounds for believing the Greeks to have actually created their own system, bringing it by successive steps of elaboration to the form in which Ptolemy finally presents it. In the history of the science among the Greeks, everything is clear and open; they tell us what they owed to the Egyptians, what to the Chaldeans: we trace the conceptions which were the germs of their scheme of epicycles, the observations on which it was based, the inductive and deductive methods by which it was worked out and established. In the Hindu astronomy, on the other hand, all is groundless assumption and absurd pretense: we find, as basis for the system, neither the conceptions—for these are directly or impliedly denied or ignored—nor the observations—for not a mention of an actual observation is anywhere to be discovered—nor the methods: the whole is gravely put forth as a complete and perfect fabric, of divine origin and immemorial antiquity. On the agreement of the two sciences in point of numerical data we will not lay any stress, since it might well enough be supposed that two nations, if once set upon the same track toward the discovery of truth, would arrive independently at so near an accordance with nature and with one another. We will look for other evidences, of a less ambiguous character, to sustain our main argument. The division of the circle, into signs, degrees, minutes, and seconds, is the same in both systems, and, being the foundation on which all numerical measurements and calculations are made, is an essential and integral part of both. Now the names of the first subdivisions, the signs, are the same in Greece and in India (see note to i. 58): but with the Greeks they belong to certain fixed arcs of the ecliptic, being derived from the constellations occupying those arcs; with the Hindus they are applied to successive arcs of 30°, counted from any point that may be chosen: this is an unambiguous indication that the latter have borrowed them, and forgotten or neglected their original significance. But farther, the ordinary Hindu name of that division of the circle which is in most frequent use, the minute, is no Sanskrit word, but taken directly from the Greek, being *liptâ*, which is λεπτόν. Again, the planets are ordinarily named in the Siddhântas in the order in which they succeed one another as regents of the days of the week; and not only has it been shown above that the week is no original Hindu institution, but it has even appeared that, on tracing it to its very foundation, we find there another Greek word, ὥρα, represented by *horâ*. Once more, in the cardinal operation of finding by means of the system of epicycles the true place of a planet, we see that one of the most important data, the mean anomaly, is called by another name of Greek origin, namely *kendra*, which is κέντρον. These three words, occurring where they do, not upon the outskirts of the Hindu science, but in its very centre and citadel, amount of themselves almost to full proof of its Greek origin: taken in connection with the other concurrent evidences, they form an argument which can neither be set aside nor refuted. Of those other evidences, we will only mention farther here that Hindu treatises and commentaries of an early date often refer to the *yavanas*, "Greeks" or "westerners," and to *yavanâcâryâs*, "the Greek (or western) teachers," as authorities on astronomical subjects—that astronomical treatises are found bearing names which come more or less distinctly from the West (note to i. 4–6)—and that floating traditions are met with, to the effect that some of the Siddhântas were revealed to their human promulgators in Romaka-city, that is to say, at Rome. Farther witness to the same truth, deducible from other coincidences of the two systems, we pass unnoticed here, since it is not our object to discuss the question exhaustively, but only to bring forward the main grounds of our opinions.

The question next arises, when, and in what manner the knowledge of astronomy was communicated from Greece to India. In reply to this, only probabilities offer themselves, yet in some points the indications are pretty distinct. It is, in our own view, altogether likely that the science came in connection with the lively commerce which, during the first centuries of our era, was carried on by sea between Alexandria, as the port and mart of Rome, and the western coast of India. Two considerations especially favor this supposition: first, that the chief site of the Hindu science is found to be the city which lay nearest to the route of that commerce (note to i. 62): secondly, that Rome is the only western city or country which is distinctly mentioned in the astronomical geography (xii. 39), and the one with which, as above noticed, the astronomical traditions connect themselves. Had the Hindus derived their knowledge overland, through the Syrian, Persian, and Bactrian kingdoms which stood under Greek government, or in which Greek influence was predominant, and Greek culture known and prized, the name of Rome would have been vastly less likely to stand forth with such prominence, and the capitals of Hindustan proper would more probably have been the cradles of the new science. The absence from the Hindu system of any of the improvements introduced by Ptolemy into that of the Greeks (note to ii. 43–45) tends strongly to prove that the transmission of the principal groundwork of the former took place before his time: nor can we think it likely that the numerical elements adopted by the Hindus would vary so much as in many cases they are found to

do from those of the Syntaxis, if the latter had been already in existence, and acknowledged as the principal and most authoritative exponent of Greek astronomy. Whether the information was transmitted through the medium of Hindus who visited the Mediterranean, or of learned Greeks who made the voyage to India, or by the translation of Greek treatises, or by what other methods, we would not at present even offer a conjecture; and the point is one of only subordinate consequence.

Whatever may have been the date of the first communication of the elements out of which the Hindu system was elaborated, there is good reason to suppose that its final reduction to its present form did not take place until some time during the fifth and sixth centuries. That period is distinctly pointed out by the choice of the equinox of A. D. 570 as the initial and principal point of the fixed sphere (note to i. 27), by the definition of position of the junction-stars of the asterisms (p. 211), and by the Hindu traditions which refer to that time the names of greatest prominence and authority in the early history of the science. It is evident that the elaboration of the system must have been a work of time, probably of many generations: what were the forms which it wore in the interval we do not know; here, as in many other departments of the Hindu literature, all record of the steps of development appears to be lost, only the final and fully formed product being preserved and transmitted to us: yet more light upon this point may still be hoped for, from the careful examination of all documents now accessible, or of such as may hereafter be discovered. The process of assimilation and adaptation to Hindu conceptions and Hindu methods was thoroughly and completely performed. Among the changes of method introduced, the most useful and important was the substitution of sines for chords (p. 66); the general substitution of an arithmetical for a geometrical form also deserves particular notice. That no great amount of geometrical science is implied in any part of the system, is very evident: it is distinguished by the constant and dexterous application of a few simple principles: the equality of the square of the hypothenuse to the sum of the squares of the base and perpendicular—the comparison of similar right-angled triangles—the formation and combination of proportions, the rule of three—are the characteristic features of the early Hindu mathematical knowledge, as displayed in the Sûrya-Siddhânta. Of other treatises, of an earlier or later period, as those of Brahmagupta and Bhâskara, which (see Colebrooke's Hindu Algebra) give evidence of knowledge more profound in arithmetic and algebra, we cannot at present speak; but we hope at some future time to be able to revert to the subject of the Hindu astronomy, in connection with these or other of the text-books by which it is represented.

Rev. Mr. Burgess, having placed his translation and notes in the hands of the Committee of Publication for farther elaboration, has very liberally allowed them entire freedom in their work, even where their deductions, and the views they expressed, did not accord with his own opinions. The most important point at issue between us is that discussed in the next preceding pages, or the originality of the Hindu astronomy; upon this, then, he is desirous of expressing independently his dissenting views, as in the following note.

Concluding Note by the Translator.

It may not be improper for me to state, in a closing note, that I had prepared a somewhat extended and elaborate essay on the history of astronomy among the Hindus, to be published in connection with the preceding translation. But the length of this essay is such—the subject matter of it not being material to the illustration of the Siddhânta, and the translation and notes having already occupied so much space—that it was not thought advisable to insert it here.

Yet as my investigations have led me to adopt opinions on some points differing from those advanced by Prof. Whitney in his very valuable additions to the notes upon the translation, truth and consistency seem to require me to present at least a brief summary of the results at which I arrived in that essay in reference to the points in question. By so doing, I free myself from any embarrassment under which I should labor, if hereafter—as I now intend—I shall wish to express the grounds for my opinions on these points, in this Journal or elsewhere.

The points to which I allude bear upon the claims of the Hindus to the honor of original invention and discovery in astronomical science—especially, their claims to such an honor in comparison with the Greeks.

Prof. Whitney seems to hold the opinion, that the Hindus derived their astronomy and astrology almost bodily from the Greeks—and that what they did not borrow from the Greeks, they derived from other people, as the Arabians, Chaldeans and Chinese (see pp. 34, 204, 206, et al.). I think he does not give the Hindus the credit due to them, and awards to the Greeks more credit than they are justly entitled to. In advancing this opinion, however, I admit that the Greeks, at a later period, were the more successful cultivators of astronomical science. There is nothing among the Hindu treatises that can compare with the great Syntaxis of Ptolemy. And yet, from the light I now have, I must think the Hindus original in regard to most of the elementary facts and principles of astronomy as found in their systems, and for the most part also in their cultivation of the science; and that the Greeks borrowed from them, or from an intermediate secondary source, to which these facts and principles had come from India. I might perhaps so far modify this statement as to admit the supposition that neither Greeks nor Hindus borrowed the one from the other, but both from a common source. But with my present knowledge, I cannot concur in the opinion that the Hindus are, to any great extent, indebted to the Greeks for their astronomy, or that the latter have any well grounded claims to the honor of originality in regard to those elementary facts and principles of astronomical science which are common to their own and other ancient systems, and which are of such a nature as indicates for them a single origin, and a transmission from one system to another. For the sake of clearness, it is well that I should state more specifically a few of the more important facts and principles that come under the class above referred to. They are as follows:

1. The lunar division of the zodiac into twenty-seven or twenty-eight asterisms (see transl., ch. viii). This division is common, with slight modifications, to the Hindu, Arabian, and Chinese systems.

2. The solar division of the zodiac into twelve signs, with the names of the latter. These names are, in their import, precisely the same in the Hindu and Greek systems. The coincidence is such that the theory of the division and the names of the parts having proceeded from one original source is unquestionably the correct one.

3. The theory of epicycles in accounting for the motions of the planets, and in calculating their true places. This is common to the Hindu and Greek astronomies. At least, there is such a coincidence in the two systems in reference to the epicycles as almost to preclude the idea of independent origin or invention.

4. Coincidences, and even a sameness in some parts, between the systems of astrology received among the Hindus, Greeks, and Arabians, strongly indicate for those systems, in their primitive and essential elements, a common origin.

5. The names of the five planets known to the ancients, and the application of these names to the days of the week (see notes, i. 52).

In regard to these specifications I remark in general:

First, in reference to no one of them do the claims of any people to the honor of having been the original inventors or discoverers appear to be better founded than those of the Hindus.

Secondly, in reference to most of them, the evidence of originality I regard as clearly in favor of the Hindus; and in regard to some, and those the more important, this evidence appears to me nearly or quite conclusive.

I have not space for detail, nor is it the design of this note to enter into the details of argument on any point whatever. A brief remark, however, for the sake of clearness, seems called for in reference to each of the above five specifications of facts and principles common to some or all of the ancient systems of astronomy and astrology.

1. As to the lunar division of the zodiac into twenty-seven or twenty-eight asterisms. The undoubted antiquity of this division, even in its elaborated form, among the Hindus, in connection with the absence or paucity of such evidence among any other people, incline me decidedly to the opinion that the division is of a purely Hindu origin. This is still my opinion, notwithstanding the views advanced by M. Biot and others in favor of another origin.

2. As to the solar division of the zodiac into twelve parts, and the names of those parts. The use of this division, and the present names of the signs, can be proved to have existed in India at as early a period as in any other country; and there is evidence less clear and satisfactory, it is true, yet of such a character as to create a high degree of probability, that this division was known to the Hindus centuries before any traces can be found in existence among any other people.

As corroborative of this position in part, or at least as strongly favoring the idea of an eastern origin of the division of the ecliptic in question, I may be allowed to adduce the opinions of Ideler and Lepsius, as quoted by Humboldt (Cosmos, Harper's ed., iii. 120, note): "Ideler is inclined to believe that the Orientals had names, but not constellations, for the Dodecatomeria, and Lepsius regards it as a natural assumption 'that the Greeks, at the period when their sphere was for the most part unfilled, should have added to their own the Chaldean constellations from which the twelve divisions were named.'" Whether Ideler meant by "Orientals" the Chaldeans, or some other eastern people, the application of the term in this connection to the Hindus exactly suits the supposition of the Indian origin of the division in question, since in Indian astronomy the names of the signs are merely names of the twelfth parts of the ecliptic, and are never applied to constellations. Humboldt's opinion is, that the solar divisions of the ecliptic, with the names of the signs, came to the Greeks from Chaldea. I think the evidence preponderates in favor of a more eastern, if not a Hindu, origin.

3. The theory of epicycles. The difference in the development of this theory in the Greek and Hindu systems of astronomy precludes the idea that one of these people derived more than a hint respecting it from the other. And so far as this point alone is concerned, we have as much reason to suppose the Greeks to have been the borrowers as the contrary; but other considerations seem to favor the supposition that the Hindus were the original inventors of this theory.

4. As regards astrology, there is not much honor, in any estimation, connected with its invention and culture. The coincidences that exist between the Hindu and Greek systems are too remarkable to admit of the supposition of an independent origin for them. But the honor of original invention, such as it is, lies, I think, between the Hindus and the Chaldeans. The evidence of priority of invention and culture seems, on the whole, to be in favor of the former; the existence of three or four Arabic and Greek terms in the Hindu system being accounted for on the supposition that they were introduced at a comparatively recent period. In reference, however, to the word *horâ*, Greek ὥρα (see notes to i. 52; xii. 78–79), it may not be inappropriate to introduce the testimony of Herodotus (B. II, ch. 109): "The sun-dial and the gnomon, with the division of the day into twelve parts, were received by the Greeks from the Babylonians." There is abundant testimony to the fact that the division of the day into twenty-four hours existed in the East, if not actually in India, before it did in Greece. In reference, farther, to the so-called Greek words found in Hindu astronomical treatises, I would remark that we may with entire propriety refer them to that numerous class of words common to the Greek and Sanskrit languages, which either came to both from a common source, or passed from the Sanskrit to the Greek at a period of high antiquity; for no one maintains, so far as I am aware, that the Greek is the parent of the Sanskrit, to the extent indicated by this numerous class of words, and by the similarity of grammatical inflections in the two languages.

5. As to the names of the planets, I remark that the identity of all of them in the Hindu and Greek systems is not to my mind clearly made out. However this may be, I think the present names of the planets in Greek astronomy originated at least as far east as Chaldea. Herodotus says (B. II, ch. 52) . . . "the names of the gods came into Greece from Egypt." The names of the planets are names of gods. Herodotus's opinion indicates the belief of the Greeks in reference to the origin of these names. Other considerations show for them, almost beyond a question, an origin as far east, to say the least, as Chaldea.

As to the application of the names of the planets to the days of the week, it is impossible to determine definitely where it originated. Respecting this matter, Prof. H. H. Wilson expresses his opinion—in which I concur—in the following language: "The origin of this arrangement is not very precisely ascertained, as it was unknown to the Greeks, and not adopted by the Romans until a late period. It is commonly ascribed to the Egyptians and Babylonians, but upon no very sufficient authority, and the Hindus appear to have at least as good a title to the invention as any other people" (Jour. Roy. As. Soc., ix. 84).

One word on the claims of the Arabians to the honor of original invention in astronomical science. And first, they themselves claim no such honor. They confess to having received their astronomy from India and Greece. They had at an early period some two or three of the first Hindu treatises of astronomy. "In the reign of the second Abbasside Khalif Almansúr . . . (A. D. 773), as is related in the preface to the astronomical tables of Ben-Al-Ádamí, published . . . A. D. 920, an Indian astronomer, well versed in the science which he professed, visited the court of the Khalif, bringing with him tables of the equations of planets according to the mean motions, with observations relative to both solar and lunar eclipses, and the ascension of the signs; taken, as he affirmed, from tables computed by an Indian prince, whose name, as the Arabian author writes it, was PHÍGHAR" (Colebrooke's Hindu Algebra, p. lxiv). That the Arabians were thoroughly imbued with a knowledge of the Hindu astronomy before they became acquainted with that of the Greeks, is evident from their translation of Ptolemy's Syntaxis. It is known that this great work of the Greek astronomer first became known in Europe through the Arabic version. In the Latin translation of this version, the ascending node (Greek *ἀναβιβάζων σύνδεσμος*) is called *nodus capitis*, "node of the head," and the descending node (Greek *καταβιβάζων σύνδεσμος*), *nodus caudæ*, "node of the tail"—which are pure Hindu appellations (see Latin Translation of Almagest, B. iv, ch. 4; B. vi, ch. 7, et al.). This fact, with other evidence, clearly shows the influence of Hindu astronomy on that of the Arabians. In fact, this latter people seem to have done little more in this science than work over the materials derived from their eastern and western neighbors.

Another fact showing the belief of the Arabians themselves respecting their indebtedness, in matters of science, to the Hindus, should be mentioned here. They ascribe the invention of the numerals, the nine digits (the credit of whose invention is quite generally awarded to the Arabians), to the Hindus. "All the Arabic and Persian books of arithmetic ascribe the invention to the Indians" (Strachey, on the Early History of Algebra, As. Res., xii. 184; see likewise Colebrooke's Hindu Algebra, pp. lii–liii, where the same is shown from a different authority. Strachey's article was published subsequently to the work of Colebrooke).

The above facts and considerations, showing the indebtedness of the Arabians to the Hindus in regard to mathematical and astronomical science, clearly have an important bearing on the question of priority of invention in regard to the lunar division of the zodiac into twenty-eight asterisms, at least so far as the Arabians are concerned. Taking all the facts into account, the supposition that this people were the inventors is altogether untenable.

I close this note—already longer than I intended—with a quotation from that distinguished orientalist, H. T. Colebrooke. In a very valuable essay entitled "On the Notions of the Hindu Astronomers concerning the Precession of the Equinoxes and Motions of the Planets," having stated with some detail some of the more striking peculiarities of the Hindu systems, and likewise coincidences existing between them and that of the Greeks, with the evidence of communication from one people to the other, he says: "If these circumstances, joined to a resemblance hardly to be supposed casual, which the Hindu astronomy, with its apparatus of eccentrics and epicycles, bears in many respects to that of the Greeks, be thought to authorize a belief, that the Hindus received from the Greeks that knowledge which enabled them to correct and improve their own imperfect astronomy, I shall not be inclined to dissent from the opinion" (As. Res., xii. 245–6; Essays, ii. 411).

This is all that so learned and cautious a writer could say in favor of the opinion that the Hindus derived astronomical knowledge from the Greeks. More than this I certainly could not say. After the solar division of the zodiac, with the names of its parts, it is evident, I think, that only hints could have passed from one people to the other, and that at an early period; for on the supposition that the Hindus borrowed from the Greeks at a later period, we find it difficult to see precisely what it was that they borrowed; since in no case do numerical data and results in the systems of the two peoples exactly correspond. And in regard to the more important of such data and results—as for instance, the amount of the annual precession of the equinoxes, the relative size of the sun and moon as compared with the earth, the greatest equation of the centre for the sun—the Hindus are more nearly correct than the Greeks, and in regard to the times of the revolutions of the planets they are very nearly as correct: it appearing from a comparative view of the sidereal revolutions of the planets (p. 24), that the Hindus are most nearly correct in four items, and Ptolemy in six. There has evidently been very little astronomical borrowing between the Hindus and the Greeks. And in relation to points that prove a communication from one people to the other, with my present knowledge on the subject, I am inclined to think that the course of derivation was the opposite to that supposed by Colebrooke—from east to west rather than from west to east; and I would express my opinion in relation to astronomy, in the language which this eminent scholar uses in relation to some coincidences in speculative philosophy and religious dogmas, especially the doctrine of metempsychosis, found in the Greek and Hindu systems, which indicate a communication from one people to the other: "I should be disposed to conclude that the Indians were in this instance teachers rather than learners" (Transactions of the Roy. As. Soc., i. 579). This opinion is expressed in the last essay on oriental philosophy that came from the pen of Colebrooke. E. B.

Boston, May, 1860.

SANSKRIT INDEX.

The following Index contains all the Sanskrit words, excepting proper names, which have been cited in the text and notes, in connection with their translation or more detailed explanation. It includes many terms of trivial importance, but we prefer to err upon the side of fullness, if upon either. All the cases of occurrence of each word are not given, but it is referred to a characteristic passage, or to the note where it is explained. The references by Roman and Arabic figures are to chapter and verse, and an added *n* denotes the note next following the verse given: Arabic figures when used alone refer to pages.

GENERAL INDEX.

The references are as in the preceding Index.